AUDITORY PATHWAY

Structure and Function

AUDITORY PATHWAY
Structure and Function

Edited by
Josef Syka
Czechoslovak Academy of Sciences
Prague, Czechoslovakia

and
R. Bruce Masterton
Florida State University
Tallahassee, Florida

PLENUM PRESS • NEW YORK AND LONDON

Library of Congress Cataloging in Publication Data

Auditory Pathway-Structure and Function Satellite Symposium (1987: Prague, Czechoslovakia)
  Auditory pathway: structure and function / edited by Josef Syka and R. Bruce Masterton.
    p.   cm.
  "Proceedings of the Auditory Pathway-Structure and Function Satellite Symposium to the World Congress of Neurosciences, held August 23-26, 1987, Prague, Czechoslovakia" — T.p. verso.
  Bibliography: p.
  Includes index.
  ISBN 0-306-42994-2
  1. Auditory pathways — Congresses. I. Syka, Josef. II. Masterton, R. Bruce. III. World Congress of Neurosciences (1987: Prague, Czechoslovakia) IV. Title.
QP460.A95   1987   88-22372
596′.01825 — dc19   CIP

Proceedings of the Auditory Pathway — Structure and Function
Satellite Symposium to the World Congress of Neurosciences,
held August 23-26, 1987, in Prague, Czechoslovakia

© 1988 Plenum Press, New York
A Division of Plenum Publishing Corporation
233 Spring Street, New York, N.Y. 10013

Printed in the United States of America

PREFACE

    Since the last symposium on "Neuronal Mechanisms of Hearing" held
in Prague in 1980 and published in the volume of the same name (J. Syka
and L. Aitkin, Eds., Plenum Press, 1981), remarkable progress has been
achieved in the understanding of the auditory system. A variety of new
ideas and new methods have emerged.

    This progress can be easily documented by comparing the volume based
on the 1980 Symposium with the program for the 1987 Symposium. For example,
there were 45 contributions to auditory physiology in each symposium but
there were 27 contributions focusing on anatomy in 1987 as compared to 7
in 1980, and perhaps most telling, there were 12 contributions to the
neurochemistry of the system in 1987 while there were only 3 in 1980. In
terms of percentages of contributions, neuroanatomy rose from 13% to
32% and neurochemistry (or chemical anatomy) rose from 5% in 1980 to 14%
in 1987. These increases in the numbers and proportions of anatomical and
neurochemical contributions undoubtedly reflects the increasing availabil-
ity and rising expertise in the new neuroanatomical and biochemical
techniques most notably, tract-tracing by exploitation of axonal transport
or by intracellular micro-injection methods, and neurotransmitter identification by use of immunocytochemistry or receptor-binding techniques.

    New ideas have emerged on the function of cochlear hair cells
particularly in connection with olivocochlear bundle stimulation and
supported by findings of contractile proteins in outer hair cells. Further,
the distribution of neurotransmitters and their associated enzymes,
localized with antibodies or other microchemical methods, have enriched
understanding of the cellular physiology as well as the tissue-level
organization of the entire auditory pathway.

    Auditory system structure and function have now become the object of
investigation in several nonmammalian vertebrates, as well as in many new
species of mammals, adding valuable comparative, phylogenetic, and onto-
genetic dimensions.

    Although the complexity of the auditory pathway has been known for
many years, only recently have anatomists with modern tracing methods
(tritiated amino acids, horseradish peroxidase) started to unravel the
interconnections and neurochemical interactions of auditory nuclei.
Promising data have also been obtained with tomography enabling the
mapping of changes in the functional activity of auditory centers in the
human brain.

    More traditional topics in auditory physiology such as the neuro-
physiological basis of sound localization and the coding of communication
and speech signals also appeared in the 1987 Symposium as well as the

previous one. Furthermore, the structure of the efferent auditory pathway from forebrain to cochlea is now traceable. The new methods for studying the response characteristics of single identified neurons of the olivo-cochlear bundles have brought new insight to this old and persistent problem.

The 1987 Symposium occupied three days and was organized mostly along anatomical levels, with separate sessions on Cochlea, Auditory Nerve, Auditory Brainstem Nuclei, Inferior Colliculus, Auditory Cortex, and Efferent Pathway. But recent discoveries made it necessary to expand this outline along physiological dimensions -- thus there were also sessions on Sound Localization and on Coding of Complex Stimuli. Approximately 110 scientists registered for the Symposium and provided 84 separate papers or posters. The present volume contains 49 of these contributions.

We would like to acknowledge support for the symposium provided by the Institute of Experimental Medicine, Czechoslovak Academy of Sciences. We thank our many colleagues who contributed to the success of the meeting, particularly Jiří Popelář, Rostislav Druga, Ivo Melichar, Alexandra Vlková, Milan Jilek, Zbyněk Kubec and Jarmila Ullschmiedová. Their efforts are greatly appreciated.

Josef Syka and Bruce Masterton
November, 1987

CONTENTS

EFFERENT AUDITORY SYSTEM

PROCESSING OF COMPLEX ACOUSTIC STIMULI

AUDITORY LOCALIZATION

COCHLEA

NEW ASPECTS OF COMPARATIVE PERIPHERAL AUDITORY PHYSIOLOGY

Geoffrey A. Manley, Jutta Brix, Otto Gleich, Alexander Kaiser,
Christiane Köppl and *Graeme Yates

Institute of Zoology
Technical University Munich
Lichtenbergstr. 4, 8046 Garching, F.R.G.
*Department of Physiology
University of Western Australia
Nedlands, 6009, Australia

Although the study of the auditory system of nonmammals has a long
history, it has only been in the last ten years that there has been
a greatly increased interest in comparative studies. There are two main
reasons for this upsurge in interest. Firstly, the hearing organs of
amphibians, reptiles and birds display a structural variety not found in
the cochlea of mammals, offering the chance to investigate structure-
function relationships without interfering with the normal structure.
Secondly, it has been recognized that the sensory papillae of many nonmammals are mechanically and physiologically relatively robust, which
allows extensive and detailed investigation of hair-cell function in
isolated organs.

These and other advantages have led to a significant increase
in the quantity of data derived from nonmammalian preparations, data
which has made it possible to distinguish between those functional
features which are also found in mammals and those features which are
apparently observable only in nonmammals. Thus, we are increasingly
in a better position to judge which aspects of comparative auditory
physiology are directly comparable to mammalian inner ear function
and provide an alternative means of investigation, if not necessarily
a 'simple' model of the mammalian ear.

A number of features of the organization of the auditory receptors of
terrestrial vertebrates are common to both mammals and nonmammals. Taken
together with a comparison to other hair-cell systems such as the auditory
receptors of fish, and vestibular and lateral-line receptors, this fact
suggests that many of these features are the result of the conservation
throughout evolution of fundamental, primitive hair-cell properties
(Manley, 1986). Thus, the receptor cells themselves and the genetic
programmes which determine their ontogeny into an organized mosaic are the
primary determinants of many basic features of the function of the inner
ear.

There are, on the other hand, several different mechanisms by which
individual hair cells and hair-cell arrays achieve their frequency
selectivity, and there are variations in the expression of different

3

mechanisms in the various vertebrate groups (Manley, 1986) and perhaps even in hair cells of different frequency ranges within one papilla. It is useful to begin by summarizing some of the published data on important similarities and differences between nerve-fibre activity in the various groups of terrestrial vertebrates.

Similarities in Auditory-Nerve Fibre Activity between Mammals and Nonmammals

1) In the equivalent frequency range (amphibians up to about 3 kHz, reptiles and most birds up to 5-6 kHz), the frequency selectivity of nonmammalian auditory-nerve tuning curves is generally as high as or even higher than that of mammalian primary fibres (Manley, 1981; Manley et al., 1985; Sachs et al., 1974; Turner, 1987).

2) All vertebrate auditory papillae are strongly tonotopically organized, although in certain cases in lizards, the distribution of CFs is not monotonic as in mammals (Manley, 1981).

3) The general patterns in the distribution of intervals in the spontaneous activity of nerve fibres are similar; that is, a modified Poisson distribution underlies the activity in both mammals and nonmammals.

4) Otoacoustic emissions (OAE) have been demonstrated in frogs (Palmer and Wilson, 1981), in the caiman (Klinke and Smolders, 1984) and in the starling (Manley et al., 1987b), which in most respects resemble those reported from mammals.

Aspects in which Auditory-Nerve Fibre Activity Differs between Mammals and Nonmammals

1) Tuning curve symmetry seldom shows the typical form of the asymmetry seen in mammalian fibres of CF >1 kHz. There are often differences between different frequency ranges, however, as in mammals (Manley, 1981; Manley et al., 1985; Turner, 1987). An explanation of these findings assumes an understanding of the mechanisms of frequency selectivity in each case.

2) Spontaneous activity in a number of nonmammals shows systematic deviations from a Poisson distribution (red-eared turtle, Crawford and Fetti-place, 1980; a lizard, Eatock and Manley, 1981; Manley, 1979; the starling, Manley and Gleich, 1984; Manley et al., 1985; the pigeon, Temchin, 1985). Low-CF (characteristic frequency) cells (seldom above CF 1.5 kHz) show preferred intervals in their spontaneous activity, intervals which are roughly the same as or multiples of the CF period. In the turtle, it has been shown that these intervals reflect the presence of oscillations of the hair-cell membrane potential, oscillations whose main energy lies near the cell's CF (Crawford and Fettiplace, 1981). These oscillations seem to be intimately related to an electrical filter mechanism in the properties of the ionic channels of the cell membrane. Such preferred intervals have not been reported in mammalian auditory-nerve fibres.

3) With the exception of basilar-papilla units in the amphibia (Moffat and Capranica, 1976), all nonmammalian primary auditory-nerve fibres investigated show a systematic temperature sensitivity (Gekko, Eatock and Manley, 1976, 1981; Caiman, Smolders and Klinke, 1984; pigeon, Schermuly and Klinke, 1985). Although the most obvious effect is a reversible rise of CF with a rise in temperature (at about 0.06 octaves/°C), the spontaneous activity may also be affected. In mammals, there is no temperature sensitivity over a non-lethal range of temperatures (Gummer and Klinke, 1983). The temperature sensitivity of nonmammalian fibres can be explained in terms of an effect on cell membrane channels involved in an electrical tuning mechanism (Eatock and Manley, 1981).

4

4)   The primary fibre discharge patterns in nonmammals are not always "primary-like" (Manley, 1981, Manley et al., 1985). For example, evidence is accumulating that many reptile and bird primary fibres display primary suppression, that is, the spontaneous activity is depressed by certain single tones (Gross and Anderson, 1976; Manley et al., 1985; Temchin, 1985). This effect is probably related to a firm connection between most hair cells in nonmammals and the tectorial membrane and the apparent absence of such a connection in mammalian inner hair cells.

Differences in activity patterns between mammalian and nonmammalian auditory-nerve fibres are traceable to features at two different levels. The first level consists of cellular and biochemical features (such as membrane ion channels) which are not amenable to normal anatomical analysis. At the second level are features which can be described in normal anatomical terms.

Morphological Substrate as a Potential Origin of Differences in Activity Patterns

Between mammals and nonmammals there are differences in the size of the papilla, which ranges from little over 100 μm up to a few cm, associated, of course with enormous differences in the total number of hair cells from about 100 to many thousands. Some papillae are even divided into two subpapillae. The hair cells may or may not be overlaid by a tectorial membrane, whose size and form can vary widely between species and even within one papilla. The hair cells in primitive reptilian papillae (e.g., turtles, Tuatara) all resemble each other, but in other groups of reptiles they are divided into two or more types recognizable by their location and their cytological features (Manley, 1981; Miller and Beck, 1988). Lizards show a division of the two hair-cell types between the two different frequency areas. The low-CF area is covered by unidirectional-type hair cells, the high-CF area by bidirectional-type hair cells (Miller and Beck, 1988). In the Archosaurs (Crocodilia and birds) and mammals, two or more hair-cell types are generally found in all frequency areas.

Additionally, the innervational pattern of the hair cells plays an important role. Recent data indicates that there are systematic innervational differences between different hair-cell types, and between species in reptiles, birds and mammals. It is apparent in mammals that only the inner hair-cell population plays a significant role in stimulus transformation and transmission of information to the brain (Liberman, 1982).

In view of these differences, one might well be surprised that superficially, the response patterns in auditory-nerve fibres of different vertebrates are so similar. As outlined above, however, it should be remembered that many similarities are based on common hair-cell properties. In addition, the possibility of convergent or parallel evolution, this is, the independent evolution of a similar solution to the same problem in different groups, should be kept in mind. Some of our recent data from reptiles and birds indicate that there are greater resemblances between activity patterns which can be observed in mammals and nonmammals than some of us had anticipated. The presence of otoacoustic emissions in nonmammals has already been mentioned above. Here, we will briefly discuss three further recent findings from our research groups, which indicate unexpected similarities in (1) the response patterns of different hair-cell populations, in (2) the mechanical resonance properties which contribute to tuning and in (3) the tonotopic distribution on the papillae of mammals and nonmammals.

Functional Differentiation of Hair-Cell Populations in Birds

The basilar papilla of birds contains a continuum of hair-cell forms, from the columnar tall hair cells on the neural edge to the bowl-shaped short hair cells on the abneural edge. The actual distribution of the hair cells (Smith, 1985) and of various aspects of hair-cell morphology (Gleich and Manley, 1988) are species-specific. For descriptive purposes, different hair-cell types (tall, intermediate, short and lenticular) have been defined (Smith, 1985; Takasaka and Smith, 1971). In the related Crocodilia, the tall and short hair-cells are much more sharply differentiated and easily distinguishable from one another (Leake, 1977).

The tall hair cells are the less specialized of the two. This fact, the pattern of innervation and the fact that the tall hair cells are found on the neural (inner) side of the papilla has led to the supposition that the tall and short hair cells are directly comparable to the inner and outer hair cells of mammals, respectively. It should, however, be kept in mind that mammals and birds are derived from different groups of ancestral reptiles and have a very long history of separate evolution.

The ancestors of the different groups of modern reptiles, of birds and of mammals diverged from each other during permiantriassic times (about 200 million years ago). Reptiles regarded as having primitive ears (e.g., Tuatara, turtles) do not show a division into more than one hair-cell population. Assuming this represents the primitive (triassic) condition, it is unlikely that the reptiles ancestral to birds and mammals already possessed two equivalent hair-cell populations. We are thus forced to conclude that the mammals and birds have evolved neural and abneural hair-cell groups independently. We will describe evidence that there are not only anatomical parallels, but also at least one physiological function common to tall hair cells of birds and inner hair cells of mammals. This is a remarkable case of parallel evolution.

In studies of the tonotopic organization of the basilar papilla and its ontogeny in birds, we have used both HRP and cobalt staining techniques to label physiologically characterized primary nerve fibres in the starling and the chick (Gleich, in prep.;Manley et al., 1987a). With only two exceptions, all characterized, unambiguously-labelled fibres made contact with tall hair cells (usually only one cell, Fig. 1a, b). Two broadly-tuned, insensitive cells in the starling with CF near 100 Hz innervated abneurally-lying cells (Gleich, in prep.) in an area described in the pigeon by Klinke and Schermuly (1986) as responding to infrasound. Thus, as in the cat (Liberman, 1982), only the neurally-lying (tall) hair cells of birds seem to transmit information to the brain. These are the first physiological data which support the working hypothesis that a functional equivalence exists between hair-cell populations of birds and mammals.

A Simple Mechanical Resonance Phenomenon as the Basis of Sharp Tuning in the Auditory Nerve of the Bobtail Lizard

An interactive (hair cell - tectorial membrane) mechanical resonance phenomenon has been suggested to underly the sharp tuning of mammalian nerve fibres (e.g., Neely and Kim, 1983; Strelioff and Flock, 1984; Zwislocki, 1979, 1985). Our study of the tuning properties of the basilar membrane and of nerve fibres in the bobtail lizard suggested that a similar resonance could be involved (Manley et al., 1987c).

The basilar papilla of this skink is about 2mm long and contains almost 2000 hair cells. It is divided into a shorter apical area, where the hair cells are covered by a huge, helmet-like tectorial structure

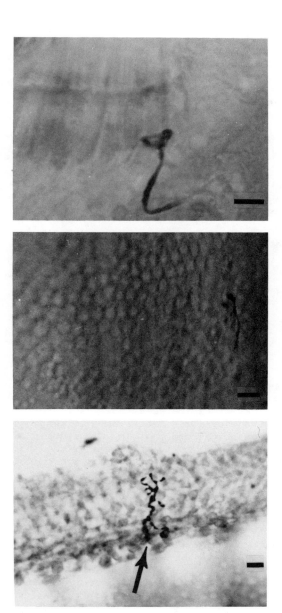

Fig. 1. a) HRP-stained primary nerve fibre in a cross-section of the
papilla of a young chick. This fibre appears to make a cup-shaped
synapse on a single tall hair cell. b) A primary fibre in the
starling papilla seen in surface view. This fibre was stained
with cobalt and makes a button-shaped synapse on a hair cell on
the neural side of the papilla. c) An auditory-nerve fibre of
the bobtail lizard, stained with cobalt during the recording
and seen in a wholemount of the papilla. The arrow indicates
the fibre's point of entry into the papilla. In sectioned
material, this fibre was seen to synapse with at least five
different hair cells. The calibration bar is 10 ʌm in all
cases.

and a long basal area, where the hair cells are covered by a delicate chain of individual "sallets".

These sallets each cover a strip of hair cells running across the papilla and are connected to each other by a thin strip of tectorial material running along the midline of the papilla (Köppl, in preparation).

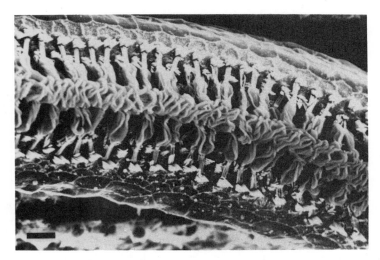

Fig. 2. Scanning-electron-microscope photograph of a small portion of the upper surface of the basal area of the papilla of the bobtail skink. It can be seen that the tectorial membrane consists of a chain of sallets, each of which joins a series of hair cells across the papilla. The bundles of marginal hair cells are distorted due to shrinkage of the tectorial structures caused by the histological procedure. The sallets are themselves linked together by a rope-like tectorial structure. The calibration bar is 10 μm.

Unlike the tuning of the basilar membrane, the tuning of primary nerve fibres was strictly place-specific, revealing a clear tonotopic organization. Our measurements also showed that at one and the same location in the same papilla, the neural tuning was substantially sharper than that of the basilar membrane, each neural curve having a sensitive tip up to about 40 dB deep. Subtracting the basilar-membrane tuning from the neural tuning in each case revealed a tuning component subsequent to the basilar-membrane motion which resembled a sharply-tuned high-pass filter.

Our model of this tuning assumed coupling between neighbouring, sharply-tuned elements each represented as a simple, sharply-tuned high-pass filter. Adding the modelled resonance curves to a typical basilar-membrane function produced a family of tuning curves for various CFs which bears a close resemblance to those of the measured neural elements. It was suggested that the sharp tuning in the basal area of the bobtail skink papilla is the result of the resonance of local groups of hair cells (stereovillar bundles) and their associated tectorial sallet. The shapes of the tuning curves suggest some form of interaction between adjacent sallet groups, since they are simultaneously relatively broad and deep. This interaction could be a mechanical one between adjacent sallets, or it could be due to the parallel connection of several nearby hair cells to one nerve fibre, or be due to both factors

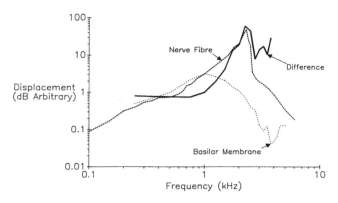

Fig. 3. A comparison of basilar-membrane (dotted line) and single nerve-
fibre (dashed line) tuning measured for the same location in one
papilla of the bobtail lizard. The neural tuning curve has been
inverted and its placement on the y axis was made using the best
match for the low-frequency flanks of the basilar-membrane and
neural curves. The continuous line represents the calculated
difference curve, which strongly resembles a high-pass filter
characteristic.

(Manley et al., 1987c). Labelled single auditory fibres in this species
innervated more than five or six neighbouring hair cells. It could be
that the sallets are present as semi-independent tectorial units in
order to increase the isolation of the mechanical resonances of neighbouring elements. In this way, sharp tuning can be achieved over a wider
frequency range in a short papilla.

<u>Similarities in the Tonotopic Pattern in the Auditory Papillae in a
Reptile, a Bird and a Mammal</u>

In discussions of tonotopic organization in unspecialized mammals,
it is often assumed that octaves are distributed linearly on the receptor
mosaic. As the amount of space devoted to each octave is roughly
equivalent to the number of receptors responsible for that octave,
variations in the amount of space per octave in specialized mammals has
been taken as an indicator of the relative "importance" of that octave
to the animal. Thus it seems obvious that the "acoustic fovea" of some
bats indicates that a particular frequency region is of special importance
(Vater et al., 1985). Since we can no longer assume that the frequency
analysis mechanisms are the same in all frequency ranges and in all
animal groups, however, the assumption that the amount of space is a
reflection of "importance" may not always be true. If the analysis
mechanisms in different cases are not known, then neither are the
constraints which determine the amount of space devoted to each part
of the frequency range.

A comparison of the frequency distribution on the hearing organs
of the bobtail lizard (Manley and Köppl, in prep.), chick (Manley et al.,
1987a) and cat (Libermann, 1982) revealed that in all three animals,
much less space is devoted to octaves below about 1 kHz than to octaves
above this. The question arises as to whether this means that the lower
frequencies are less "important" to all these animals.

The frequency map in the bobtail was studied both by direct recording of fibres as they emanated from the papilla and by tracing single
fibres labelled with cobalt (Fig. 1c; Köppl and Gleich, 1987). In lizards,
there is a discontinuity between a low-CF hair-cell area (where the hair

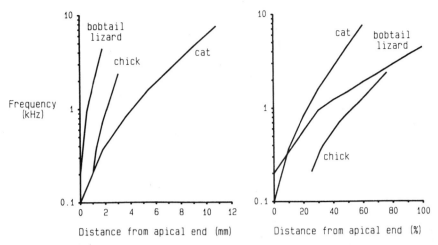

Fig. 4. A comparison of the tonotopic organization of the hearing organs
of the cat (after Liberman, 1982), chick and bobtail lizard.
The left half of the figure shows the distribution of characteristic frequencies in terms of the absolute distance of the
innervated cells from the apical end of the organs. On the right,
the values are given as a percentage of each papillar length,
in order to make the comparison easier. Less space is devoted
to octaves below about 1 kHz in all three cases.

cells are oriented in the same direction) and one or more bidirectionally-
oriented areas with CFs above about 1 kHz. More space is devoted to each
octave in the higher-CF area than in the low-CF area. The fact that in
mammals and in the chick this general pattern is preserved suggests that
the division in reptiles is not coincidental, but perhaps correlates
with a change in frequency selectivity mechanisms towards higher
frequencies.

In nonmammals there are two mechanisms of frequency tuning in which
an individual cell is the functional unit and does not greatly influence
its neighbours. At lower frequencies, there is evidence for an electrical
tuning mechanism. Depending on the required resolution, such a mechanism
does not require a large number of receptor cells spread over a large
area. At higher CFs, some lizards have a mechanical selectivity mechanism
in which individual cells can function independently of their neighbours
because they are not attached to them via a tectorial membrane (i.e., the
so-called "free-standing" hair cells of iguanid-agamid-anguid lizards,
e.g., Weiss et al., 1976).

Where a tectorial membrane is present in an area with CF above 1 kHz,
the lateral linking of the hair cells may lead to a reduced selectivity
of tuning, unless more space is devoted to each octave. In the latter
case, neighbouring cells would have a very similar CF. During the evolution of the some reptile groups, birds and mammals, there was a strong
tendency to devote more space to these octaves (Manley, 1986).

What about the low-CF area in mammals? In spite of the apparent
absence of electrical tuning in mammals and the fact that a tectorial
membrane covers all frequency areas, the cat still devotes less space
to lower CF regions than to higher. Perhaps the large brain of mammals
(and birds?) is able to use other information in the stimulus at low
frequencies (e.g. time information from phase-locked responses). This
could function with fewer labelled lines and reduce the necessity for a

large physical separation of frequencies in the low-CF periphery. The resemblance between the tonotopic distributions in these three groups may thus not necessarily point to common mechanisms.

REFERENCES

Crawford, A. C. and Fettiplace, R. 1980, The frequency selectivity of auditory nerve fibres and hair cells in the cochlea of the turtle, J. Physiol., 306:79-125.

Crawford, A. C. and Fettiplace, R., 1981, An electrical tuning mechanism in turtle cochlear hair cells, J. Physiol., 312:377-412.

Eatock, R. A. and Manley, G. A., 1976, Temperature effects on single auditory nerve fiber responses, J. Acoust. Soc. Amer., 60:S80.

Eatock, R. A. and Manley, G. A., 1981, Auditory nerve fibre activity in the tokay gecko: II, temperature effect on tuning, J. Comp. Physiol. A, 142:219-226.

Gleich, O. and Manley, G. A., 1988, Quantitative morphological analysis of the sensory epithelium of the starling and pigeon basilar papilla, (Submitted).

Gross, N. B. and Anderson, D. J., 1976, Single unit responses recorded from the first order neuron of the pigeon auditory system, Brain Res., 101:209-222.

Gummer, A. W. and Klinke, R., 1983, Influence of temperature on tuning of primary-like units in the guinea pig cochlear nucleus, Hearing Res., 12:367-380.

Klinke, R. and Schermuly, L., 1986, Inner ear mechanics of the crocodilian and avian basilar papillae in comparison to neuronal data, Hearing Res., 22:183-184.

Klinke, R. and Smolders, J., 1984, Hearing mechanisms in caiman and pigeon, in: "Comparative Physiology of Sensory Systems", L. Bolis, R. D. Keynes, S. H. P. Maddrell, eds., Cambridge Univ. Press, 195-211.

Köppl, C. and Gleich, O., 1987, Cobalt labelling of single primary auditory neurones: an alternative to HRP, (Submitted).

Leake, P. A., 1977, SEM observations of the cochlear duct in Caiman crocodilus, Scan. Electr. Micros., 2:437-444.

Lieberman, M. C., 1982, The cochlear frequency map for the cat: Labeling auditory-nerve fibers of known characteristic frequency, J. Acoust. Soc. Amer., 72:1441-1449.

Manley, G. A., 1979, Preferred intervals in the spontaneous activity of primary auditory neurones, Naturwiss. 66:582.

Manley, G. A., 1981, A review of the auditory physiology of the reptiles, Progr. Sens. Physiol., 2:49-134.

Manley, G. A., 1986, The evolution of the mechanisms of frequency selectivity in vertebrates, in: "Auditory Frequency Selectivity", B. C. J. Moore, R. D. Patterson, eds., Plenum Press, New York, London, 63-72.

Manley, G. A. and Gleich, O., 1984, Avian primary auditory neurones: the relationship between characteristic frequency and preferred intervals, Naturwissenschaften, 71:592-594.

Manley, G. A., Gleich, O., Leppelsack, H.-J. and Oeckinghaus, H., 1985, Activity patterns of cochlear ganglion neurones in the starling, J. Comp. Physiol. A., 157:161-181.

Manley, G. A., Brix, J. and Kaiser, A., 1987a, Developmental stability of the tonotopic organization of the chick's basilar papilla, Science, 237:655-656.

Manley, G. A., Schulze, M. and Oeckinghaus, H., 1987b, Otoacoustic emissions in a song bird, Hearing Res., 26:257-266.

Manley, G. A., Yates, G. and Köppl, C., 1987c, Auditory peripheral tuning: evidence for a simple resonance phenomenon in the lizard Tiliqua, (Submitted)

Miller, M. R. and Beck, J., 1988, Auditory hair cell innervational patterns in lizards, (Submitted).

Moffat, A. J. M. and Capranica, R. R., 1976, Effects of temperature on the response properties of auditory nerve fibers in the american toad (Bufo americanus), J. Acoust. Soc. Amer., 60:S580.

Neely, S. T. and Kim, D. O., 1983, An active cochlear model showing sharp tuning and high sensitivity, Hearing Res., 9:123-130.

Palmer, A. and Wilson, J. P., 1981, Spontaneous evoked emissions in the frog Rana esculenta, J. Physiol., 324:66P.

Sachs, M. B., Young, E. D. and Lewis, R. H., 1974, Discharge patterns of single fibers in the pigeon auditory nerve, Brain Res., 70:431-447.

Schermuly, L. and Klinke, R., 1985, Change of characteristic frequencies of pigeon primary auditory afferents with temperature, J. Comp. Physiol. A., 156:209-211.

Smolders, J. W. T. and Klinke, R., 1984, Effects of temperature on the properties of primary auditory fibres of the spectacled caiman, Caiman crocodilus (L), J. Comp. Physiol., 155:19-30.

Smith, C. A., 1985, Inner ear, in: "Form and Function in Birds", A. S. King, J. McLeland, eds., Vol 3. Academic Press, London, 273-310.

Strelioff, D. and Flock, A., 1984, Stiffness of sensory-cell hair bundles in the isolated guinea pig cochlea, Hearing Res., 15:19-28.

Takasaka, T. and Smith, C. A., 1971, The structure and innervation of the pigeon's basilar papilla, J. Ultrastruct. Res., 35:20-65.

Temchin, A. N., 1985, Acoustical reception in birds, in: "Acta XVIII Congressus Internat Ornithol.", V. D. Ilyichev, V. M. Gavrilov, eds., Vol. 1 Moskow Nauka, 275-282.

Turner, R. G., 1987, Neural tuning in the granite spiny lizard, Hearing Res., 26:287-299.

Vater, M., Feng, A. S. and Betz, M., 1985, An HRP-study of the frequency-place map of the horseshoe-bat cochlea: morphological correlates of the sharp tuning to a narrow frequency band, J. Comp. Physiol.A., 157:671-686.

Weiss, T. F., Mulroy, M. J., Turner, R. G. and Pike, C. L., 1976, Tuning of single fibres in the cochlear nerve of the alligator lizard: relation to receptor morphology, Brain Res. 115:71-90.

Zwislocki, J. J., 1979, Tectorial membrane: a possible sharpening effect on the frequency analysis in the cochlea, Acta Otolaryngol., 87:267-269.

Zwislocki, J. J., 1985, Are nonlinearities observed in firing rates of auditory-nerve afferents reflections of a nonlinear coupling between the tectorial membrane and the Organ of Corti?, Hearing Res., 22:217-221.

TRANSFER FUNCTION OF THE OUTER EAR OF THE GUINEA PIG MEASURED

BY FFT AND FFT SCAN ANALYSIS OF COCHLEAR MICROPHONIC RESPONSE

J. W. Schäfer and W. H. Fischer

Hals-Nasen-Ohren-Klinik
Bereichslabor Oberer Eselsberg, Universität Ulm
D-7900 Ulm, FRG

INTRODUCTION

Recent advances in neurophysiological research have demonstrated the important role of the outer ear for sound localization not only for echolocating bats. Knowledge of the influence of torso, head and pinna on the acoustic event is essential in investigations of the response of auditory neurons using a sound source under free-field conditions. Measurements of the acoustic properties of the outer ear in humans have been made in the seventies by Blauert (1974) who introduced digital signal processing into this field and developed the "impulse technique" which is now a standard procedure for measuring transfer functions of the outer ear in humans. In this study two different transfer functions of the outer ear as defined by Blauert were measured in the guinea pig: "free field transfer function": transfer function between the sound pressure at a measuring point in the ear canal of the subject and the sound pressure which can be measured at the same point in space when the object is removed. "monaural transfer function": transfer function between the sound pressure at a measuring point in the ear canal with the sound source at any given direction and distance referred to the sound pressure at the same measuring point with the sound source at zero degree azimuth and elevation to (=in front of) the subject.

We present our results of cochlear microphonic (CM) measurements in the frequency range of 5 to 20 kHz and compare them to measurements with a probe tube microphone using Blauert's impulse technique.

MATERIALS AND METHODS

All measurements were performed in a sound-shielded chamber (IAC). The animal's head was placed in the center of a hoop on which the loudspeaker (Technics EAS-10TH800A) was mounted. The distance between loudspeaker and animal was 38 cm. The loudspeaker was moved in 30° steps in the horizontal (azimuth) and vertical (elevation) plane. Six pigmented guinea pigs were anaesthetized (Ketanest/Rompun) and cochlear microphonics were recorded with a silver ball electrode placed from a retro-auricular approach at the round window. CMs were high-pass filtered, amplified and displayed with conventional techniques. The left ear was measured. For the probe tube microphone measurements a Brüel and Kjaer

4133 microphone with a short (20 mm, inner diameter 1.25 mm) tube was used. Clicks of 100, 50, 25 and 15 µs duration (Grass S-4), a sinusoidal signal swept from 5 to 30 kHz (Hewlett Packard 3314A) in 600 ms and a species-specific sound recorded on a FM-recorder (Racal 4 DS) were used for stimulation. The species-specific sound consists of a fundamental frequency rising from 1.5 to 3.3 kHz and 7 harmonically related frequency bands. Sound pressure levels for click measurements ranged from 38 to 86 dB (impulse) and was 74 dB for the swept sinusoidal signal and the species-specific sound.

Recorded CMs were sampled, read into a memory and a 1,024 point FFT was calculated. For analysis of the clicks several responses were averaged and a flat window was used. For scan analysis a Hanning window was moved along the record. Frequency resolution was 50 Hz. Read-out for the plots presented here was in 500 Hz steps, the presumed frequency discrimination ability for guinea pigs in the frequency range analyzed (Heffner et al., 1971). The results were read into a micro computer for calculation of the transfer function. The frequency range was restricted from 5 to 20 kHz because our experiments with the probe tube microphone did not yield direction dependent response below 5 kHz in the guinea pig which is in agreement with Palmer and King (1985).

RESULTS

The purpose of this study was to reveal frequency dependent sound pressure changes of a signal occuring when the direction of the sound source varies in a systematic manner. According to Blauert (1974) transfer functions of the outer ear of different individuals must not be averaged since averaging results in undue smoothing of characteristics. For this reason transfer functions of only one animal are shown in the figures.

Clicks. Independent of sound pressure level round window CM response to clicks could not be used for calculating transfer functions. In all registrations high frequencies were attenuated compared to measurements with the probe tube microphone and to other signals used for CM recording.

Sweep. Figure 1A shows our results of the monaural transfer function at 0° elevation when the sound source is moved in 30° steps in the horizontal plane. In all spectra at lateral sound directions there is a dip in amplitude at 10 to 12.5 kHz followed by a rise to a peak at 13 to 13.5 kHz with a gradual slope to higher frequencies. Below 10 kHz the amplitude diminishes when the sound source is moved laterally. At 180º the dip and peak do occur but a comb-like pattern is seen at frequencies above 13 kHz. At 30° elevation (Fig.1C) the dip at 10 to 12.5 kHz and the gradual slope to higher frequencies have disappeared. Exceptions are the angles 150° and 180° where intensity is highest at 13 and 15 kHz. At 60° (Fig. 1D) elevation the peak at 12 to 13 kHz is followed by a slope to about 16 kHz.

Moving the sound source to the contralateral side results in attenuation of the amplitude (Fig. 1B). At 30° and 60° elevation the shape of the curves is rather similar: a gradually rising amplitude with a slope above 13 kHz. The response at 60° elevation is rather uniform whereas at 30° elevation there are variations of up to 18 dB for the same frequency region.

Species-specific sound. To compare the filter function of the outer ear of the guinea pig on simple and complex (natural) sounds the transfer function was calculated from the peak amplitude of FO to F5 of the species-specific sound at 60° azimuth and 0° elevation of the sound source. The resulting transfer function is identical within $\pm$ 2.5 dB with the one calculated from the swept sinusoidal signal.

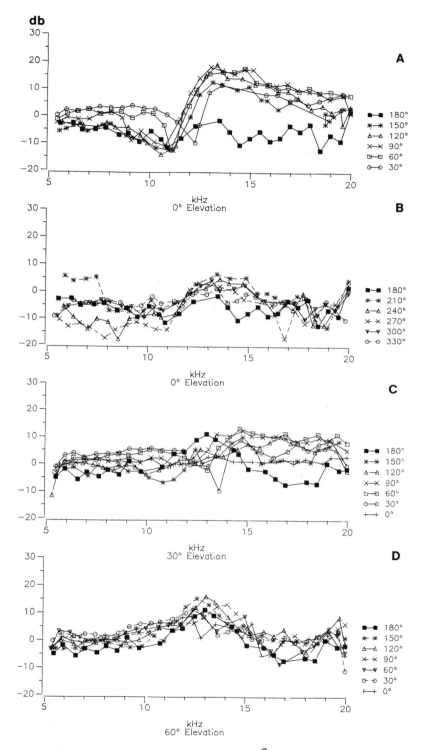

Fig. 1.   A. Monaural transfer function at 0° elevation with the sound
          source moved in 30° steps in the horizontal plane, ipsilateral
          side. Swept sinus signal, CM. B. Same as A, contralateral side.
          C. Same as A, 30° elevation. D. Same as A, 60° elevation.

CM response vs. probe tube microphone measurement. In our probe tube microphone measurements the difference between measurements with a 15 µs click and CM response to the swept sinusoidal signal described was less than ± 2 dB. This is far better than Blauert's results in the human outer ear (Blauert, 1974) and is probably due to the short ear canal in the guinea pig.

CONCLUSIONS

Probably due to its non-linearity CM response to short clicks recorded from the round window could not be used to measure the influence of the outer ear of the guinea pig on the spectral content of the signal. However, our measurements with a swept sinusoidal signal and a species-specific sound demonstrate that in guinea pig as well as in other animals the outer ear is responsible for dramatic changes in sound pressure at the eardrum which are dependent on the azimuth and elevation of the sound source. This is valid not only for simple but also for complex signals. Probably due to the short ear canal of the guinea pig differences between probe tube microphone and CM measurements of the monaural transfer function are small.

REFERENCES

Blauert, J., 1974, Raumliches Hören, S. Hirzel Verlag, Stuttgart.
Heffner, R., Heffner, H. and Masterton, B., 1971, Behavioral measurements of absolute and frequency-difference thresholds in guinea pig, J. Acoust. Soc. Amer., 49:1888-1895.
Palmer, A. R. and King, A. J., 1985, A monaural space map in the guinea pig superior colliculus, Hearing Res., 17:267-280.

ELASTIC PROPERTIES OF THE TECTORIAL MEMBRANE IN VIVO

J. J. Zwislocki, N. B. Slepecky, S. C. Chamberlain

and L. K. Cefaratti
Institute for Sensory Research
Syracuse University
Syracuse, NY 13244, USA

The subject of this paper concerns the role of the tectorial membrane in cochlear hair-cell stimulation. According to our measurements this role in the mammalian cochlea was misunderstood for the past three quarters of this century.

It must be clear by now that the cochlear hair cells are stimulated by the deflection of their stereocilia in the radial cochlear direction. Since the tips of the stereocilia are coupled to the tectorial membrane and the hair cells are held in the reticular lamina, the deflection is only possible through shear motion between the two structures.

According to the classical concept, first proposed by ter Kuile in 1900 and modified somewhat by H. Davis in 1956, the shear motion arises because the tectorial membrane and the reticular lamina move in cross section like two stiff beams rotating around two parallel but offset axes - the margin of the spiral limbus and the margin of the spiral lamina. As illustrated in Fig. 1, such a rotation leads to shear motion and deflection of the stereocilia. In panel A, the cochlear partition is at

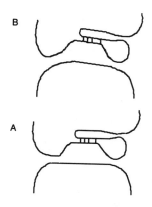

Fig. 1.   Schematic representation of the shear motion between the tectorial membrane and the reticular lamina according to the classical concept. A. Cochlear partition at rest; B. displaced toward scala vestibuli.

rest, and the stereocilia bundles are shown for simplicity as standing vertically on the reticular lamina. In panel B, the partition is displaced toward scala vestibuli, and the stereocilia bundles lean relative to the reticular lamina toward Hensen's cells - this is their excitatory position.

To deflect the stereocilia toward Hensen's cells, mechanical force is required. This force must be substantial since Flock (1977) and others have demonstrated that the stereocilia are stiff and stiffly anchored in the hair cells (Tilney et al., 1983). The force must act at a right angle to the stereocilia. It is anatomically clear that such a force can only be exerted by the tectorial membrane and must be tangential to the membrane in the radial cochlear direction. To exert it effectively, the tectorial membrane must be stiffer in this direction than the aggregate of the stereocilia acting together. Unless this condition is satisfied, the classical concept cannot be valid.

Because of the critical importance of the radial stiffness of the tectorial membrane, it is surprising that it was not measured in the past. The key importance of the measurement compelled us to undertake it in spite of anticipated difficulties, compounded by the sensitivity of the tectorial membrane to its chemical environment. To obtain chemical conditions as close to natural as possible, we decided to perform the measurement _in situ_ and _in vivo_.

Our Mongolian-gerbil preparation (Schmiedt and Zwislocki, 1977) appeared particularly suitable for the task. The angle between the basilar membrane and the tectorial membrane is wide in this animal, particularly in the second cochlear turn, as can be seen in Fig. 2. This makes it possible to access the tectorial membrane through the lateral wall of scala media without injuring Reissner's membrane. Such injury opens a flood gate for sodium ions which are known to have deleterious effects on the tectorial membrane. According to Tonndorf et al. (1962) and Kronester-Frei (1978), the tectorial membrane shrinks, lifts up from the organ of Corti, and its marginal zone curls up.

The lateral wall of the scala media was opened between Reissner's membrane and the vicinity of the basilar membrane, as indicated by the arrow and the solid line in Fig. 2. Special microsurgical techniques and instrumentation were developed for this purpose. The spiral ligament and stria vascularis were removed with great care, and intrusion of blood and sodium ions into scala media was minimized with the help of various manipulations, including rinsing with isotonic KCl solution.

In 30% of the total number of about 50 preparations, the tectorial membrane appeared to remain right on the organ of Corti, and no gap was evident between it and Hensen's cells. This is documented in Fig. 3 that shows the tectorial membrane as a wide dark band in the lower left part of the cochlear opening. The light, somewhat mottled band filling the upper right part of the opening is the mass of Hensen's cells. The cochlear apex is outside the picture to the lower left. The tectorial membrane was stained selectively with Alcian blue 8Gx in isotonic KCl solution. Janus green B, also in isotonic KCl solution, was used as a counterstain to make Hensen's cells visible. Since Alcian blue was found to affect the mechanical properties of the tectorial membrane, only Janus green in small concentrations (.03 to .1%), or no stain at all were used in subsequent experiments. In the absence of staining, the position of the tectorial membrane was marked with 9-um latex microspheres. No mechanical effects of light Janus-green staining were observed. This was not true for heavy staining, such as used by Békésy (1953) or Tonndorf

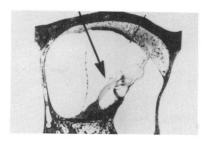

Fig. 2. Cross section of the second cochlear turn in the Mongolian gerbil. The cochlea was opened between the arrow and the straight line to the right of it. The arrow shows the location of one of the two micropipette placements used.

Fig. 3. View of the tectorial membrane (dark band between two light bands) through the opening in the lateral wall of the scala media. The wide light band is the mass of Hensen's cells.

Fig. 4. View of a micropipette inserted frontally through the marginal zone into the tectorial membrane and pulling it away from the organ of Corti. The tectorial membrane appears as a dark band indented by the micropipette to the left and separated from the dark band of Hensen's cells by the unstained band of the organ of Corti. The micropipette enters the picture from the top.

et al. (1962). In the majority of preparations, the tectorial membrane was lifted up only slightly. This did not noticeably affect its mechanical properties.

The stiffness of a tectorial membrane was measured by inserting into it a flexible micropipette and moving it laterally so as to produce a desired amount of deformation of the membrane. The force required for the deformation produced a bend in the micropipette. The micropipettes used in the measurements were calibrated on an extremely sensitive instrument developed especially for the purpose. The calibration allowed

19

us to convert the amount of bend of a micropipette into the force exerted on its tip by the tectorial membrane. Division of this force by the membrane's displacement gave us the membrane's stiffness. In all the measurements, the micropipette was inserted 30 μm deep. This is only slightly less than the thickness of the tectorial membrane at the experimental location, so that nearly all the layers of the tectorial membrane contributed to the measured stiffness.

A micropipette inserted into a tectorial membrane frontally, through the marginal zone, and pulling the membrane away from the organ of Corti is shown in Fig. 4. This direction of pull allowed us to measure the transversal stiffness of the membrane. A micropipette inserted through the top surface of the same membrane as perpendicularly as possible and pulling the membrane toward Hensen's cells is shown in Fig. 5. Note the exponential deformation of the membrane on either side of the micro-pipette. It indicates the presence of shear rather than bending forces. The same pattern observed on another tectorial membrane in the presence of lighter staining is shown in Fig. 6. The exponential lines of strain can be followed all the way to the spiral limbus.

Fig. 5. A micropipette inserted into the tectorial membrane through its top surface and indenting it toward Hensen's cells which appear as a wide, mottled band at the upper edge of the opening. Note the stretching of the membrane indicated by lighter appearing stain.

Fig. 6. Similar to Fig. 5 but with lighter staining. Note the exponential lines of strain within the tectorial membrane, extending all the way to the spiral limbus. They are particularly clear to the lower right of the maximum membrane indentation.

Since it was not possible to insert a micropipette entirely perpendicularly through the top surface of a tectorial membrane, the lateral movement of the micropipette caused the membrane to be displaced not only radially but also transversally. As a result, the micropipette was bent by a vector sum of the radial and transversal forces. To obtain the pure radial force, the transversal force was vectorially subtracted. Division of

the resulting radial force by the radial component of the displacement
gave us the radial stiffness of the membrane. From the exponential shape
of the membrane deformation and its space constant, which amounted to only
about 100 μm, the stiffness of the membrane per cm of its length was
calculated.

The following numerical results were obtained: 9000 dyne/cm^2 for the
transversal stiffness and only about 3000 dyne/cm^2 for the radial stiffness. For the corresponding stiffness of the stereocilia of the outer
hair cells, Flock's group (Strelioff et al., 1985) obtained 60000 dyne/cm^2
at the matching characteristic-frequency location for guinea-pig cochleas
- this is 20 times more. It is unlikely that the stiffness of the gerbil
stereocilia is smaller, and the in vitro technique used by the Flock
group could hardly increase the stiffness - the opposite would be expected.

It appears safe to conclude that the tectorial membrane has a much
smaller radial stiffness than the corresponding stiffness of the hair-cell
stereocilia. This is in direct contradiction of the classical concept of
cochlear hair-cell stimulation. The tectorial membrane cannot act as a
stiff anchor for the stereocilia, it can only provide a mass load.

ACKNOWLEDGMENT

Work supported by NIH Javits grant NS 03950.

REFERENCES

Békésy, G. v., 1953, Description of some mechanical properties of the
    organ of Corti, J. Acoust. Soc. Am., 25:770-785.
Davis, H., 1956, Initiation of nerve impulses in the cochlea and other
    mechano-receptors, in: "Physiological Triggers and Discontinuous
    Rate Processes", T. H. Bullock, ed., American Physiological Society,
    Washington, DC, 60-71.
Flock, A., 1977, Physiological properties of sensory hairs in the ear,
    in: "Psychophysics and Physiology of Hearing", E. F. Evans and
    J. P. Wilson, eds., Academic Press, London, 15-25.
Kronester-Frei, A., 1978, Sodium dependent shrinking properties of the
    tectorial membrane, in: "Scanning Electron Microscopy", Vol. II,
    Sem Inc., AMF O'Hare, Illinois 60666, 943-948.
Kuile, E. ter, 1900, Die Uebertragung der Energie von der Grundmembran
    auf die Haarzellen, Pflueg. Arch. ges. Physiol., 79:146-157. Cit.
    Békésy (1960).
Schmiedt, R. A. and Zwislocki, J. J., 1977, Comparison of sound-trans-
    mission and cochlear-microphonic characteristics in Mongolian
    gerbil and guinea pig, J. Acoust. Soc. Am., 61:133-149.
Strelioff, D., Flock, A., and Minser, K. E., 1985, Role of inner and
    outer hair cells in mechanical frequency selectivity of the cochlea,
    Hearing Res., 18:169-175.
Tilney, L. G., Egelman, E. H., De Rosier, D. J., and Saunders, J. C.,
    1983, Actin filaments, stereocilia and hair cells of the bird
    cochlea. II. Packing of actin filaments in the stereocilia and in
    the cuticular plate and what happens to the organization when the
    stereocilia are bent, J. Cell. Biol., 96:822-834.
Tonndorf, J., Duvall, A. J., III, and Reneau, J. P., 1962, Permeability
    of intracochlear membranes to various vital stains, Ann. Otol.
    Rhinol. Laryngol., 71:801-842.

CYTOSKELETAL ORGANISATION IN THE APEX OF COCHLEAR HAIR CELLS

David N. Furness and Carole M. Hackney

Department of Communication and Neuroscience
University of Keele
Keele, Staffs. ST5 5BG England

The cytoskeleton is an essential component of cells, performing biomechanical functions ranging from support to motility and intracellular transport. Two major cytoskeletal proteins which occur ubiquitously in eukaryotic cells are actin and tubulin. Actin occurs with actin-binding and actin-bundling proteins, where it functions primarily to provide support or stiffening, and with myosin and tropomyosin, a combination often associated with motility, in both muscle and non-muscle cells. Tubulin occurs mainly in the form of microtubules which perform functions ranging from cell shape determination to motility in a wide variety of mammalian cells, as well as in cells in lower vertebrates and invertebrates (for reviews on the cytoskeleton see Stebbings and Hyams, 1979; Tucker, 1979).

Cochlear hair cells are sensory receptors which are structurally adapted to transduce mechanical stimuli into electrical events. The apical region and the stereocilia of hair cells have a complex cytoskeleton which may be involved in the transduction process (Hudspeth, 1983). Moreover, isolated outer hair cells (OHCs) have motile properties (Brownel et al., 1985, Zenner et al, 1985) some of which, though not all (Ashmore, 1987), have been attributed to interactions based on cytoskeletal components, in particular, actin and myosin (Drenkhahn et al., 1985).

The distributions of actin, actin-associated proteins, myosin and tropomyosin have been investigated in hair cells using immunocytochemistry (see for example Macartney et al, 1980; Flock et al, 1982; Drenkhahn et al., 1985; Slepecky and Chamberlain, 1985) and electron microscopy (e.g. Tilney et al, 1983). However, immunocytochemical techniques have so far been comparatively unsuccessful in determining the distribution of microtubules in hair cells (Slepecky and Chamberlain, 1985) although they have been observed in electron microscopic studies (e.g. Engström and Engström, 1978). We have therefore used transmission electron microscopy of serial sections to determine microtubule organisation in relation to other apical cytoskeletal structures. In this paper we shall concentrate on the shape of the cuticular plate and the distribution of microtubules within inner hair cells (IHCs), with some data from OHCs presented for comparison. In addition, some new observations have been made on the stereocilia, which represent a significant departure from the established view of their ultrastructure.

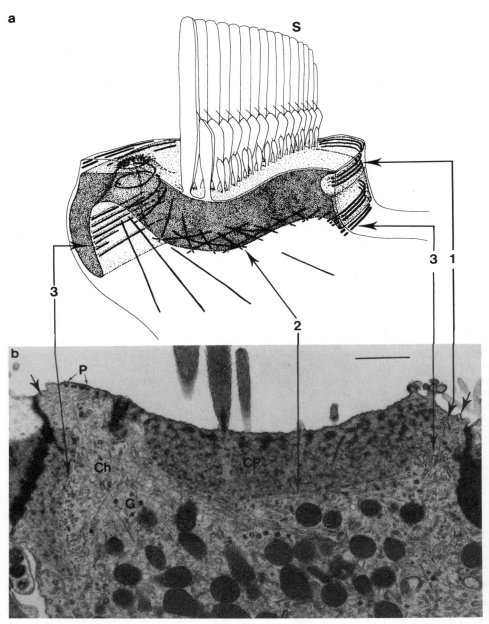

Fig. 1.   (a) Schematic diagram of an IHC apex showing the shape of the
cuticular plate and distribution of microtubules (dark lines).
The microtubular groups are numbered as in the text, the cuticular
plate is stippled and stereocilia shown (S). Not to scale. (b)
Radial section of an IHC apex showing the cuticular plate (CP)
with cytoplasmic belt (arrows) and channel (Ch), microtubular
groups numbered as in the text, dense granules (G), apical submembranous dense patches (P) and many apical vesicles. Note that
circumferential microtubules (e.g. (3)) are sectioned transverse-
ly in this plane so that they tent to appear as characteristic
circular profiles. Scale bar = 1 ∧um.

Cochleae from pigmented guinea pigs were fixed with 2.5% glutaraldehyde (in 0.1M sodium cacodylate and 2mM calcium chloride buffer) for
2h, and postfixed in 1% osmium tetroxide (in the same buffer) for 1h, then
dehydrated and embedded in Spurr resin as described previously (Furness
and Hackney, 1985). Serial sections were cut parallel to the reticular
lamina for three-dimensional (3-D) reconstruction of IHCs. All micro-
graphs from the series were printed at the same magnification and for each
section outlines of the cell, cuticular plate and positions of microtubules were traced onto an acetate sheet. The outlines were superimposed
serially to reveal the 3-D shape of the cuticular plate and the distribution of microtubules in the apex, as shown schematically in Fig. 1a. The
distribution was confirmed in other IHCs, in sections cut either in the
same plane or in a plane radially (perpendicular to the reticular lamina)
as in the hair cell shown in Fig. 1b, and compared in both section planes
with OHCs.

Inner Hair Cells

The upper surface of the cuticular plate is frequently grooved,
though sometimes flat, and lies adjacent to the apical plasma membrane
except for a belt of cytoplasm around the periphery of the apex. The belt
is widest on the side of the cell facing the strial wall of the scala
media, where it is continuous with a channel between the upper and lower
surface of the plate (Fig. 1a, b). The lower surface of the cuticular
plate follows the contours of the upper surface centrally but at the
edges is extended downwards, to the greatest extent on the strial side
though considerably less on the modiolar-facing side of the cell.

There are many microtubules in the cytoplasm surrounding the cuticular
plate, and following its contours. Three main groups can be identified
on the basis of position and orientation (Fig. 1a, b). These are: (1)
occupying the cytoplasmic belt and lining the channel, from where they
extend into the cell body; (2) below the central lower surface of the
plate; (3) in a ring just inside the lateral lower extensions of the
plate. A basal body is often found in the channel.

Other features of this area include many 70nm diameter dense granules
interspersed with the microtubules of the meshwork, and submembranous
electron dense patches located below the apical membrane in the channel
(Fig. 1b).

Outer Hair Cells

The cuticular plate of OHCs has some similarities with that of IHCs,
as it is partly surrounded by a cytoplasmic belt at its upper edges,
continuous with a channel through the plate on the strial side of the
area of the stereociliary rootlets (Fig. 2a). Moreover, the lower lateral
edges extend some way down on all sides of the cell. However, the central
lower surface is irregular and there is an extensive protrusion downward
into the cell body, as first described by Kimura (1975) and which may
correspond with the actin-containing feature described by Carlisle et al.,
(1987) (Fig. 2a). The protrusion contains many, highly-ordered microfilaments arranged longitudinally and in parallel, with associated
dense areas (Fig. 2b) and is particularly prominent in OHCs from the
cochlear apex.

Microtubules are found in orientations similar to those of IHCs,
although there are fewer in the apex of OHCs. They occur in the cytoplasmic belt and channel, and in a ring and central meshwork at the
underside of the main body of the plate. Many more microtubules are found
around the protrusion in the cell body, most of which are orientated

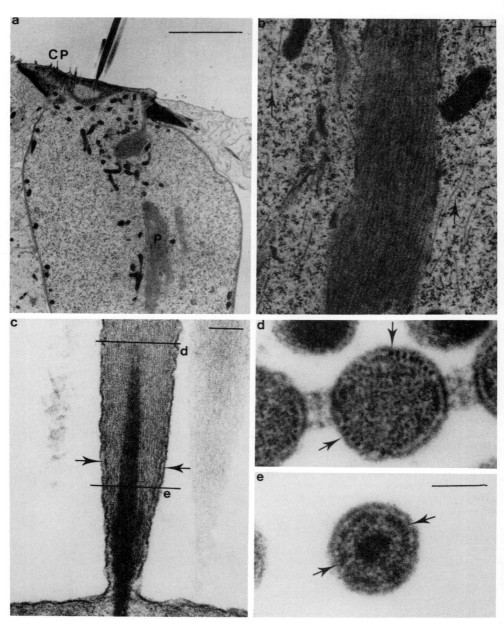

Fig. 2. (a) Radial section of an OHC apex showing cuticular plate (CP)
with the downward protrusion into the cell body (P). Scale bar =
5 um. (b) Detail of protrusion showing longitudinally orientated
parallel microfilaments (arrows) and dense material. Note the
large numbers of associated microtubules (arrowheads). Scale bar=
1 um. (c) LS of an OHC stereocilium showing the dense filament at
the edge of the actin filament bundle, extending distally from the
ankle region (arrows). Approximate section planes for 2d, e are
shown. Scale bar = 200 nm. (d, e) TS of OHC stereocilia in shaft
and ankle regions respectively. Note the dense band around the
actin bundles in the ankle region which can be resolved into
individual thick filaments more distally (arrows). Scale bar=100nm.

longitudinally, in directions similar to the microfilaments of the protrusion (Fig. 2b).

Functions of the Microtubular Systems

The microtubular organisations described here are found consistently in hair cells from different cochlear regions and from different animals. Their functions are not known, yet several possibilities are consistent with their roles in other cells (Stebbings and Hyams, 1979). They could provide structural or mechanical support at the apex, where the events associated with transduction are thought to occur. They may have a contractile or motile role, for example modifying the shape and distribution of apical structures. They may provide a means of intracellular transport, for example, of vesicles and organelles. Different groups of microtubules may have separate functions, or may be integrated to alter the biomechanical properties of hair cells and thus influence their function.

Stereociliary Ultrastructure: Some New Observations

A previously unreported feature has been observed in stereocilia (Fig. 2c-e). In transverse sections the actin filament bundle is enclosed by a peripheral dense zone in the ankle region (Fig. 2e). More distally, in the main shaft, individual profiles of 10-12 nm diameter filaments occur (Fig. 2d). In longitudinal sections (LS) dense strands occur at the edges of the actin bundle which appear to be correlated in position, staining intensity and dimensions with the profiles observed in TS and which extend from the ankle region some way along the shaft (Fig. 2c).

Thus there appears to be a ring of longitudinally-orientated filaments which differ markedly in appearance from the adjacent actin filaments and which encloses the actin filament bundle, particularly in the proximal part of the stereocilium. Similar structures appear to be present in both IHC and OHC stereocilia. These thicker filaments could represent a modified form of actin filament, for example with an associated protein such as tropomyosin which has been reported from the rootlet region (Flock, 1985; Slepecky and Chamberlain, 1985). An alternative possibility for the composition of these filaments is myosin. This latter protein has been detected in stereocilia by some authors, using immunofluorescence, (Macartney et al., 1980) although this report was considered controversial, on steric grounds, and explained as non-specific staining or a light piping effect (Drenkhahn et al., 1982). The observation of the thick filaments reopens the possibility that myosin is present in stereocilia, a suggestion with implications for the functions of stereociliary bundles in transduction and tuning processes. We are currently developing immunocytochemical techniques to determine the composition of these filaments.

ACKNOWLEDGEMENTS

This work was supported by the Wellcome Trust and the MRC.

REFERENCES

Ashmore, J. F., 1987, A fast motile response in guinea pig outer hair cells, J. Physiol., 388:323-347.
Brownell, W. E., Bader, C. R., Bertrand, D. and de Ribaupierre, Y., 1985, Evoked mechanical responses of isolated cochlear hair cells, Science, 227:194-196.
Carlisle, L., Thorne, P. R., Zajic, G., Schacht, J., and Altschuler, R. A., 1987, F-actin in the cochlea: differential distribution

between outer hair cells of different rows and turns, Abstracts of the 10th Midwinter Meeting of the ARO.

Drenkhahn, D., Kellner, J., Mannherz, H. G., Groschel-Steward, U., Kendrick-Jones, J. and Scholey, J., 1982, Absence of myosin-like immunoreactivity in stereocilia of cochlear hair cells, Nature, 300:531-532.

Drenkhahn, D., Schafer, T. and Prinz, M., 1985, Actin, myosin, and associated proteins in the vertebrate auditory and vestibular organs: immunocytochemical and biochemical studies, in: "Auditory Biochemistry", Drescher, D., ed., Charles C. Thomas, Springfield, Illinois.

Engström, H. and Engström, B., 1978, Structure of the hairs on cochlear sensory cells, Hearing Res., 1:49-66.

Flock, A., 1985, Contractile and structural proteins in the auditory organ, in: "Auditory Biochemistry", Drescher, D., ed., Charles C. Thomas, Springfield, Illinois.

Flock, A., Bretscher, A. and Weber, K., 1982, Immunohistochemical localisation of several cytoskeletal proteins in inner ear sensory and supporting cells, Hearing Res., 6:75-89.

Furness, D. N. and Hackney, C. M., 1985, Cross-links between stereocilia in the guinea pig cochlea, Hearing Res., 18:177-188.

Hudspeth, A. J., 1983, Mechanoelectrical transduction by hair cells in the acousticolateralis sensory system, Ann. Rev. Neurosci., 6:187-215.

Kimura, R. S., 1975, The ultrastructure of the organ of Corti, Int. Rev. Cytol., 42:173-222.

Macartney, J. C., Comis, S. D. and Pickles, J. D., 1980, Is myosin in the cochlea a basis for active motility? Nature, 288:491-492.

Slepecky, N. and Chamberlain, S. C., 1985, Immunoelectron-microscopic and immunofluorescent localisation of cytoskeletal and muscle-like contractile proteins in inner ear sensory hair cells, Hearing Res., 20:245-260.

Stebbings, H. and Hyams, J. S., 1979, Cell Motility, Longman, London.

Tilney, L. G., Egelman, E. H., De Rosier, D. J. and Saunders, J. C., 1983, Actin filaments, stereocilia and hair cells of the bird cochlea II. Packing of actin filaments in the stereocilia and in the cuticular plate and what happens to the organisation when the stereocilia are bent, J. Cell Biol., 96:822-834.

Tucker, J. B., 1979, Spatial organization of microtubules, in: "Microtubules", Roberts, K, Hyams, J. S., eds., Academic Press, London.

Zenner, H. P., Zimmerman, U. and Schmitt, U., 1985, Reversible contraction of isolated mammalian cochlear hair cells, Hearing Res., 18:127-133.

FURTHER INVESTIGATION ABOUT THE FUNCTION OF INNER

EAR MELANIN

A. M. Meyer zum Gottesberge and K. Schallreuter*

Dept. of Exp. Otology and Physiology
ENT-Clinic, University of Düsseldorf
*Dept. of Dermatology, University of Hamburg

INTRODUCTION

Clinical observation on various genetic combinations of sensory neural hearing impairments with pigmentation disorders (Schuknecht, 1974) as well as animal mutants (Schrott and Spoendlin, 1987; Steel et al., 1987) indicate a crucial position of the melanocytes onto the sensory areas. The audiological findings of minor susceptibility to noise exposure by pigmented animals and humans suggest an influence of melanin on inner ear physiology and pathophysiology (Conlee et al., 1986; Hood et al., 1976; Westerstrom, 1984; Tota and Bocci, 1967).

It is striking that melanin in the inner ear is located in well vascularized areas in close relation to the cells responsible for the maintenance of the unique ionic composition of the endolymph.

It has been suggested that the melanin pigments of the central nervous system and retinal pigment (Dräger, 1985) may play some active functional role in the tissues where they occur. Several investigators have postulated that melanin may have a cytoprotective function because of its capacity to trap (or prevent formation of) potentially toxic free radicals (Benedetto et al., 1981; Korytowski et al., 1985; Peters, 1986). Moreover, the melanin polymer has cationic exchange properties (Sealy et al., 1980) which may give it capacity to influence the micro-environment. Melanin is involved in Ca^{++} homeostasis in the inner ear (Meyer zu Gottesberge - Orsulakova and Kaufman, 1986) by mechanism not yet known.

The present study was designed to clarify:1) the transfer of calcium through the melanocyte membrane. The experiments were performed in vitro with or without Ca-ionophore A 12187. Enriched ^{44}Ca was used instead of naturally occuring ^{40}Ca and determined by laser induced mass spectrometry (LAMMA). 2) an enzyme involved in free radical intoxication - thioredoxin reductase - was examined in the inner ear tissue by electron spin resonance spectroscopy (ESR). These results will be published separately elsewhere.

MATERIAL AND METHODS

The inner ear tissue of 8-dark pigmented guinea pigs was dissected in oxygen - saturated 0.32 M sucrose (Negishi and Obiha, 1985). To exclude individual differences in quality of melanin (coat color) the tissue of each experimental animal was divided in two equal parts and incubated - in vitro - in modified Hank's medium (Gibco, Hank's without Mg^{++} and Ca^{++}) supplemented by 1.8 mM $^{44}CaCl_2$ and 0.89 mM $MgCl_2^+$, $MgSo_4$ of final concentration. Enriched $^{44}CaCl_2$ was used for this study (gift of Dr. Schröder, Jülich, FRG) with purity of 96.4% (Fain and Schröder, 1987). The tissue was incubated for 1, 5, 15, 30 min. in bathing medium with/without Ca-ionophore A 23187 (Sigma), pH 7.2 at 37°C. It was reported that this concentration causes an irreversible dispersion of melanophores (Negishi and Obiha, 1985). Additionally inner ear tissue of 4 animals was taken as a control without in vitro incubation.

Each sample of the inner ear tissue supported by a piece of cotton was shock-frozen by quick immersion in melting propane and then transferred for storage to liquid nitrogen.

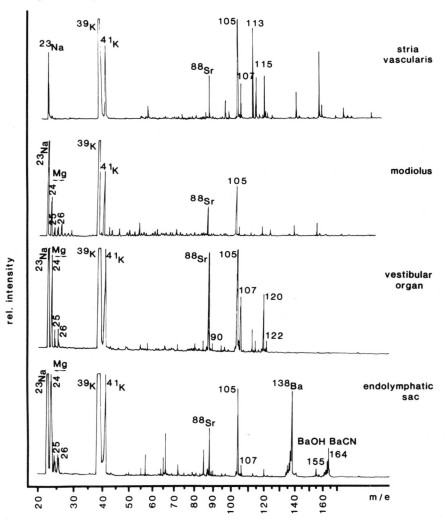

Fig. 1. Representative LAMMA spectra of melanin granules of the melanocytes in different locations of the labyrinth.

Freeze-drying procedure was performed in Balzer's freezedryer by warming the samples up to -110°C at 10^{-6} - 10^{-7} mbar. Semithin sections

were cut without floating liquid. Laser microprobe analyses were
performed with Laser Mass Microprobe Analyser - LAMMA - 500 (Leybold
Heraeus, Köln, FRG).

RESULTS AND DISCUSSION

LAMMA spectra obtained by measurements of melanin in different locations of the labyrinth indicate a variable mass distribution of ions and
organic fractions dependent on the location. Melanin mainly differs in
content of divalent ions such as Zn, Mg, Ca, Sr, Mn, Cd and Ba (measurements partly supplemented by x-ray analysis). The mass spectra of vestibular organ, modiolus, and endolymphatic sac melanin resemble each other
in their high amount of magnesium, one exception is the presence of
barium in the endolymphatic sac melanin. These spectra differ considerably
from the spectra of stria vascularis melanin (Fig. 1).

It has been suggested (Hilding and Ginzberg, 1977; Schrott and
Spoendlin, 1987) that the intermediate cells of the stria vascularis are
also melanocytes. However, the LAMMA measurements showed clear differences
in the ion content of the melanin in the functionally various parts of
the labyrinth. According to the hypothesis that melanin acts as a biolo-
gical reservoir of divalent ions (Meyer zum Gottesberge-Orsulakova, 1986),
melanin may have a specific function in each location.

Melanin in the inner ear may act as an ion exchanger (Meyer zum
Gottesberge-Orsulakova, 1986). In vivo (Meyer zum Gottesberge-Orsulakova
and Kaufmann, 1986) studies showed that melanin binds calcium and
simultaneously releases magnesium. Additionally an increase of Ca^{++} in
the melanocytes has at the cellular level several initiating effects
resulting in melanogenesis, in cell motility, and in endocytotic/exocytotic
activities (Meyer zum Gottesberge, 1988a). Therefore in series of in vitro
experiments we studied the transport of calcium through the melanocyte
and melanosome membrane.

We bathed the vestibular tissue in Hank's medium supplemented by
labelled ^{44}Ca in the presence of 5 μm Ca-ionophore or without. Only 1 min.
of exposure to 1,8 mM $^{44}Ca^{++}$ in the bathing medium leads to the presence
of ^{44}Ca in melanin granules independently of Ca-ionophore (Fig. 2). These
results indicate that the transmembrane movement of calcium is not necessary a "shuttle" type process in a form of voltage or receptor dependent
chanel (Peters, 1986). A transport directly through the plasma-lemmal
lipid bilayer (so called calcium leak) has to be considered. No time
dependent increase of calcium in melanosomes occurs. A process of adaptation of Ca-uptake by melanosomes takes place.

A high affinity of ^{45}Ca to the ocular pigment tissue of mice,
associated with a high concentration of melanin, was observed by Dräger
(1985). The binding of calcium to the melanin containing tissue was pH
dependent. This author suggests that the affinity of calcium to the
melanin might be ubiquitous. Our previous LAMMA measurements performed
on the pathological condition of experimental hydrops (Meyer zum Gottesberge and Kaufmann, 1986) and our present - in vitro - experiments with
^{44}Ca confirm this suggestion. However, the selective affinity of divalent
ions on melanin in the inner ear as well as in the ocular tissue (Meyer
zum Gottesberge, 1988b) may play an important role by binding calcium,
especially if the binding depends on ions such as Ba (Dräger, 1985).

It is documented that hypopigmentation is associated with defects
in the ocular system (Leventhal and Vitek, 1985; Creel et al., 1986) as
well as in the auditory system (Conlee et al., 1986). Misrouting of the
nerve fibres in the optical and acoustical pathway, especially in the

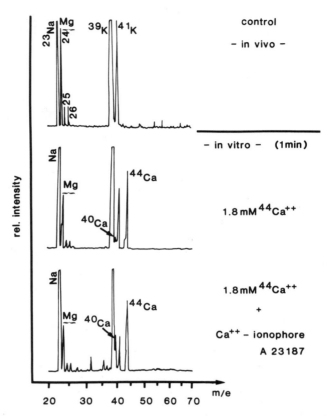

Fig. 2. Representative LAMMA spectra of melanin granules obtained after 1 min. of incubation of vestibular tissue in Hank's medium (1.8 mM $^{44}Ca^{++}$) with/without 5 μm Ca-ionophore A 23187

superior olive (Conlee et al., 1984), as well as poor development of the vascularisation pattern of the stria vascularis (Moorjani et al., 1986) have been described. Dräger (1985) discussed the possibility that melanin especially during synthesis may represent a temporary calcium sink, hence causing a lowering of calcium in adjacent cells during development. On the other hand a lack of such a kind of buffering system by hypopigmentation might be significant for receptor function and might influence the physiology of the inner ear, especially under pathological conditions such as experimental hydrops.

SUMMARY

Selective activity of divalent ions to inner ear melanin dependent on its location was determined by means of laser induced mass spectrometry (LAMMA). The same instrument was used to study the transport of calcium through the melanocyte membrane. The experiments were performed on utricular and ampullar wall under in vitro conditions with/without Ca-ionophore A 23187. Instead of natural ^{40}Ca an enriched ^{44}Ca was used. The transport of calcium across the membrane and the uptake by melanin takes place within one minute and seems to be independent of Ca-ionophore presence. The possible buffering capacity of melanin for calcium and its physiological relevance are discussed.

ACKNOWLEDGEMENTS

This research was supported by the Ministry of Science NRW (West-Germany) grant no. 2709325934086/32/84-254. The technical assistance of Mrs. S. Hertel is gratefully acknowledged.

REFERENCES

Benedetto, J.-P., Ortonne, J. P., Voulot, C., Khatchadourian, C., Prota, G. and Thivolet, J., 1981, Role of thiol compounds in mammalian melanin pigmentation: part I. Reduced and oxidized glutathione, J. of Investigative Dermatology, 77:402-405.
Conlee, J. W., Khader, J. Addul-Baqi, McCandless, G. A. and Creel, D. J., 1986, Differential susceptibility to noise-induced permanent threshold shift between albino and pigmented guinea pigs, Hearing Res., 23:81-91.
Conlee, J. W., Parks, T. N., Romero, C. and Creel, D. J., 1984, Auditory brainstem anomalies in albino cats: II. neuronal atrophy in the superior olive, J. Comp. Neurol., 225:141-148.
Creel, D. J., Bendel, C. M., Wiesner, G. L., Wirtschfter, J. D., Arthur, D. O. and King, R. A., 1986, Abnormalities of the central visual pathway in Prader-Willi syndrome associated with hypopigmentation, New England Journal of Medicine, 314:1606-1609.
Dräger, U. C., 1985, Calcium binding in pigmented and albino eyes, Proc. Natl. Acad Sci. USA, 82:6716-6720.
Fain, G. L. and Schröder, W. H., 1987, Calcium in dark-adapted toad rods: Evidence for pooling and cyclic-guanosine-3'-5'-monophosphate-dependent release, J. Physiol., (in press).
Hilding, D. A. and Ginzberg, R. D., 1977, Pigmentation of the stria vascularis, Acta Otolaryngol. (Stockh) 84:24-37.
Hood, J. D., Poole, P. J. and Freedman, L., 1976, Eye color and susceptibility to TTS, J. Acoust. Soc. Am., 59:706-707.
Korytowski, Hintz, W. R., Sealy, R. C. and Kalzanaraman, B., 1985, Mechanism of dismutation of superoxide produce during autooxidation of melanin pigments, Biochem. Biophys. Res. Comm., 131:659-665.
Leventhal, A. H. and Vitek, D. J., Abnormal visual pathways in normally pigmented cats that are heterozygous for albinism, Science, 229.
Moorjani, P. A., Glenn, N. and Steel, K. P., 1986, Strial vasculature in mice with cochlea-saccular abnormalities, XXIII Workshop on Inner Ear Biology, Berlin
Meyer zum Gottesberge-Orsulakova and Kaufmann, R., Is an imbalanced calcium-homeostasis responsible for the experimentally induced endolymphatic hydrops? Acta Otolaryngol. (Stockh) 102:92-98.
Meyer zum Gottesberge-Orsulakova, A. M., 1986, Melanin in the inner ear: micromorphological and microanalytical investigations, Acta histochemica, Suppl. Band XXXII:245-253.
Meyer zum Gottesberge, A. M., 1988a, Modulation of melanocytes during experimentally induced endolymphatic hydrops, in press.
Meyer zum Gottesberge, A. M., 1988b, Microanalytical investigations of the inner ear, uveal tract and retina pigment epithelium melanin, Adv. of Bioscience, in press.
Negishi, S. and Obiha, M., 1985, The role of calcium and magnesium on pigment translocation in melanophores of Oryzions latipes. Pigment Cell, 235-259.
Orsulakova, A., Morgenstern, C., Kaufmann, R. and D'Haese, M., 1982, The laser microprobe mass analysis technique in the studies of the inner ear, Scanning Electron Microsc., 4:1763-1766.
Peters, T., 1986, Calcium in physiological and pathological cell function, European Neurology, 25:Suppl. 1, 27-44.
Schallreuter, K. U., Pittelkow, M. R., Gleason, F. K. and Wood, J. M., 1986, The role of calcium in the regulation of free radical reduction

by thioredoxin reductase at the surface of the skin, <u>J. of Inorganic Biochemistry</u>, 28:227-238.

Schrott, A. and Spoendlin, H., 1987, Pigment anomaly-associad inner ear deafness, <u>Acta Otolaryngol.</u> (Stockh), 103:451-457.

Schuknecht, H. F., 1974, <u>in</u>: "Pathology in the inner ear".

Sealy, R. C., Felix, C. C., Hyde, J. S. and Swartz, H. M., 1980, Structure and reactivity of melanin: Influence of free radicals and metal ions, <u>Free Radicals in Biology</u>, Vol. IV: 209-259.

Steel, K. P., Barkway, C. and Bock, G. R., 1987, Strial dysfunction in mice with cochleo-saccular abnormalities, <u>Hearing Res.</u> 27:11-26.

Tota, G. and Bocci, G., 1967, Importance of the color of the iris in the evaluation of resistance of hearing to fatigue, <u>Riv. Otoneurooftalmol.</u>, 43:183-192.

Wästerström, S. A., 1984, Accumulation of drugs on inner ear melanin, <u>Scand. Audiol.</u>, Suppl. 23:1-40.

THE THROMBOXANE/PROSTACYCLIN BALANCE IN GUINEA PIG COCHLEA

A. Ernst, Ch. Taube*, P. Lotz and H.-J. Mest*

ENT Clinic and *Dept. Pharmacology and Toxicology
Martin Luther University
Halle/Saale, 4020, Leninallee 12, G. D. R.

INTRODUCTION

Thromboxane and prostacyclin as metabolites of arachidonic acid
are substances with opposite effects. While the first acts vasoconstric-
tively and is accumulated in impaired tissue, the latter is attributed
as cytoprotective, vasodilative substance. They represent the opposite
pole of the same homeostatic mechanism and their ratio reflects the
viability and responsiveness of the corresponding tissue to stimuli, such
as hypoxia, vascular impairment or metabolic changes (Oates and Fitz-
Gerals, 1985). The balance between thromboxane and prostacyclin can be
selectively influenced by drugs favouring the last (Deckmyn et al., 1983).
This is also possible in the vital cochlea where these substances can
cross the blood-labyrinth barrier (Ernst et al., 1987). Although the
involvement of prostanoids in drug ototoxicity affecting the hair cells
is hardly credible (Schacht, 1986), it remains tempting to speculate on
changes of eicosanoids linked to alterations of the non-sensory tissue
caused by either noise or e.g. loop diuretics (McFadden, 1986; Rybak et
al., 1986). It was demonstrated by Tachibana and Nishimura (1986) that
precursors of arachidonic acid are preferably incorporated in non-sensory
cochlear structures, i.e. stria vascularis and spiral ligament. These
findings correspond to what was found in the CNS (Keller et al., 1985).
However, in vitro studies of strial prostaglandins and leukotrienes could
not reveal any decisive influence of these substances on furosemide ototoxicity (Tran Ba Huy et al., 1987), but it is admitted that the in vitro
conditions might be in part responsible for the results. The determined
substances are mainly formed in response to stimuli in the physiological
situation and have very short half-lifes (thromboxane around 37 sec and
prostacyclin 3 min). The present study was aimed at investigating the
basal levels of the stable metabolites of thromboxane and prostacyclin
in the in vivo cochlea (perilymphatic perfusion experiments) as well as
in the cochlear lateral wall. The latter clearly differed from Escoubet
et al. (1985) since no blood wash-out and homogenizing were performed
(Fig. 1).

MATERIALS AND METHODS

Pigmented guinea pigs, with intact Preyer's reflex, weighing between
250-300 g were used. In a first group of 7 animals, under Ketamine
anaesthesia (50 mg/kg) the cochleas were perfused with a physiologically

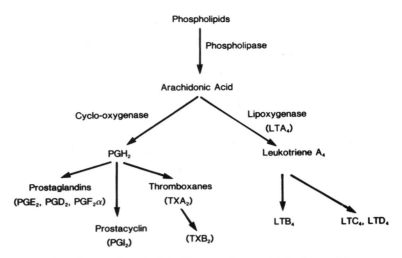

Fig. 1. Major metabolites of arachidonic acid

adequate, artificial perilymph. The inlet was in scala tympani, the outlet in scala vestibuli. The perfusate flow was about 3 μl/min. Subsequently, the samples were collected at 10-min intervals. The detailed arrangement is described by Nuttall et al. (1982) and Lotz et al. (1986). CM, HF, ECG, and rectal temperature were registered. The samples were analyzed due to TXB_2 as the stable metabolite of thromboxane and 6-keto-PGF_1 as the stable metabolite of prostacyclin (see below).

In a second group of 7 animals, the animals were decapitated, the bullae removed, opened, and their cochleas transferred to ice-cold Tyrode's solution. Under stereomicroscopical observation, the 2 LW of each animal were prepared and pooled. Until assay they were stored at -27° C. After thawing, they were incubated in a shaking water bath with 1 ml Tyrode's which has already been added before freezing. After centrifuging, the supernatant was used for RIA. The two metabolites TXB_2 and 6-keto were determined. The RIA procedure is described in detail by Taube et al. (1982). In the figures, the data are presented as mean values with SEM. Finally, the ratio TXB_2/6-keto was calculated in each experimental series.

RESULTS

Within the perfusate samples, a level of 20-22 pg TXB_2/100 μl perfusate and of about 18-20 pg 6-keto/100 μl perfusate was determined (Ernst et al., 1987). The time course is depicted in Fig. 2. The calculated ratio of TXB_2/6-keto varied around 1.0.

In tissue experiments, the observed levels showed a smaller inter-individual variability. The absolute values, given in pg/animal (i.e. for 2 LW) were as means 350 ($\pm$ 25) pg for TXB_2 and 380 ($\pm$ 45) pg for 6-keto-PGF_{1a}. Hence, the ratio TXB_2/6-keto amounted to 0.92.

DISCUSSION

It could be demonstrated that there exists a more or less inter-individually varying balance between thromboxane and prostacyclin in various compartments of the cochlea. The ratios are in the same order of magnitude though absolute levels are hardly comparable. Additionally, it was shown that this balance can be shifted by influencing the endogenous

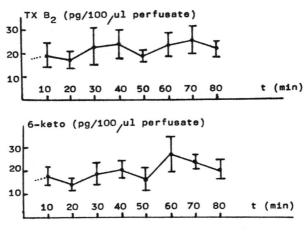

Fig. 2. Time course of the levels of TXB$_2$ and 6-keto-PGF$_1$ when perfussing the perilymphatic spaces of the guinea pig cochlea

balance without administering exogenous analogues (Ernst et al., 1987). The reported results in literature are based on a different methodology, including the blood wash-out before sacrifice, a homogenizing of the material, and a partial adding of substrate to the incubative medium (Escoubet et al., 1985; Tran Ba Huy et al., 1987). Thus, the attained levels can not easily be compared. In his furosemide experiments, Tran Ba Huy (1987) gave prior to measuring the EP which declined after the diuretic, either aspirin or indomethacin. Both drugs completely block the whole cyclo-oxygenase pathway, hence, abolishing both the synthesis of thromboxane and prostacyclin as well. Probably, a selective blocking of a target enzyme (e.g. TX synthase) might produce different synthesis results. However, infusing a synthetic prostacyclin analogue did not exert, too, a decisive influence on the furosemide-induced changes of EP. With respect to reports from the CNS where a changed thromboxane/prosta-cyclin balance càn counter-act against edema, enzyme release as a substrate of metabolic impairment (Renkawek et al., 1986; Schror, 1986), the structural aspect of a pharmacologically changed balance under conditions of noise or drugs remains to be elucidated.

Gratacap et al. (1985) and Forge (1986) demonstrated the strial changes after drug or noise exposure, e.g. edema, uncoupling of cells, mitochondrial alterations. Forge (1986) explains his aminoglycoside findings in accordance with Schacht (1986) by an interfering with poly-phosphoinositides in membranes. As arachidonic-acid precursors they are attributed as target of the observed OHC lesions, too (Schacht, 1986). One should bear this in mind when considering Rybak and Whitworth's suggestion (1987) that an organic-acid transport mechanism might be involved in loop diuretic-induced strial changes. The uncoupling of cells observed (Forge, 1986) and the subsequent altered ionic composition (Anniko and Wroblewski, 1986) favor the hypothesis of an eicosanoid involvement in a regulation of changed cellular functions (Majerus et al., 1986). This is valid as well for potassium as calcium ions (Ikeda et al., 1987 and Yoshihara and Igarashi, 1987). A changed calcium content in LW tissue would also be responsible for different mechanical properties of the basilar membrane linked to protein alterations of the spiral ligament (Henson et al., 1985).

Morphological investigations characterizing the strial vessels (Watanabe, 1986) outline the lack of nervous control which suggests that these capillaries might be the site of action for prostanoids. This becomes relevant considering the TTS caused by salicylates and the effects of noise or a combination of both (McFadden, 1986). Rybak et al. (1986) investigated the influence of salicylate on EP changes, contrasting the findings of Tran Ba Huy obtained under in vitro conditions (1987). The rapid changes of cochlear morphology and the very short half-lifes of arachidonic acid metabolites introduce the question whether the experimental findings - as conflicting as they are with respect to eicosanoids actually reflect the cochlea's response to any challenge. At these sites, i.e. the non-sensory structures, the cyclo-oxygenase derivatives are under discussion. This is of importance with respect to the so-called "blood-labyrinth barrier" (Juhn et al., 1985).

Recently, the metabolites of the lipoxygenase pathway have been documented in sensory structures. This alternative metabolic way contrasting the prostanoids is suggested to act as a second messenger in sensory structures (Beven and Wood, 1987). It was reported that they even might mimic the effect of other second-messenger systems such as cyclic AMP-dependent phosphorylation ones. These substances, mainly hydroxyeicosatetraenoic acid (HETEs and HPETEs), will attract further attention for their signalling effects. At least, they are also released after forming arachidonic acid, from phospholipids. This item will deserve our attention, too, since they are targets for aminoglycosides (Schacht, 1986). Additionally, for the first time it was evidenced that arachidonic acid metabolites might be involved in excitatory structures, too (Seregi et al., 1987; Bevans and Wood, 1987).

REFERENCES

Anniko, M. and Wroblewski, R., 1986, Ionic environment of cochlear hair cells, Hearing Res., 22:279-293.

Bevan, S. and Wood, J. N., 1987, Arachidonic-acid metabolites as second messengers, Nature (London), 328:20.

Deckmyn, H., van Houtte, E., Verstraete, M. and Vermylen, J., 1983, Manipulation of the local thromboxane and prostacyclin balance in vivo by the antithrombotic compounds Dazoxiben, Acetylsalicylic acid, and Nafazatrom, Biochem. Pharmacol., 32:2757-2762.

Escoubet, B., Amsallem, P., Ferrary, E. and Tran Ba Huy, P., 1985, Prostaglandin synthesis by the cochlea of the guinea pig, Prostaglandins, 19:589-599.

Ernst, A., Lotz, P. and Mest, H.-J., 1987, PGI_2/TXA_2 balance in guinea pig perilymph, in: "Cochlea Symposion 87", L.-P. Lobe and P. Lotz, eds., Halle University Press (in press).

Ernst, A., Taube, Ch., Lotz, P. and Mest, H.-J., 1988, Dazoxiben changes the thromboxane/prostacyclin balance in the lateral cochlear wall, Arch. Otorhinolaryngol. (in press).

McFadden, D., 1986, Some issues associated with interactions between ototoxic drugs and exposure to intense sound, in: "Basic and applied aspects of noise-induced hearing loss", R. J. Salvi, D. Henderson, R. P. Hamernik and V. Coletti, eds., Plenum, New York, 541-550.

Forge, A., 1986, The morphology of the normal and pathological cell membrane and junctional complexes of the cochlea, Ibid., 55-68.

Gratacap, B., Characon, R. and Stoebner, P., 1985, Results of an ultra-structural study comparing stria vascularis with organ of Corti in guinea pigs treated with Kanamycin, Acta Oto-laryngol., 99:339-342.

Henson, M. M., Burridge, K., Fitzpatrick, D., Jenkins, D. B., Pillsbury, H. C. and Henson jr., O. W., 1985, Immunocytochemical localization of contractile and contraction-associated proteins in the spiral ligament of the cochlea, Hearing Res., 20:207-214.

Ikeda, K., Kusakari, J., Takasaka, T. and Saito, Y., 1987, The Ca^{2+} activity of cochlear endolymph of the guinea pig and the effect of inhibitors, _Hearing Res._, 26:117-125.

Juhn, S. K., Rybak, L. P. and Jung, T. T. K., 1985, Transport characteristics of the blood-labyrinth barrier, _in_: "Auditory Biochemistry", D. G. Drescher and J. E. Medina, eds., Thomas, Springfield, 488-499.

Keller, M., Jackisch, R., Seregi, A. and Hertting, G., 1985, Comparison of prostanoid forming capacity of neuronal and a stroglial cells in primary cultures, _Neurochem. Int._, 7:655-665.

Lotz, P., Posse, D., Haberland, E.-J., Kuhl, K.-D. and Ernst, A., 1986, The metabolic reaction of the cochlea to unphysiological noise exposure, _Acta Oto-laryngol._, 102:20-26.

Majerus, P. W., Connolly, T. M., Deckmyn, H., Ross, T. S., Bross, E. E., Ishii, H., Bansal, V. S. and Wilson, D. B., 1986, The metabolism of phosphoinositide-derived messenger molecules, _Science_, 234:1519-1526.

Nuttall, A. L., LaRouere, M. J. and Lawrence, M., 1982, Acute perilymphatic perfusion of the guinea pig cochlea, _Hearing Res._, 6:207-221.

Oates, J. A., FitzGerals, G. A., 1985, Endogenous Biosynthesis of prostacyclin and thromboxane A_2 and the prevention of Vascular Occlusion in man, _in_: "Prostacyclin - Clinical Trials", R. J. Gryglowski, ed., Raven, New York, 115-123.

Renkawek, K., Herbaczynska-Cedro, K. and Mossakowski, M. J., 1986, The effect of prostacyclin on the morphological and enzymatic properties of CNS cultures exposed to anoxia, _Acta Neurol. Scand._, 73:111-118.

Rybak, L. P., Santiago, W. and Whitworth, C., 1986, An experimental study using sodium salicylate to reduce cochlear changes induced by furosemide, _Arch. Otorhinolaryngol._, 243:180-182.

Rybak, L. P. and Whitworth, C., 1987, Some organic acids attenuate the effects of furosemide on the endocochlear potentials, _Hearing Res._, 26:89-93.

Schacht, J., 1986, Molecular mechanisms of drug-induced hearing loss, _Hearing Res._, 22:297-304.

Schror, K., 1986, Prostaglandins and other fatty acid peroxidation products in cerebral ischemia, _in_: "Pharmacology of cerebral ischemia", J. Krieglstein, ed., Elsevier, Amsterdam, 199-209.

Seregi, A., Keller, M. and Hertting, G., 1987, Are cerebral prostanoids of astroglial origin?, _Brain Res._, 404:113-120.

Tachibana, M. and Nishimura, H., 1986, Incorporation of inositol in the cochlear tissues as determined by autoradiography, _Hearing Res._, 24:105-109.

Taube, Ch., Hoffmann, P. and Forster, W., 1982, Enhanced thromboxane production in the aorta of spontaneously hypertensive rats in vitro, _Prostagl. Leukotr. Med._,9:411-414.

Tran Ba Huy, P., Ferrary, E., Escoubet, B. and Sterkers, O., 1987, Strial Prostaglandins and leukotrienes, _Acta otolaryngol._, 103:558-567.

Watanabe, K., 1986, Ultrastructural characteristics of capillaries entering and leaving the stria vascularis, _Ann. Otol. Rhinol. Laryngol._, 95:309-312.

Yoshibara, T. and Igarashi, M., 1987, Cytochemical localization of Ca^{2+}-ATPase activity in the lateral cochlear wall of the guinea pig, _Arch. Otorhinolaryngol._, 243:395-400.

PERIPHERAL AUDITORY DEVELOPMENT IN THE CAT AND RAT

L. P. Rybak, E. Walsh, and C. Whitworth

Southern Illinois University School of Medicine

P.O. Box 19230, Springfield, Illinois 62794-9230

INTRODUCTION

The cat and the rat are altricious animals. They are well-suited for studying the development of hearing because their cochleae are still immature at birth. However, other investigators have reported differences in the age of onset for cochlear microphonic (CM) potentials and compound action potentials (AP) in these two species. In the cat, CM appears two days before birth (Romand, 1971), whereas the rat's onset of CM is reported to be 8-9 days after birth with the onset of AP occurring 11 days after birth (Crowley and Hepp-Raymond, 1966; Uziel et al., 1981).

The purpose of the present study was to compare the development of endocochlear potentials (EP) and AP's in kittens and rats and to determine the effects of the cochlear inhibitor, furosemide, on these potentials at various ages.

METHODS

Kittens and rat pups of various ages were anesthetized with ketamine HCl (30 mg/kg) followed by pentobarbital (30 mg/kg IM) and supplemented with smaller doses as needed. After anesthesia was achieved, a tracheostomy was performed. The ear canal was transected to allow direct delivery of the sound stimulus to the tympanic membrane. The bulla was opened and a microelectrode was inserted through the round window and advanced until the scala media was entered, producing a positive EP. AP was recorded simultaneously with a tungsten wire placed on the surface of the round window membrane. Alternate condensation and rarefaction clicks were presented through the closed acoustic system. AP responses were amplified 1000-fold by a DAM-60 preamplifier (WPI, Inc.) and the output of a Tektronix oscilloscope amplifier (50 mV/div) served as the input to a Nicollet 1170 microprocessor. A total of 512 trials were averaged and plotted on a standard x-y plotter. Thresholds were defined as the intensity producing a just detectable and reproducible negative deflection of the baseline. Following the acquisition of control levels, 35 mg/kg furosemide was injected intravenously and potentials were monitored for 1-2 hours.

RESULTS

EP values in the kitten and the rat undergo an initial rapid rate of increase with increasing postnatal age, followed by a more gradual rate of increase until adult values are reached (Fig. 1). AP thresholds and latencies undergo a correspondingly steep exponential decline, followed by a more gradual reduction to adult values (Fig. 2). AP input-output curves show adultlike behavior by 26 days postnatally.

The sensitivity of developing animals to furosemide varied with the stage of development. Animals 45 days old or younger exhibited a 15 milli-volt or less reduction of EP, whereas adults exhibited a larger EP reduc-tion, five-fold greater in magnitude (75 millivolts).

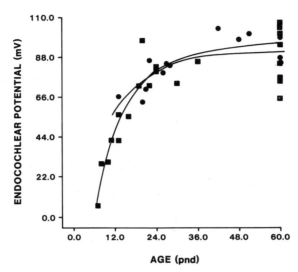

Fig. 1. Endocochlear potentials (EP) measured at various ages in kittens (squares) and rat pups (circles). Points are connected by computer best fit of exponential functions.

DISCUSSION

In the present study we found that EP development in the rat and cat followed similar patterns. A steep growth phase was noted from days 9-15 with a subsequently more gradual rate of increase of EP with age. These data agree with other reports in the kitten (Fernandez and Hinojasa, 1974) and the rat (Bosher and Warren, 1971). A rapid improvement in AP thresholds in the kitten (Romand, 1971) and rat pup (Uziel et al., 1981) have been previously reported. The biphasic adultlike input-output func-tions noted in rats 26 days old and older have been explained by an initial activation of the outer hair cells and their efferents, followed by activation of inner hair cells and their synaptic connections (Carlier and Pujol, 1978).

Inner and outer hair cells behave similarly at their best frequency, i.e. the amplitude-level functions exhibit relatively gradual saturation in the depolarizing direction, contrasted with sharp limiting for the hyperpolarizing phase. Below their best frequency, however, outer hair cells exhibit a markedly different response pattern. At high sound levels, the pattern resembles that seen at best frequency, but at low levels the response asymmetry is reversed (Dallos, 1986).

The differential responses to furosemide at various ages may be due

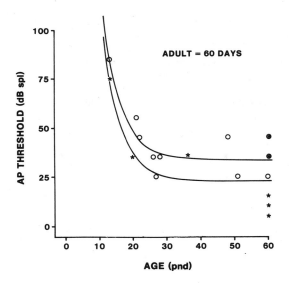

Fig. 2. The action potential of the eighth nerve (AP). Threshold is
plotted at various ages in the rat (circles) and kitten
(asterisks). With increasing age, an improvement in AP threshold
occurs, as demonstrated by the detection of AP response at
increasingly greater attenuation of the auditory stimulus.
These data were fit to an exponential curve.

to developmental changes in the stria vascularis. Others have noted that
the EP is less sensitive to anoxia prior to 16 days of age in the gerbil
(Woolf et al., 1986). The marginal cells increase in size with development
and display more mitochondria and fewer glycogen particles, with greater
cochlear extensions. Basal cells become smaller (Fernandez and Hinojosa,
1974) and move away from the strial capillaries. Perhaps this allows
greater access of furosemide to the marginal cells. (Supported by NIH
and Deafness Research Foundation)

REFERENCES

Bosher, S. K. and Warren R. L., 1971, A study of the electrochemistry
    and osmotic relationships of the cochlear fluids in the neonatal
    rat at the time of the development of the endocochlear potential,
    J. Physiol. (London), 212:739-761.
Carlier, E. and Pujol, R., 1978, Role of inner and outer hair cells in
    coding sound intensity:an ontogenetic approach, Brain Res., 147:
    174-176.
Crowley, D. E. and Hepp-Reymond, M-C., 1966, Development of cochlear
    function in the ear of the infant rat, J. Comp. Physiol. Psychol.,
    62:427-432.
Dallos, P., 1986, Response amplitude sound level functions of cochlear
    hair cells, Abstracts of the Ninth Midwinter Research Meeting,
    Association for Research in Otolaryngology, Feb. 2-6, Clearwater
    Beach, Fl, 67.
Fernandez, C. and Hinojosa, R., 1974, Postnatal development of endo-
    cochlear potential and stria vascularis in the cat, Acta Otolaryngol.,
    78:173-186.
Romand, R., 1971, Maturation des potentiels cochleaires dans la periode
    perinatale chez le chat et le cobaye, J. Physiol. (Paris), 63:763-
    782.

Uziel, A., Romand, R. and Marot, M., 1981, Development of cochlear potentials in rats, Audiology, 20:89-100.

Woolf, N. K., Ryan, A. F. and Harris, J. P., 1986, Development of mammalian endocochlear potential: normal ontogeny and effects of anoxia, Am. J. Physiol., 250:R493-498.

NOISE INDUCED HAIR CELL LOSS DURING THE SENSITIVE PERIOD OF DEVELOPING

RAT PUP COCHLEA

Martin Braniš

Institute of Experimental Medicine
Czechoslovak Academy of Sciences
Prague 2, Czechoslovakia

The mammalian inner ear is very susceptible to acoustic overstimulation, especially around the time when cochlear elements enter their
final stages of maturation. During the past decade, a supra-normal period
of sensitivity to acoustic trauma during development has been reported
for several mammalian species, such as hamsters, guinea pigs, mice, rats
(Falk et al., 1974; Bock and Saunders, 1977; Lenoir et al., 1979; Henry,
1984;and others) and was also hypothesized for humans (Douek et al., 1976).
However, there is no study which deals with numerous age graded series
of developing subjects from the morphological aspect. Known experimental
data are based mainly on electrophysiological methods. In this report we
present a light microscopic study of numerical hair cell loss during the
sensitive period in low frequency pure tone stimulated immature rats
compared with young adults 110 days old at the time of sound exposure.
White Sprague-Dawley rats were utilized in this experiment, since the
identical strain was chosen for verifying the sensitive period by recording
the compound action potential and by means of a different histological
technique by Lenoir et al. (1979) and Lenoir and Pujol (1980). Timing of
the sensitive period, quantitative aspects, and distribution of hair cell
degeneration are discussed in this paper.

METHODS

Thirteen groups of awake Sprague-Dawley rats were exposed to a
continuous pure tone noise of 3.15 kHz, 126 dB SPL for one hour. Animals
were exposed at 13, 15, 17, 19, 21, 23, 25, 27, 29, 31, 33, 35 and 110
days of age. In each group, four animals were exposed (except for 110 day
old group where five individuals were utilized). Animals were exposed in
body-size wire mesh cages in a sound proof and anechoic box. Intensity
levels were measured through a Brüel and Kjaer microphone No. 4134. After
surviving for 30 days the animals were killed, the auditory bullae were
removed and fixed in ten per cent neutral formaldehyde solution. The bony
cochlear wall was then opened from the apex, the basilar membrane with
the organ of Corti was stained in situ by toluidine blue and Ehrlich's
haematoxylin and removed from the cochlea coil after coil under a dissecting microscope. Separated coils of the basilar membrane were mounted in
glycerine and observed under a light microscope. The method is described
in detail elsewhere (Ûlehlová and Voldřich, 1987). The inner and outer
hair cells as well as phalangeal scars in the reticular lamina representing
missing hair cells were counted. In order to compare cochleae of different

length, each basilar membrane was divided into ten segments of equal
length (each segment represented ten per cent of the whole basilar
membrane length). In the final evaluation the ratio between the number of
hair cells (scars and nonaffected cells) and the number of missing hair
cells (only scars) was expressed as a percentage. Mean value and standard
deviation were calculated for each group of animals. Individuals exhibiting
severe otitis media or otitis interna were excluded from further evaluations.

RESULTS AND DISCUSSION

All values are summarized in Tables 1, 2, 3, and illustrated graphically on Fig. 1. Histocochleograms (Fig. 2) demonstrate the variability and
hair cell loss distribution along the whole organ of Corti in four 21-day-

Table 1. Percentages of outer hair cell loss (ears of one animal in
parentheses). NE = not evaluated due to otitis media or otitis
interna.

day				
13	(0.54 - 0.39)	(0.31 - 0.92)	(0.19 - 0.19)	(NE - NE)
15	(1.84 - 4.09)	(5.28 - 1.69)	(0.40 - 0.16)	(NE - NE)
17	(0.98 - 8.39)	(2.01 - 3.45)	(5.01 - 1.61)	(5.73 - 3.02)
19	(51.84 - 13.03)	(7.17 - 2.74)	(9.14 - 0.67)	(0.44 - 0.38)
21	(26.70 - 16.74)	(11.52 - 11.49)	(4.29 - 2.47)	(42.57 - 2.57)
23	(25.90 - 56.80)	(7.16 - 0.64)	(0.22 - 0.50)	(0.17 - NE)
25	(0.16 - 0.47)	(0.37 - 0.34)	(0.28 - 0.30)	(0.28 - 2.11)
27	(33.22 - 7.83)	(0.03 - 0.14)	(8.27 - 2.91)	(0.53 - 0.32)
29	(11.66 - 0.42)	(0.74 - 0.08)	(2.28 - 0.43)	(6.90 - 0.48)
31	(27.88 - 7.47)	(8.95 - 4.62)	(5.76 - 0.64)	(20.47 - 0.30)
33	(22.73 - 0.61)	(0.41 - 0.69)	(0.27 - 0.31)	(1.44 - 0.42)
35	(2.45 - 1.16)	(10.46 - 1.05)	(NE - NE)	(NE - NE)
110	(0.40 - 12.08)	(0.69 - 5.12)	(8.47 - 3.22)	(7.88 - 12.74)
	(0.72 - 0.61)			

Table 2. Percentages of inner hair cell loss (captions - see Tab. 1)

day				
13	(0.00 - 0.00)	(0.00 - 0.00)	(0.00 - 0.00)	(NE - NE)
15	(0.00 - 0.00)	(0.00 - 0.00)	(0.00 - 0.00)	(NE - NE)
17	(0.00 - 0.21)	(0.00 - 0.00)	(0.11 - 0.00)	(0.21 - 0.00)
19	(41.21 - 2.24)	(0.93 - 0.00)	(4.95 - 0.31)	(0.00 - 0.00)
21	(7.42 - 7.96)	(3.49 - 0.87)	(0.00 - 0.00)	(17.55 - 2.98)
23	(8.46 - 34.40)	(1.88 - 0.00)	(0.32 - 0.00)	(0.11 - NE)
25	(0.00 - 0.00)	(0.11 - 0.00)	(0.21 - 0.00)	(0.00 - 0.00)
27	(9.53 - 1.38)	(0.00 - 0.00)	(2.02 - 0.54)	(0.11 - 0.00)
29	(5.18 - 0.11)	(0.21 - 0.00)	(0.54 - 0.10)	(0.85 - 0.11)
31	(9.61 - 1.58)	(3.48 - 0.42)	(0.21 - 0.00)	(7.65 - 0.21)
33	(8.69 - 0.21)	(0.00 - 0.43)	(0.00 - 0.11)	(0.00 - 0.00)
35	(0.52 - 0.20)	(0.00 - 0.00)	(NE - NE)	(NE - NE)
110	(0.42 - 6.16)	(0.00 - 0.32)	(3.22 - 0.65)	(7.14 - 2.15)
	(0.22 - 0.22)			

Table 3. Mean values (MEAN) and standard deviations ($\pm$ SD) of percentual outer (OHC) and inner (IHC) hair cell loss after one hour of 3.15 kHz, 126 dB SPL sound exposure. DAY = day of exposure; A/C = number of animals/cochleae evaluated.

DAY	A/C	OHC loss MEAN	$\pm$ SD	IHC loss MEAN	$\pm$ SD
13	3/6	0.4	0.3	0.0	0.0
15	3/6	2.2	2.0	0.0	0.0
17	4/8	3.8	2.5	0.1	0.1
19	4/8	10.7	17.3	6.2	14.2
21	4/8	14.8	13.9	5.0	5.9
23	4/7	13.1	21.4	6.5	12.7
25	4/8	0.6	0.6	0.04	0.1
27	4/7	6.7	11.3	1.8	3.2
29	4/8	2.9	4.2	0.9	1.8
31	4/8	9.5	9.7	2.9	3.8
33	4/8	3.4	7.8	1.2	3.0
35	2/4	3.8	4.5	0.2	0.3
110	5/10	5.2	4.8	2.0	2.6

old animals representing the most affected group. Results of the present study confirm the existence of a sensitive period for acoustic impairment as previously described by Lenoir et al. (1979) for the same (Sprague-Dawley) strain of laboratory rats by means of electrophysiological methods. The peak of this period - 22 days post partum - reported by Lenoir et al. (1979) agrees with the 21 days ascertained during our investigation. Quantitative assessment of numerical hair cell loss indicates that the beginning of the sensitive period can be dated on day 13 post partum. It is the first day when a slight outer hair cell degeneration occurs. Before this day no damage to cochlear sensory epithelim has been found (unpublished observations). The end of a period of enhanced susceptibility to acoustic overstimulation cannot be, however, rigidly defined. Functional damage to inner ear structures (threshold loss) measured electrophysiologically showed very low variability (Lenoir et al., 1979). Contrarywise, in the course of our histopathological study we found an unexpected variability not only between individuals of the same age group but also between cochleae of the same individual. Since this high variability range cannot be satisfactorily explained by the low frequency stimulation used in our experiment (the lower perception limit for rats is, according to Gourewitch and Hack, 1966 approximately 0.5 kHz, and 3.15 kHz should be perceived near the end of the first apical coil - with helicotrema being the "zero" point - /Burda and Voldřich, 1980/) we assign it most likely to individual susceptibility (vulnerability) to noise exposure. The controversy between physiological (Lenoir et al., 1979) and our histopathological findings may be partly explained by the fact that functionally important damage to the cochlea is usually more extensive than indicated by topographical abnormalities in the organ of Corti (Thorne and Gavin, 1985). Another important finding is the distribution of hair cell degeneration along the spiral organ of Corti. Generally, ears with moderate or severe hair cell loss were damaged mostly in the basal portion of the cochlea. Pathological changes within regions corresponding tonotopically to the sound frequency used for stimulation in our study were found only sporadically, accompanyig usually more damaged base of the cochlea. Saunders and Chen (1985) report that in the course of the sensitive period there is a massive outer hair cell damage against relatively no effect on inner hair cells. Our findings refer both to outer and inner hair cells.

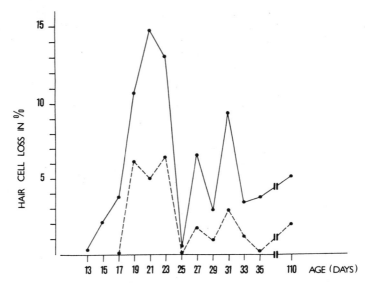

Fig. 1. Percentals of outer (solid line) and inner (dashed line) hair
cell loss in noise exposed animals. Numerical values are given
in Table 3.

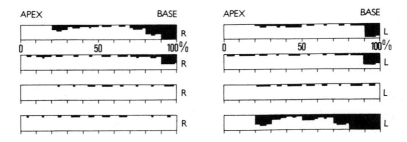

Fig. 2. Histocochleograms of eight cochleae of animals exposed on day
21 showing loss of outer hair cells (black area). Note high
intra-/and interindividual variability. R = Right cochlea,
L = left cochlea of the same animal. The whole basilar membrane
is divided into ten segments of equal length.

Percental hair cell loss was, however, about two times lower within the
inner hair cell population. It is commonly accepted that apart from severe
mechanical lesions caused by very intense noise - the true acoustic trauma
as defined by Voldřich and Úlehlová (1982) - three important factors may
contribute to cochlear hair cell damage:
1)   intoxication by endolymph via small ruptures in the reticular lamina,
2)   enzymatic exhaustion within sensory cells due to functional fatigue,
3)   vascular disorders (Bohne, 1976). We assume that enzymatic depletion
and intoxication by penetrating endolymph may be rather the result of
a direct impact on certain tonotopically related regions of the cochlea
(which in our experiment correlates with slightly prominent damage in
the third segment - end of the apical coil of the basilar membrane).
Vascular disorders, however, may well represent topographically nonspecific

effects of very intense noise, which can influence different parts of the
whole cochlea; possibly those parts of it with heightened susceptibility
to noise stress - the base. Interestingly, in a similar experiment with
wide band noise, Lenoir et al. (1979) also found the high frequency
percepting basal regions to be the most affected part of the whole cochlea.

SUMMARY

Forty eight immature and five adult Sprague-Dawley rats were exposed
to a continuous noise of 3.15 kHz, 126 dB SPL for one hour. Young animals
were exposed in groups of four individuals from the 13th to the 35th
postnatal day. Degeneration of cochlear hair cells was assessed by means
of the standard surface specimen technique. Subjects exposed between days
19 and 23 showed drastic hair cell loss (mean value for day 21 was 14.8
$\pm$ 13.9% OHC loss and 5.0 $\pm$ 5.9% IHC loss). In some cochleae the maximum
loss was higher than 40% of the total outer and inner hair cell population.
After this period hair cell losses were smaller but showing, however,
greater variability. In five control animals exposed to sound of equal
characteristics on day 110 a moderate hair cell loss was found with lower
variability. The sensitive period ascertained in our experiment with low
frequency pure tone stimulation corresponds in timing to previously
published data. However, according to our results, in most noise exposed
individuals, the hair cell population is not necessarily affected equally
within the same group or even within one subject (i.e. high variability
exists between cochleae of the same animal). The distribution of sensory
cell degeneration along the organ of Corti does not fully correspond with
the frequency used for stimulation. The most affected part of the basilar
membrane was its basal portion.

REFERENCES

Bock, G. R. and Saunders, J. C., 1977, A critical period for acoustic
    trauma in the hamster and its relation to cochlear development,
    Science, 197:396-398.
Bohne, B., 1976, Mechanisms of noise damage in the inner ear, in: "Effects
    of noise on hearing", D. Henderson, R. Hamernik, D. S. Dosanjh, J. H.
    Mills, eds, Raven Press, New York.
Burda, H. and Voldřich L., 1980, Correlation between the hair cell density
    and the auditory threshold in the white rat. Hearing Res., 3:91-93.
Douek, E., Dodson, H. C., Bannister, L. H., Aschcroft, P. and Humphries,
    K. N., 1976, Effect of incubator noise on the cochlea of the newborn,
    Lancet, 20:1110-1113.
Falk, S. A., Cook, R. O., Haseman, J. K. and Sanders G. M., 1974, Noise
    induced inner ear damage in newborns and adult guinea pigs,
    Laryngoscope, 84:444-453.
Gourewitch, G. and Hack, M. H., 1966, Audibility in the rat, J. Comp.
    Physiol. Psychol., 62:289-291.
Henry, K. H., 1984, Noise and the young mice: Genotype modifies the
    sensitive period for effects on cochlear physiology and audiogenic
    seizures, Behav. Neurosci., 6:1073-1082.
Lenoir, M., Bock, G. R. and Pujol, R., 1979, Supra-normal susceptibility
    to acoustic trauma of the rat pup cochlea, J. Physiol. Paris, 75:
    521-524.
Lenoir, M. and Pujol, R., 1980, Sensitive period to acoustic trauma in
    the rat pup cochlea, Histological findings, Acta Otolaryngol.,
    89:317-322.
Saunders, J. C. and Chen, C.-S., 1985, Developmetal periods to auditory
    trauma in laboratory animals, in: "Toxicology of the eye and ear
    and other special senses", A. W. Hayes, ed., Raven Press, New York.
Thorne, P. R. and Gavin, J. B., 1985, Changing relationships between
    structure and function in the cochlea during recovery from intense

sound exposure, <u>Ann. Otol. Rhinol. Laryngol.</u>, 94:81-86.

Úlehlová, L. and Voldřich, L., 1987, Modified staining technique for surface preparation of the organ of Corti, <u>Hearing Res.</u>, 26:221-224.

Voldřich, L. and Úlehlová, L., 1982, Correlation of the development of acoustic trauma to the intensity and time of acoustic overstimulation, <u>Hearing Res.</u>, 6:1-6.

INTERAURAL DIFFERENCE OF PSEUDOTHRESHOLD OF STIMULATED

OTOACOUSTIC EMISSION IN UNILATERAL INNER EAR IMPAIRMENTS

Y. Tanaka, T. O-Uchi, H. Shimada and Y. Koseki

Dept. Otolaryng, Dokkyo Univ. Sch. Med. Koshigaya Hosp.

Koshigaya, Saitama, Japan 343

INTRODUCTION

Stimulated otoacoustic emissions (OAEs) described by Kemp (1978) are acoustic responses which could be detected at the external ear canal with a delay of several milliseconds after tonal stimulation. Anderson and Kemp (1979) have shown with animal experiments that OAE relates to physiologically vulnerable inner ear function. It has been reported by Mountain (1980) and Siegel and Kim (1982) that acoustic distortion products in the external ear canal can be changed by electrical stimulation of the crossed olivo-cochlear bundle. These observations lead to a hypothesis that the OAE originates from the cochlear micromechanics linked with outer hair cells.

Many investigators have attempted to make use of OAEs clinically for patients with inner ear impairments (Johnsen and Elberling, 1982; Hinz und Wedel, 1984; Kemp et al., 1986; Tanaka et al., 1987). In this paper we conducted a study to determine whether or not OAEs would be reliable indicator for the evaluation of inner ear functions.

MATERIALS AND METHODS

The OAEs were examined in normal hearing ears of human subjects with type A tympanogram and in cases of unilateral sensorineural deaf ears. The pseudothreshold of the OAE (the OAE threshold) was measured in 119 normal hearing ears and the interaural difference of the OAE threshold (IAD-OAE) was measured in 56 subjects with normal hearing, bilaterally. The subjects with unilateral deaf ear consist of 7 cases with congenital inner ear anomaly, 22 examples with mumps deafness, 125 from sudden deafness, 6 functional deafness and finally 18 patients having sensorineural hearing loss of unknown cause.

An acoustic probe was prepared for measuring OAEs. The probe was made from a double barrelled plastic tube, a subminor earphone (Danavox, SMW-68) and a miniature microphone (Knowles, EA-1843). The plastic tube of the barrel was 40 mm long and 5 mm inside diameter at its tip. The subminor earphone was connected to the basal end of one plastic tube and the miniature microphone was inlaid into the basal side of the other tube. The tip of the probe was inserted into the external ear canal where it was secured using a dental mold or an air cuff depending on the

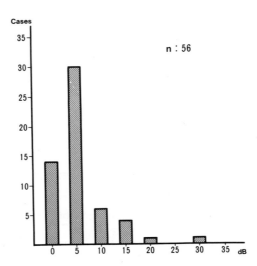

Fig. 1. Distribution of the interaural difference of OAE pseudothreshold (IAD-OAE) in 56 subjects with bilateral normal hearing ears.

shape of the canal of the patient tested. The sine wave signals generated with a function generator (NF Electronic Instruments, FG-143) were amplified and transmitted to drive the earphone. The stimulation was applied as 3 ms tone burst with 1 ms rise/fall time. The outputs of the microphone were averaged with a signal processor (San-ei, 7S11) that was connected to a X-Y recorder (San-ei, 8U16). Average values of 256 acoustic sequences for 20 ms after the onset of the sound stimulation were thus obtained.

RESULTS

The OAEs were recorded from the normal ears with mean psychoacoustic thresholds lower than 30 dB. The mean psychoacoustic threshold in the present study means the average of four hearing threshold values of pure tone audiometry, a + 2b + c at the frequencies of 500 Hz (a), 1000 Hz (b) and 2000 Hz (c). The lowest level of sound stimulation required to evoke the OAEs with 1 mVp-p in the microphone output at each best frequency, was defined as OAE threshold. The thresholds of 119 normal hearing ears were in the range of -15 to 30 dB nHL. Because of this large intersubject variation, the OAE threshold was inadequate for normal control. Accordingly, the IAD-OAE was examined. The range of the IADs was between zero and 30 dB in 56 subjects whose OAEs were measured simultaneously from both ears. The IAD values were within 10 dB in 89.3% (50 out of 56) as shown in Fig. 1.

The IAD-OAE was investigated in subjects with various unilateral sensorineural hearing losses, such as congenital inner ear anomaly, mumps deafness, sudden deafness and functional deafness. In 7 inner ear anomalies studied (4: definite, 3:suspected), the affected ears were either totally deaf or near totally deaf. Mean value of the psychoacoustic thresholds was 106.3 dB and their OAE thresholds were in the range of 25 to 55 dB nHL. The IADs-OAE the 7 inner ear anomalies were ranged between 3o dB and 55 dB; the mean value of the IADs was 37.1 dB ± 6.0. The relation between the IAD of psychoacoustic threshold (IAD of mean threshold value in pure tone audiometry:IAD-PTA) and the IAD-OAE are shown in the top of Fig. 2. There was a positive correlation between the IADs (r=0.86).

52

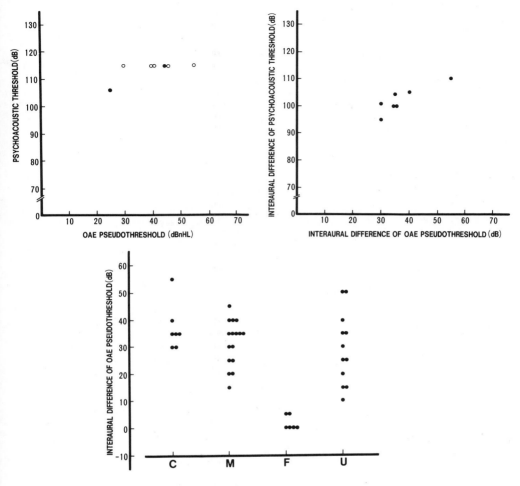

Fig. 2. Top, relation between psychoacoustic threshold and OAE
threshold in 7 unilateral inner ear anomalies. Left, values
in threshold, closed circle: actual value, open circle: value
higher than 115 dB; right value in IADs (r=0.86). Bottom,
comparison of IADs-OAE among unilateral profound sensorineural
deaf ears. C: congenital inner ear anomaly, M: mumps deafness,
F: functional deafness, U: deafness of unknown cause.

The IAD-OAE in the inner ear anomalies were compared with those in
the ears of unilateral mumps deafness, functional deafness and
sensorineural deafness of unknown cause. The IADs-OAE were examined in
16 subjects with mumps deafness whose IADs-PTA of affected ears were
ranged from 81.2 dB to 110.0 dB, excepting 6 subjects with the OAE
threshold being higher than 30 dBnHL in the unaffected side. The mean
value of the IADs-OAE was 31.6 dB $\overset{+}{-}$ 4.1. The IADs in 12 subjects of the
deafness of unknown cause excepting 6 cases with high OAE threshold of
unaffected side, ranged from 36.2 dB to 107.5 dB in psychoacoustic
threshold and were from 15 dB to 50 dB in OAE threshold, respectively.
In contrast with the IADs- OAE of such sensorineural deaf subjects, those
of 6 unilateral functional deaf subjects were small in range and were
very low in value. The IAD values were 5 dB or less. The distribution
of IAD-OAE values in the unilateral sensorineural hearing losses were
compared as seen in the bottom of Fig. 2. The data of the functional

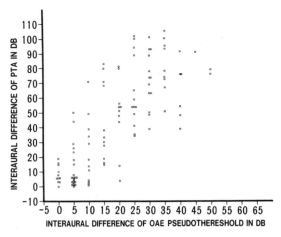

Fig. 3. Relation between the IAD-PTA (threshold in pure tone audiometry)
and the IAD-OAE in 109 subjects of unilateral sudden deafness.

deafness differed significantly from those obtained in other sensorineural deaf subjects.

    The OAEs were measured in 125 subjects with unilateral sudden
deafness which were thought to be induced by inner ear impairments. The
type and grade of hearing loss in these subjects varied. For measurement
of IAD-OAE 16 subjects were excluded since the OAE thresholds in
unaffected side of them were higher than 30 dBnHL. The relation of the
IAD-OAE to the IAD-PTA in 109 of sudden deafness are illustrated in Fig.
3 is 0.77. Of these patients with sudden deafness, about 38% of them
recovered from the deafness, whose IADs-OAE was noted to be reduced as
the recovery of the IADs-PTA.

DISCUSSION

    Despite many facts implying that the OAE is an acoustic phenomenon
of inner ear origin, Kemp et al. (1986) have suggested that acoustic
emission cochleography, as a clinical test, does not quantify hearing
loss, but detects its presence. In the present study the intersubject
variation of the OAE threshold in normal hearing subjects is large whereas
the variation of the IAD-OAE is small. Furthermore, the IAD-OAE of unilateral inner ear anomalies is related to the IAD-PTA. Based on the
results of IAD measurements in unilateral profound hearing losses the OAE
threshold elevates clearly in inner ear impairments but it does not in
sensorineural deaf ears without organic change of the inner ear.

    Hearing loss due to sudden deafness is characterized by its sudden
onset and by affecting usually only one ear in varying degree. Sudden
deafness has been regarded as induced inner ear impairment and Schuknecht
(1974) reported degenerative changes of hair cells in a case of sudden
deafness at autopsy. The notable results obtained from IAD-OAE
measurements in the sudden deafness cases suggest that the IAD-OAE
is a reliable indicator to determine whether or not organic changes
exist in the inner ear and furthermore to evaluate the degree of inner
ear impairments.

SUMMARY

    1) The interaural difference (IAD) of OAE threshold was investigated
in subjects with normal hearing and with unilateral sensorineural hearing
loss. 2) The intersubject variation of the OAE threshold was large but
that of IAD-OAE was small in normal subjects. 3) The IADs-OAE were large
in inner ear impairments. 4) The IAD-OAE was very small in functionally
deaf subjects. 5) There was a positive correlation between the IAD-PTA
and the IAD-OAE in 109 subjects of sudden deafness. 6) The results
support the hypothesis that the OAE originates from the inner ear. The
IAD-OAE is a useful indicator for the evaluation of inner ear impairments.

ACKNOWLEDGEMENTS

    The authors express their thanks to Dr. Nobuhiro Tokita for his
editing of the English. This work was supported in part by grants from
the Ministry of Education, Science and Culture of Japan.

REFERENCES

Anderson, S. D. and Kemp, D. T., 1979, The evoked mechanical response in
    laboratory primates, Arch Otorhinolaryngol., 224:47-54.
Hinz, M. and Wedel, H. V., 1984, Otoakustische Emissionen bei Patienten
    mit Hörsturz, Arch Otorhinolaryngol., Suppl II:128-130.
Johnsen, N. J. and Elberling, C., 1982, Evoked acoustic emissions from
    the human ear, Scand Audiol., II:3-12.
Kemp, D. T., 1978, Stimulated acoustic emissions from within the human
    auditory system, J. Acoust. Soc. Am., 64:1386-1389.
Kemp, D. T., Bray, P., Alexander, L. and Brown, A. M., 1986, Acoustic
    emission cochleography,Practical aspects, Scand. Audiol. Suppl. 25:
    71-95.
Mountain, D. C., 1980, Changes in endolymphatic potential and crossed
    olivocochlear bundle stimulation alter cochlear mechanics,
    Science, 210:71-72.
Schuknecht, H., 1974, Pathology of the ear. Harvard Univ. Press,
    Cambridge, Massachusetts.
Siegel, J. H. and Kim, D. O., 1982, Efferent neural control of cochlear
    mechanics? Hearing. Res., 6:171-182.
Tanaka, Y., O-Uchi, T., Arai, Y. and Suzuki, J., 1987, Otoacoustic
    emission as an indicator in evaluating inner ear impairments,
    Acta Otolaryngol., 103:644-648.

RESPONSES OF COCHLEAR AFFERENTS TO LOW-FREQUENCY TONES:

INTENSITY DEPENDENCE

Mario A. Ruggero and Nola C. Rich

Department of Otolaryngology, University of Minnesota

Minneapolis, MN 55414 U. S. A.

INTRODUCTION

It is remarkable that the adequate stimulus for the inner hair cell (IHC) of the mammalian cochlea has not yet been fully specified in terms of the vibration of the cochlear partition. There is little doubt that the proximal stimulus for depolarization is displacement of the stereocilia toward the longer stereocilia (Russell et al., 1986). What is not known is the manner in which this stimulus arises from interactions of the basilar membrane (BM), tectorial membrane and surrounding fluid spaces. In an effort to resolve this issue, we have been studying the responses of cochlear afferents and of the BM to the same stimuli -- low-frequency tones -- under comparable experimental conditions in a single species, the chinchilla. Previously we have discussed the response phases of the basal region of the BM (Ruggero et al., 1986b) and the dependence on site of innervation of near-threshold cochlear-afferent response phases (Ruggero and Rich, 1983 and 1987). Here we focus on the intensity dependence of IHC response phases relative to BM displacement, as reflected in cochlear afferent excitation.

METHODS

Single unit recordings were made in the cochlear nerves of anesthetized chinchillas using conventional microelectrode techniques. Spike trains were analyzed by means of dot displays and period histograms (Fig. 1). The dot displays encode response phase in their ordinates (one stimulus period), stimulus intensity in their abscissas (2-dB steps) and response magnitudes in terms of dot (spike) density. Period histograms are constructed by binning the spikes in a given vertical slice of the dot display (i.e., responses at a particular stimulus intensity).

RESULTS

Fig. 1 represents the responses of a low-best frequency (BF) cochlear afferent to 100 Hz tones at intensities 0-100 dB SPL. Above 40 dB SPL the spikes tend to cluster at times leading rarefaction by some 90-180 degrees. At 80 dB SPL a second preferred phase appears, lagging rarefaction by 90 degrees. In order to use cochlear-afferent phase data measured relative to eardrum pressure to derive IHC response phases relative to the displacement of the adjacent region of BM, it is necessary to compensate for the delays

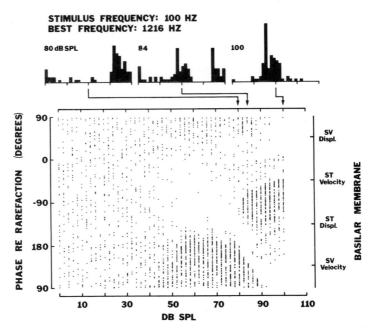

Fig. 1. Responses to 100 Hz tones of a low-BF cochlear afferent. The lower panel presents the spike discharges as a dot display. Three period histograms at the top show selected responses at intensities above and below that at which "peak splitting" (Kiang and Moxon, 1972) occurred.

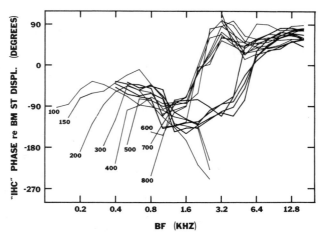

Fig. 2. Average near-threshold "IHC" phases of responses to 30-1000 Hz tones, as functions of log-BF (or equivalently, innervation site).

imposed on signal transmission by the middle ear, the travel time along the BM and neural axons, and the IHC-neural synapses (Ruggero et al., 1986b; Ruggero and Rich, 1983 and 1987). Subtracting the phase lags corresponding to these delays from cochlear-afferent phases re: eardrum pressure (Fig. 1, left ordinate) yields "IHC" response phases relative to BM displacement (right ordinate). Thus, at stimulus levels below 82 dB SPL

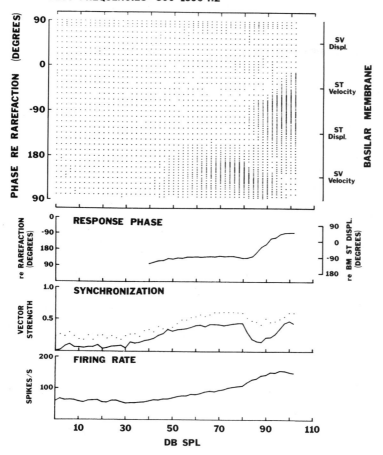

STIMULUS FREQUENCY: 100 HZ
BEST FREQUENCIES: 800-1600 HZ

Fig. 3.   Intensity dependence of responses to 100 Hz tones for 52 low-BF
          cochlear afferents from 8 chinchillas.

the responses of this low-BF neuron occur preferentially synchronous with
velocity of the BM toward scala vestibuli (SV); at higher levels, responses
switch to somewhere between displacement and velocity toward scala
tympani (ST).

      In order to investigate the BF and stimulus-frequency dependences of
"IHC" response phases it is convenient to focus on near-threshold
responses. Fig. 2 presents averages of phase re: BM displacement from
large samples of near-threshold phases in many animals, as functions of
BF and of stimulus frequency. For frequencies 30-80 Hz (thick lines,
unlabelled) and 100-1000 Hz (thin lines), "IHC" responses occur with two
distinct phases relative to BM displacement. Basal "IHCs" (high BFs)
respond some 40-90 degrees leading maximum BM displacement toward ST.
Apical "IHCs" (low BFs), with BFs as low as twice stimulus frequency,
respond approximately in phase with maximum BM velocity toward SV (i.e.,
lagging ST displacement by some 90 degrees). The derived IHC responses for
cochlear locations with BFs lower than 2 kHz match the phases of depolarization measured in intracellular IHC recordings in guinea pig cochleas,
both at the base (Nuttall et al., 1981; Russell and Sellick, 1983) and
at the apex (Dallos, 1985). However, derived IHC responses from cochlear

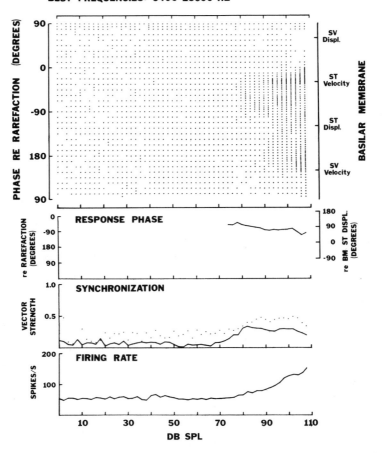

STIMULUS FREQUENCY: 100 HZ
BEST FREQUENCIES: 6400-25600 HZ

Fig. 4. Intensity dependence of responses to 100-Hz tones for 51 high-BF
         cochlear afferents from 8 chinchillas.

locations with BFs higher than 6 kHz are almost out-of-phase with intracellular IHC recordings.

    Fig. 3 illustrates the averaged responses to 100 Hz tones of many
neurons with BFs 800-1600 Hz. Three measures extracted from the dot display
are plotted against stimulus intensity: average phase, vector strength
(Ruggero and Rich, 1983) and average rate. Vector strength is plotted in
two ways: the solid line is the vector strength for the pooled data,
whereas the dotted line indicates the mean of 52 individual vector
strengths. To the extent that these lines overlap, response phases are
uniform across neurons. The pooled results are similar to those described
for the single afferent in Fig. 1. In particular, a phase jump takes place
above 80 dB SPL. Results for other low-BF ranges, down to twice stimulus
frequency, are also similar to Fig. 1 <u>if the phases are referred to the BM.</u>

    Fig. 4 represents the pooled responses to 100 Hz tones of many basal
neurons (BFs exceeding 6400 Hz). In contrast to the low-BF responses of
Figs. 1 and 3, at near-threshold levels (70 dB SPL) responses slightly
lag BM ST velocity; at higher intensities a second peak of activity appears
somewhat leading BM SV velocity.

60

Although we do not yet have as extensive a data sample for higher stimulus frequencies as is available for 100 Hz, it seems that the intensity dependence of derived IHC response phases illustrated for 100-Hz stimuli in Figs. 1, 3 and 4 holds true for stimulus frequencies at least as high as 500 Hz, <u>as long as phases are referred to BM vibration.</u>

DISCUSSION

Our recordings from cochlear afferents suggest that, depending on the intensity of low-frequency stimuli, IHCs in the chinchilla depolarize with two distinct phases, some 130 degrees apart. One phase is synchronous with BM SV velocity, while the other falls somewhere between BM ST displacement and velocity. The near-threshold phase of each of the two regions of the cochlea (low-BF, apical, and high,BF, basal) is the same as the high-intensity phase of the other (Fig. 5). At "peak splitting" intensities (80-95 dB SPL for low-BF neurons, 5-15 dB higher for high-BF neurons) the same two distinct phases coexist.

Until recently there were no intracellular recordings providing IHC counterparts of "peak splitting". Possible counterparts have now been reported by two laboratories (Dallos et al., 1986 and Cody et al., 1986). In addition, there is a suggestion that outer hair cells also undergo abrupt intensity-dependent response phase changes (Zwislocki, 1986). Nevertheless, the near-threshold response phases of basal cochlear afferents still remain without any clear basis in hair cell potentials or in

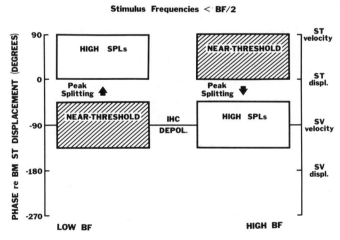

Fig. 5. Schema for the BF and intensity dependences of IHC responses relative to BM displacement, as reflected in the responses of chinchilla cochlear afferents (rectangles). Also shown (horizontal line) is the IHC phase according to intracellular recordings (referenced in text).

cochlear mechanics. In the past, we advocated a hypothesis of Sellick et al. (1982) which postulates an electrical effect of outer hair cells upon basal IHCs (Ruggero and Rich, 1983; Ruggero et al., 1986a). More recent data, showing that the nearly out-of-phase responses of basal and apical cochlear afferents persist for frequencies as high as 500 Hz, lead us to conclude that the postulated interaction between outer and inner hair cells is probably micromechanical in nature (Ruggero and Rich, 1987). In any case, an incompatibility appears to exist between cochlear afferent data and IHC recordings for basal cochlear regions; is it possible that the latter are somehow flawed (Zwislocki, 1985)?

ACKNOWLEDGEMENT

Supported by NIH (NS12125 and NS 25012) and NSF (BNS-834587).

REFERENCES

Cody, A. R., Mountain, D. C. and Russell, I. J., 1986, Acoustically evoked potentials in the basal turn of the guinea pig cochlea, in: "IUPS Satellite Symposium on Hearing", Univ. of California, San Francisco, 24.

Dallos, P., 1985, Response characteristics of mammalian cochlear hair cells, J. Neurosci., 5:1591-1608.

Dallos, P., Cheatham, M. A. and Oesterle, E., 1986, Harmonic components in hair cell response, in: "Auditory Frequency Selectivity", B. C. J. Moore and R. D. Patterson, eds., Plenum, London, 73-80.

Kiang, N. Y. S. and Moxon, E. C., 1972, Physiological considerations in artificial stimulation of the inner ear, Ann. Otol. Rhinol. Laryngol., 81:714-730.

Nuttall, A. L., Brown, M. C., Masta, R. I. and Lawrence, M., 1981, Inner hair cell responses to the velocity of basilar membrane motion in the guinea pig, Brain Res., 211:171-174.

Ruggero, M. A. and Rich, N. C., 1983, Chinchilla auditory-nerve responses to low-frequency tones, J. Acoust. Soc. Am., 73:2096-2108.

Ruggero, M. A. and Rich, N. C., 1987, Timing of spike initiation in cochlear afferents: dependence on site of innervation, J. Neurophysiol., 58:379-403.

Ruggero, M. A., Robles, L. and Rich, N. C., 1986a, Cochlear microphonics and the initiation of spikes in the auditory nerve:correlation of single unit data with neural and receptor potentials recorded from the round window, J. Acoust. Soc. Am., 79:1491-1498.

Ruggero, M. A., Robles, L. and Rich, N. C., 1986b, Basilar membrane mechanics at the base of the chinchilla cochlea. II. Responses to low-frequency tones and relationship to microphonics and spike initiation in the VIII-nerve, J. Acoust. Soc. Am., 80:1375-1383.

Russell, I. J., Richardson, G. P. and Cody, A. R., 1986, Mechanosensitivity of mammalian auditory hair cells in vitro, Nature, 321:517-519.

Russell, I. J. and Sellick, P. M., 1983, Low-frequency characteristics of intracellularly recorded receptor potentials in guinea-pig cochlear hair cells, J. Physiol., Lond. 338:179-206.

Sellick, P. M., Patuzzi, R. and Johnstone, B. M., 1982, Modulation of responses of spiral ganglion cells in the guinea pig cochlea by low frequency sound, Hear. Res., 7:199-221.

Zwislocki, J. J., 1985, Cochlear function - an analysis. Acta Otolaryngol., 100:201-209.

Zwislocki, J. J., 1986, Comments on "Basilar membrane motion and spike initiation in the cochlear nerve" by M. A. Ruggero, L. Robles, N. C. Rich and J. A. Costalupes, in: "Auditory Frequency Selectivity", B. C. J. Moore and R. D. Patterson, eds., Plenum, London, 197.

AUDITORY BRAINSTEM NUCLEI

ANATOMY OF THE MAMMALIAN COCHLEAR NUCLEI; A REVIEW

Kirsten K. Osen

Anatomical Institute
University of Oslo
Karl Johansgt. 47, N-0162 Oslo 1, Norway

Nearly half a century after his first publications on the anatomy of the eighth nerve, Lorente de Nó (1981) portrayed "the primary acoustic nuclei" as "a miniature brain that has a cerebellum of its own, the tuberculum acousticum". Here, selected aspects of this "brain" are discussed on the basis of recent literature. Included are also some preliminary immunocytochemical data from our institute (Osen et al., 1987), obtained by postembedding staining of 0.5 μm resin sections of glutaraldehyde-fixed tissue, using locally produced antisera against conjugated glycine and GABA (Storm-Mathisen et al., 1983; Somogyi et al., 1984; Ottersen and Storm-Mathisen, 1984; Ottersen et al., 1987). As specificity tests are used preabsorption with amino acid glutaraldehyde complexes and coincubation with test sections containing six spatially separate amino acid conjugates (Ottersen, 1987).

Optimal Planes of Sectioning

The cochlear nuclear complex in mammals is situated superficially in the brain stem. Perhaps as a consequence of this, the nuclei show considerable interspecies variation in spatial orientation. In order to compensate for this variation and to obtain sections that are optimally oriented with respect to the functional axes of the nuclei, the complex should be blocked independently of the rest of the brain stem, after removal of the cerebellum.

Cochlear Nerve Fibers

These constitute two main populations; the more numerous and thicker, myelinated inner hair cell afferents and the less numerous and thinner, unmyelinated outer hair cell afferents (Arnesen and Osen, 1978; Liberman and Simmons, 1985). The former fibers are further subdivided into three functionally distinct types according to spontaneous discharge rate (Liberman and Simmons, 1985).

As a group, the inner hair cell afferents supply the anteroventral (AVCN(, posteroventral (PVCN), and dorsal (DCN) cochlear nuclei in a regular tonotopical fashion (Fig. 1A). Their mode of termination has been described in great detail, i. e. in intracellular HRP-labeling experiments (e.g. Rouiller et al., 1986). Their ascending branches in the AVCN are distinctly fasciculated (Fig. 4B), and not arranged in sheats as could be expected

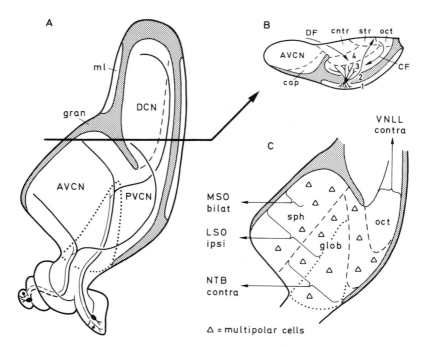

Fig. 1. Diagrams of the cat cochlear nuclei, A and C parallel to the
frequency gradient axis of VCN, B parallel to the isofrequency
planes of the DCN. A shows the bifurcation and tonotopical
organization of the inner hair cell afferents (thick lines) and
outer hair cell afferents (thin lines). B shows the relationship
of a bipolar pyramidal cell to layers 1-3 of DCN and the entrance
of the descending fibers (DF) and the cochlear fibers (CF) into
DCN. C indicates projections of AVCN according to Cant and Casseday (1986). Abbr.: cntr, central nucleus of DCN; glob, globular
cell area; gran, dense granular areas; ml, molecular layer
(=layer 1) of DCN; oct, octopus cell area; sph, spherical cell
area; str, acoustic striae.

from the lamellar tonotopical organization of the nucleus (Bourk et al.,
1981). It still remains to be shown whether the afferents from each inner
hair cell stay together in the fascicles or whether they are spread
throughout the entire mediolateral dimension of the nucleus. By means
of the HRP technique, also the outer hair cell afferents have now been
traced as far centrally as their bifurcation in the ventral cochlear
nucleus (VCN) (Brown, 1987), but their terminal parts still remain to
be demonstrated.

With reservation for the termination of these latter fibers, three
regions of the complex are without primary input, i.e., the dense granular areas, the DCN molecular layer, and the deep DCN (layer 4 and central
nucleus). There is still some disagreement concerning the absence of
cochlear input to the deep DCN. Following cochlear ablations in guinea
pig, however, I have recently found a relatively large zone free of fiber
degeneration deep to layer 3 of the DCN (Fig. 3), as previously found in
cat (Osen, 1970).

Another region that deserves special mention is the cap (= small
cell cap of Osen, 1969a; nucleus lateralis anterior and posterior by
Lorente de Nó, 1981). In contrast to the central nucleus, which is situated

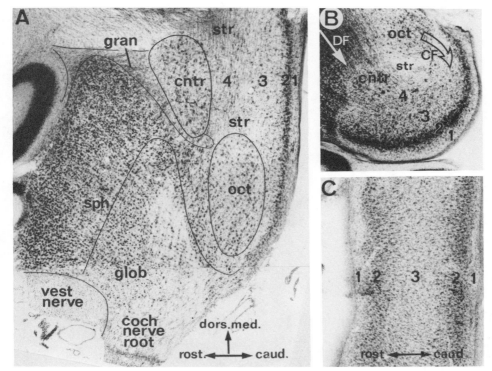

Fig. 2. Photomicrographs of 100 μm thick Nissl sections of the cat
cochlear nuclei, A and C cut parallel to the lateral surface and
the frequency gradient axis of VCN and DCN, respectively, B cut
parallel to the isofrequency planes of DCN. Note the lamination
and the distinct central nucleus of DCN. Abbr., see text Fig.1.

medially in the complex at the site of entrance of the centrifugal fiber
bundle of Lorente de Nó, the cap is situated laterally in relation to
the ventrotubercular association tract which interconnects the DCN and
VCN (Fig. 4B). The cap is supplied by long collaterals from the main
cochlear branches (Fig. 4A). It contains a variety of cells. It is present
in all mammals I have studied, including man, and it may be an integration
center worthy to focus more attention to.

Ventral Cochlear Nucleus

The VCN contains a number of anatomically and physiologically distinct
cell types packed into a small volume of tissue. It was not until research
was focused on cell types rather than subdivision boundaries that any substantial progress was made in unravelling the complicated organization of
this "brain". The four, best known cell classification systems (Lorente
de Nó, 1933, 1981; Harrison and Irving, 1965, 1966; Osen, 1969a; Brawer
et al., 1974) are based on different methods and emphasize different
morphological features of the cells. It may still be fruitful to operate
with all systems in parallel, but in the long run a synthesis should be
aimed at. Hybrid terms like "spherical bushy" and "globular bushy" are
already in common use.

In sections of the VCN, cut parallel to its lateral surface, all of
the five main cell types of the nucleus are usually represented (Figs.
1C, 2A). Three types, the spherical bushy, globular bushy, and octopus

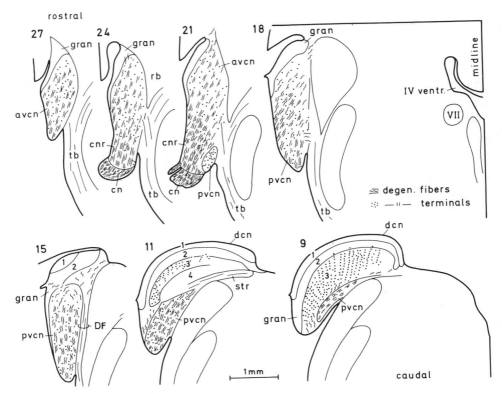

Fig. 3. Camera lucida drawings of Fink and Heimer-stained transverse
sections through the guinea pig cochlear nuclei three days
following cochlear ablation. Note absence of degeneration in
the dense granular areas and in layers 2 and 4 of DCN. Abbr.:
cn, cochlear nerve; cnr, cochlear nerve rot; tb, trapezoid body,
(see also text Fig. 1).

cells, occupy specific portions of the nucleus; the spherical cells are
found rostrally, the globular cells in the middle, and the octopus cells
caudally. The multipolar and small cells, in contrast, are present
throughout the VCN, although in less concentration rostrally and caudally.
The five types may be collected into two groups also with respect to
dendritic pattern and multitude of main axonal targets, as indicated
below.

Multipolar and small cells have tapering, moderately branched dendrites. Their cell bodies form a continuum in size from the smallest to
the largest non-granular cells of the complex. They are most numerous
peripherally where they contribute to the cap (Cant and Morest, 1979).
They receive primary afferents as small boutons, largely on their dendrites. As a group they have a widespread projection to secondary auditory
centers up to and including the mesencephalon, i.e., the periolivary
region, the nuclei of the lateral lemniscus, and the central nucleus of
the inferior colliculus (Adams, 1979). They, apparently, also supply the
motoneurons of the middle ear muscles (Itoh et al., 1986). They can all
be labeled retrogradely from the DCN (Adams, 1983), obviously by the axon
collaterals illustrated by Lorente de Nó (1981, his Fig. 3-39).

The largest multipolar cells, Harrison's h-cells (Harrison and Irving,
1965; Harrison and Feldman, 1970), have long dendrites oriented both

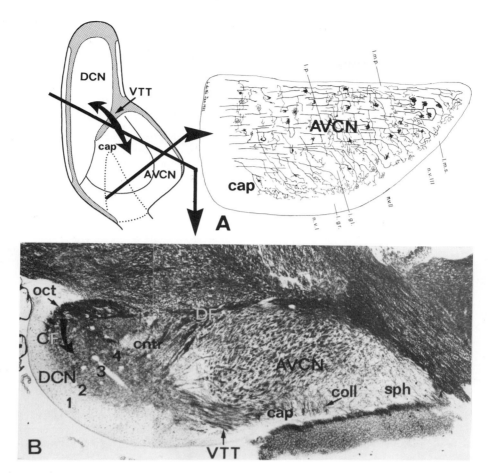

Fig. 4. A is from Lorente de Nó (1981, his Fig. 1-7) and shows a Golgi
section cut parallel to the isofrequency planes af AVCN (see
inset). Collaterals of the cochlear branches supply the cap. B
is a myelin-stained section of the cat cochlear nuclei (see inset
for sectioning plane) showing the central nucleus of DCN with
the entering descending fibers (DC), and the cap with the ventrotubercular tract (VTT) and the collaterals (coll) of the
fasciculated, cross-cut cochlear branches. Abbr., see text Fig.
1.

parallel to and across the isofrequency planes, in small rodents often
spanning the entire width of the VCN (Lorente de Nó, 1981). Wenthold (1987)
has recently shown that in the VCN of guinea pig, cells of this size have
glycine-like immunoreactivity (Gly-LI) and can be retrogradely labeled
with HRP from the contralateral cochlear nuclei. Commissural neurons with
a similar size and location have also been described by Cant and Gaston
(1982) in cat.

Large cells with Gly-LI are also scattered in the rat VCN (personal
observation). Thick, Gly-LI-positive fibers, apparently belonging to
these cells, are abundant in the rostrocaudal middle of the VCN. They
can be traced dorsally through the rostral part of the DCN to the dorsal
acoustic stria where they form a distinct fascicle of about 100-150 fibers.
The fibers can be traced to underneath the facial knee, but here they are
lost to view in between the thicker vestibular fibers, which like the

former cross the midline ventral to the olivocochlear bundle. Midline stimulation of the crossing olivocochlear fibers may also activate the presumably inhibitory, commissural auditory fibers.

Immunocytochemistry may also be used to sort out other subsets of the multipolar and small cell types, but time does not allow me to go further into this matter here.

The spherical bushy, globular bushy and octopus cells all have non-tapering dendrites ending in bush-like formations. The three types differ in number of dendritic stems and the relative length of the stem and the terminal bush. They also differ in number of afferent cochlear fibres; the spherical cells receive the smallest number, the octopus cells the largest number.

They all have a relatively restricted projection to secondary auditory nuclei below the mesencephalon; the octopus cells have their main projection to the ventral nucleus of the lateral lemniscus contralaterally, and the spherical cells to the medial superior olive (MSO) bilaterally and the lateral superior olive (LSO) ipsilaterally.

Until recently it was not clear whether the same or separate cells projected to the MSO and the LSO. I am sorry to say that I myself have contributed to this confusion. Originally I defined two types of spherical cells, large and small, with separate distribution in the VCN (Osen, 1969a). In a preliminary report (Osen, 1969b), that I never followed up, I concluded that the two cell types also had different central projections, the large variety supplying the MSO bilaterally and the small variety supplying the LSO ipsilaterally.

The group of relatively large and densely packed spherical cells in the rostroventral corner of the AVCN, although less evident in certain species (Hackney and Pick, 1986), is a feature typical for many mammals. Instead of constituting a separate type of cells, however, they could represent the one end of a gradient in cell size and packing along the frequency gradient axis of the nucleus. Gradients of this sort are found also in other tonotopically organized nuclei of the auditory pathway, like the central nucleus of the inferior colliculus (Faye-Lund and Osen, 1985). The functional implications of such gradients should be explored further.

With respect to the central projection of the spherical cells, however, my conclusions were obviously too hasty. As convincingly demonstrated in the HRP study by Cant and Casseday (1986) and schematized by me in Fig. 1C, the entire rostral part of the spherical cell area, not only its ventral part, projects to the MSO bilaterally. These same cells project to the LSO ipsilaterally together with cells in the caudal part of the sperical cell area and the rostral part of the globular cell area. The entire globular cell area projects to the NTB contralaterally, as shown before (Harrison and Irving, 1964). The retrogradely labeled cells could not be classified with certainty. If multipolar and small cells are not involved, there may be as many as four subsets of bushy cells with different central projections; a rostral set with a triple projection to the MSO bilaterally and the LSO ipsilaterally; a second set with a single projection to the LSO ipsilaterally; a third set with a double projection to the LSO ipsilaterally and the NTB contralaterally; and a fourth set with a single projection to the NTB contralaterally. Judged by Golgi material, the bushy cells could well establish a rostrocaudal gradient of dendritic shapes, but in Nissl sections no definite transitional forms between cell bodies of spherical and globular cells prevail. Thus, although the study by Cant and Casseday (1986) has brought us a large step forward in this difficult matter, the final solution may still be pending.

Dorsal Cochlear Nucleus

As mentioned, Lorente de Nó (1981) has defined the DCN as an acoustic "cerebellum". In tangential sections, it may indeed resemble a cerebellar folium (Fig. 2C), and in lower mammals its superficial layers may also be easily mistaken electron microscopically for the cerebellar cortex. On the whole, however, the DCN has a much more complicated and variable structure. It is unique both in being a primary sensory nucleus with a cortical structure and in being a cortex that is located subependymally. Whether the latter feature implies a particular influence from the cerebrospinal fluid is a question that remains to be addressed.

Many authors divide the DCN into three layers. In subprimate mammals, however, it definitely consists of four layers and a more or less distinct central nucleus (Lorente de Nó, 1981) (Figs. 1B, 2A, B). The three super-ficial layers are related to the bipolar pyramidal (fusiform) cells which project to the contralateral inferior colliculus; the apical arbor of these cells occupies layer 1, the cell body layer 2, and the basal arbor layer 3. The most important intrinsic border, thus, must be between layers 3 and 4, and not between layers 2 and 3 as usually emphasized. In rat, in which the soma and the basal dendrites of the pyramidal cells lack the radial orientation typically found in cat, layers 2 and 3 are actually more or less confluent.

As shown by computer reconstructions of Golgi sections (Blackstad et al., 1984), both dendritic arbors of the pyramidal cells are conspicuously flattened across the long, frequency gradient axis of the DCN. The highest degree of flatness and mutually parallel orientation are found in the basal arbor which occupies the same layer 3 of the nucleus as the similarly oriented, terminal arborizations of the cochlear nerve fibers. In agreement with this, electron microscopy of intracellularly HRP-labeled pyramidal cells have shown that synapses with primary afferents are located almost exclusively on the basal dendrites (Smith and Rhode, 1985).

The less flattened and less parallel apical arbors are traversed by the granule cell axons, the so-called parallel fibers, that are oriented parallel to the long axis of the DCN, i.e., across the isofrequency pla-nes.

Interneurons also have different dendritic orientation in the dif-ferent layers of the DCN. Cells with planar dendritic fields, in parallel with the isofrequency planes, are, as could be expected, found only in layer 3 (Osen, 1983). Lorente de Nó (1933) originally named these cells "corn cells", but in his latest publication (Lorente de Nó, 1981) he renamed them "vertical cells" because of their orientation normal to the nuclear surface.

Strangely enough, Lorente de Nó did not recognize the planarity of the pyramidal and vertical cells, nor the true orientation of the isofrequency planes in the DCN. On the other hand, he made very accurate observations of the vertical cell axons. The smallest of these cells have locally branching axons. The largest have axons which, after giving off collaterals to the DCN, proceed to the VCN in the mighty and tono-topically organized ventrotubercular association tract (Lorente de Nó, 1981). These fibers probably correspond to Cant and Morest's (1978) group IV of non-cochlear afferents to the AVCN.

The rat DCN contains at least one superficial and one deep group of

small-to medium-sized cells with Gly-LI (Osen et al., 1987). The superficial group represents cartwheel neurons, which are commented on below.
In agreement with the numerous, small axons with Gly-LI found in the
ventrotubercular tract, the majority of cells in the deep group must be
vertical cells. These cells, thus, may excert a tonotopically organized
inhibition both in the DCN and the AVCN.

The Cochlear Granule Cell System

Like its counterpart in the cerebellar cortex, this system represents a set of local interneurons that receives input from many different
sources and has as its target the main projection neurons, i.e., the
pyramidal cells. To the previously known sources of input can now also
be added the somatosensory system (Itoh et al., 1987). In addition to the
presumably excitatory granule cells, the system includes three types of
presumably inhibitory neurons, the Golgi cells, molecular stellate cells,
and cartwheel cells (Mugnaini et al., 1980; Wouterlood and Mugnaini,
1984; Wouterlood et al., 1984). The Golgi cells are probably intercalated
between mossy fibers and granule cells, the stellate and cartwheel cells
between parallel fibers and pyramidal cells.

It has now been proved beyond doubt that all granule cells, ir-
respective of their location, contribute parallel fibers to the DCN
molecular layer (Adams, 1983). Here they form asymetric synapses en
passage with the thick spiny dendrites of the cartwheel cells and the
pyramidal cell apical arbors (Wouterlood and Mugnaini, 1984). The thin,
electrically coupled, non-spiny dendrites of the small stellate cells are
also contacted by the parallel fibers (Wouterlood et al., 1984).

The stellate cells show a strong GABA-LI. Their axons ramify in
layer 1 (Lorente de Nó, 1981). They most probably contribute to the dense
GABA-LI positive plexus in this layer (Mugnaini, 1985; Osen et al., 1987).

As mentioned, the cartwheel cells show a strong Gly-LI. Their axons
form a dense plexus in layer 2 (Lorente de Nó, 1981). Together with the
vertical cells, therefore, they may contribute to the dense Gly-LI-positive
plexus of this layer, including the numerous terminal-like profiles out-
lining the bodies and basal stem dendrites of the pyramidal cells (Osen
et al., 1987).

In previous studies (e.g. Mugnaini, 1985) both stellate and cartwheel
cells have been regarded as GABA-ergic because of positive immunostaining
for GAD. The possibility of colocalization of glycine and GABA in the
cartwheel cells needs further exploration, but so far we have found
immunoreactivity for both amino acids only in a few medium-sized cells
which could as well represent Golgi cells. In agreement with the immunocytochemical data, iontophoretic application studies speak in favor of
GABA and glycine as inhibitory transmitters in the DCN (for a review see
Caspary et al., 1985). Physiological properties of the pyramidal cells
are also compatible with several potent inhibitory inputs.

The great interspecies variation in the structure of the DCN, particularly in the development of its granule cell system (Moore, 1980),
seems to indicate a species specific role in auditory behavior. So far,
however, practically nothing is known about the functional significance
of this challenging part of the auditory pathway.

REFERENCES

Adams, J. C., 1979, Ascending projections to the inferior colliculus,
    J. Comp. Neurol., 183: 519-538.

Adams, J. C., 1983, Multipolar cells in the ventral cochlear nucleus
    project to the dorsal cochlear nucleus and the inferior colliculus,
    Neurosci. Lett., 37:205-208.
Arnesen, A. R. and Osen, K. K., 1978, The cochlear nerve in the cat:
    topography, cochleotopy, and fiber spectrum, J. Comp. Neurol., 178:
    661-678.
Blackstad, T. W., Osen, K. K. and Mugnaini, E., 1984, Pyramidal neurones
    of the dorsal cochlear nucleus: a Golgi and computer reconstruction
    study in cat, Neuroscience, 13:827-854.
Bourk, T. R., Mielcarz, J. P. and Norris, B. E., 1981, Tonotopic organi-
    zation of the anteroventral cochlear nucleus of the cat, Hearing
    Res., 4:215-241.
Brawer, J. R., Morest, D. K. and Kane, E. C., 1974, The neuronal archi-
    tecture of the cochlear nucleus of the cat, J. Comp. Neurol., 155:
    251-300.
Brown, M. C., 1987, Morphology of labeled afferent fibers in the guinea
    pig cochlea, J. Comp. Neurol., 260:591-604.
Cant, N. B. and Casseday, J. H., 1986, Projections from the anteroventral
    cochlear nucleus to the lateral and medial superior olivary nuclei,
    J. Comp. Neurol., 247:457-476.
Cant, N. B. and Gaston, K. C., 1982, Pathways connecting the right and
    left cochlear nuclei, J. Comp. Neurol., 212:313-326.
Cant, N. B. and Morest, D. K., 1978, Axons from non-cochlear sources in
    the anteroventral cochlear nucleus of the cat. A study with the
    rapid Golgi method, Neuroscience, 3:1003-1029.
Cant, N. B. and Morest, D. K., 1979, Organization of the neurons in the
    anterior division of the anteroventral cochlear nucleus of the cat.
    Light-microscopic observations, Neuroscience, 4:1909-1923.
Caspary, D. M., Rybak, L. P. and Faingold, C. L., 1985, The effects of
    inhibitory and excitatory amino-acid neurotransmitters on the
    response properties of brainstem auditory neurons, in: "Auditory
    biochemistry", D. G. Drescher, ed., Charles C. Thomas Publ.,
    Springfield Ill.
Faye-Lund, H. and Osen, K. K., 1985, Anatomy of the inferior colliculus
    in rat, Anat. Embryol., 171:1-20.
Hackney, C. M. and Pick, G. F., 1986, The distribution of spherical cells
    in the anteroventral cochlear nucleus of the guinea pig. Brit.
    J. Audiol., 20:215-220.
Harrison, J. M. and Feldman, M. L., 1970, Anatomical aspects of the
    cochlear nucleus and superior olivary complex, in: "Contributions
    to sensory physiology", W. D. Neff, ed., Academic Press, New York
    and London.
Harrison, J. M. and Irving, R., 1964, Nucleus of the trapezoid body:
    dual afferent innervation, Science, 143:473-474.
Harrison, J. M. and Irving, R., 1965, The anterior ventral cochlear
    nucleus, J. Comp. Neurol., 124:15-21.
Harrison, J. M. and Irving, R., 1966, The organization of the posterior
    ventral cochlear nucleus in the rat, J. Comp. Neurol., 126:391-402.
Itoh, K., Nomura, S., Konishi, A., Yasui, Y., Sugimoto, T. and Muzino,
    N., 1986, A morphological evidence of direct connections from the
    cochlear nuclei to tensor tympani motoneurons in the cat: a possible
    afferent limb of the acoustic middle ear reflex pathways, Brain.
    Res., 375:214-219.
Itoh, K., Kamiya, H., Mitani, A., Yasui, Y., Takada, M. and Mizuno, N.,
    1987, Direct projections from the dorsal column nuclei and the spi-
    nal trigeminal nuclei to the cochlear nuclei in the cat, Brain
    Res., 400:145-150.
Liberman, M. C. and Simmons, D. D., 1985, Applications of neuronal
    labeling techniques to the study of the peripheral auditory system,
    J. Acoust. Soc. Am., 78:312-319.

Lorente de Nó, R., 1933, Anatomy of the eighth nerve. III. General plan
of structure of the primary cochlear nuclei, Laryngoscope, 43:327-350.
Lorente de Nó, R., 1981, "The primary acoustic nuclei", Raven Press, New
York.
Moore, J. K., 1980, The primate cochlear nuclei: loss of lamination as a
phylogenetic process, J. Comp. Neurol., 193:609-629.
Mugnaini, E., 1985, GABA neurons in the superficial layers of the rat
dorsal cochlear nucleus:light and electron microscopic immunocy-
tochemistry, J. Comp. Neurol., 235:61-81.
Mugnaini, E., Osen, K. K., Dahl, A.-L., Friedrich, V. L., jr., and
Korte, G., 1980, Fine structure of granule cells and related inter-
neurons (termed Golgi cells) in the cochlear nuclear complex of
cat, rat and mouse, J. Neurocytol., 9:537-570.
Osen, K. K., 1969a, Cytoarchitecture of the cochlear nuclei in the cat,
J. Comp. Neurol., 136:453-484.
Osen, K. K., 1969b, The intrinsic organization of the cochlear nuclei in
the cat, Acta Otolaryngol., 67:352-359.
Osen, K. K., 1970, Course and termination of the primary afferents in
the cochlear nuclei of the cat. An experimental anatomical study,
Arch. Ital. Biol., 108:21-51.
Osen, K. K., 1983, Orientation of dendritic arbors studied in Golgi
sections of the cat dorsal cochlear nucleus, in: "Mechanisms of
hearing", W. R. Webster and L. M. Aitkin, eds., Monash University
Press, Clayton.
Osen, K. K., Ottersen, O. P. and Storm-Mathisen, J., 1987, Glycine-like
immunoreactivity in the rat dorsal cochlear nucleus (DCN),
Neuroscience, Suppl. 22:S788.
Ottersen, O. P., 1987, Postembedding light- and electron microscopic
immunocytochemistry of amino acids: description of a new model
system allowing identical conditions for specificity testing and
tissue processing, Exp. Brain Res., (in press).
Ottersen, O. P., Davanger, S. and Storm-Mathisen, J., 1987, Glycine-like
immunoreactivity in the cerebellum of rat and Senegalese baboon,
Papio papio: a comparison with the distribution of GABA-like immunoreactivity and with ^{3}H glycine and ^{3}H GABA uptake, Exp. Brain
Res., 66:211-221.
Ottersen, O. P. and Storm-Mathisen, J., 1984, Glutamate- and GABA-
containing neurons in the mouse and rat brain, as demonstrated with
a new immunocytochemical technique, J. Comp. Neurol., 229:374-392.
Rouiller, E. M., Cronin-Schreiber, R., Fekete, D. M. and Ryugo, D. K.,
1986, The central projections of intracellularly labeled auditory
nerve fibers in cats: an analysis of terminal morphology. J. Comp.
Neurol., 249:261-278.
Smith, P. H. and Rhode, W. S., 1985, Electron microscopic features of
physiologically characterized, HRP-labeled fusiform cells in the
cat dorsal cochlear nucleus, J. Comp. Neurol., 237:127-143.
Somogyi, P., Hodgson, A. J., Smith, A. D., Nunzi, M. G., Gorio, A. and
Wu, J.-Y., 1984, Different populations of GABA ergic neurons in the
visual cortex and hippocampus of cat contains somatostatin- or
cholecystokinin-immunoreactive material, J. Neurosci., 4:2590-2603.
Storm-Mathisen, J., Leknes, A. K., Bore, A. T., Vaaland, J. L.,
Edminson, P., Haug, F.-M. S. and Ottersen, O. P., 1983, First
visualization of glutamate and GABA in neurones by immunocytochemistry, Nature, London, 301:517-520.
Wenthold, R. J., 1987, Evidence for a glycinergic pathway connecting the
two cochlear nuclei; an immunocytochemical and retrograde transport
study, Brain Res., 415:183-187.
Wouterlood, F. G., and Mugnaini, E., 1984, Cartwheel neurons of the dorsal cochlear nucleus: A Golgi-electron microscopic study in rat,
J. Comp. Neurol., 227:136-157.

Wouterlood, F. G., Mugnaini, E., Osen, K. K. and Dahl, A.-L., 1984, Stellate neurons in rat dorsal cochlear nucleus studied with combined Golgi impregnation and electron microscopy: synaptic connections and mutual coupling by gap junctions, J. Neurocytol., 13: 639-664.

OBSERVATIONS ON THE CYTOARCHITECTURE OF THE GUINEA PIG VENTRAL COCHLEAR

NUCLEUS

Carole M. Hackney

Dept. of Communication and Neuroscience
University of Keele
Keele, Staffs. ST5 5BG, England

All the major cell types described by Osen (1969a) in cat can be
identified in the guinea pig cochlear nucleus (Hackney and Osen, 1985).
In the ventral cochlear nucleus (VCN), granule cells, octopus cells,
globular cells, spherical cells, small cells, multipolar cells and giant
cells can all be found. There are differences between the two species in
the distribution of some of these cell types. The present account
concentrates on two of the major cell types seen in the Nissl-stained VCN
- the spherical and globular cell - and on the morphology of the bushy
cells seen in Golgi stained material in the same regions.

SPHERICAL CELLS

In cat, Osen (1969a) divided the spherical cell region in the
anteroventral cochlear nucleus (AVCN) into large and small spherical cell
areas. The large spherical cells are found more rostrally in a cap-shaped
area on the lateral side of the AVCN whilst the smaller spherical cells
are found more caudally and medially forming a broad band of cells
throughout the rest of the AVCN. This division of spherical cells into
two groups - large and small - has been suggested to have functional
implications (Osen, 1969b). The relative position of the large and small
spherical cells in cat could mean that the large cells are concerned with
low and middle characteristic frequencies whilst the small cells are
concerned with all characteristic frequencies. Osen also raised the
interesting possibility from lesion studies that the large spherical cell
area projects to the medial superior olive (MSO) bilaterally whilst the
small spherical cell area projects to the lateral superior olive (LSO)
ipsilaterally. More recently, Cant and Casseday (1986) have suggested
from dye-tracing work that whilst the MSO receives bilateral input from
the rostralmost region of the AVCN, the LSO receives an ipsilateral input
from cells throughout the AVCN. This latter division of projections does
not correspond to areas containing only one cell type - but the question
of whether the AVCN can be subdivided into regions containing predominantly large or small spherical cells and whether these project to different
olivary nuclei remains important in interpreting their function. It is
therefore of interest to locate these cells in other species used in
auditory physiology for comparison with cat.

In guinea pig Noda and Pirsig (1974) followed Osen's (1969a) scheme
for cat and divided the AVCN into large and small spherical cell areas.

On examination of Nissl stained material from guinea pig, however, I found it impossible to make a qualitative decision about whether there was a boundary between large and small spherical cell regions based on cell size. Indeed, it was not clear whether or not there was any consistent difference in the size of spherical cells at all. A quantitative investigation of this region was therefore carried out (Hackney and Pick, 1986).

Measurements were made of cell size, shape, distribution and packing density for each AVCN cell in every fifth section of series of parasagittal Nissl sections cut at 15 μm. Data was collected via camera lucida drawings using a bit pad, and analysed with a CED Alpha computer and a programme written in Fortran. Using this, the AVCN could be divided into grid squares or columns and different resolutions selected, ranging from divisions containing the whole AVCN down to divisions containing only one or two cells. All useful rotations and translations of the analysing grids could then be investigated.

Cells were placed in three categories as follows:

1. Spherical cells. A cell was only placed in this category if its nucleus was relatively centrally placed, with a distinct cap of Nissl substance and a 'necklace' of coarse Nissl granules in the cytoplasm around it.

2. Multipolar cells. Cells were placed in this category only if they had an irregular somatic outline and relatively coarse Nissl granules scattered throughout the cytoplasm.

3. Unclassified and small cells. The remaining cells were placed in this category. Many of these cells were smaller than cells in either of the other categories and would probably have been included in the small cell group by Osen (1969a). The others had some but not all the features of either the spherical or multipolar cells so could not be placed in either category with confidence.

Only the outlines of cells in which the nucleus and nucleolus had been sectioned were analysed to minimise the inclusion of cells which had not been cut across their mid region.

The data presented here is for 100 μm columns running dorsal to ventral, taken from the rostralmost point of the AVCN back to the nerve root and averaged appropriately for the reconstructed AVCN, but analyses made at angles to this (e.g. in columns running from dorsal to ventral) do not lead to a different interpretation of the results.

The most important finding was that there was no indication of the spherical cells being larger in the rostralmost portion of the AVCN. In fact, the average spherical cell area in the rostral 400 μm of the AVCN was 279.4 μm^2 (S.D. = 59.5 μm^2) compared to 294.6 μm^2 (S.D. = 61.1 μm^2) in the remaining caudal portion. The t-test showed the two populations not to be significantly different at the 5% level. Similarly if the average cell area in each 100 μm column is plotted against distance from the rostral pole, then no clear change from large to small relative to position can be seen. If anything there is an increase in cell size caudally but this is not significant (Fig. 1a).

The number of spherical cells as a percentage of the total cell the entire cell population (Fig. 1b). The packing density of all the cells is greater rostrally, as is the packing density of the spherical cells alone (Fig. 1c).

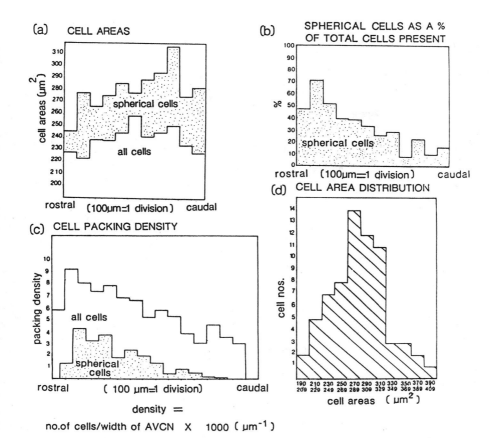

Fig. 1. (a) Histogram to show the average cell area (μm^2) in columns
100 μm wide from the rostral pole to the caudal region of the
AVCN just above the nerve root for all the cells and for the
spherical cells alone. (b) Histogram showing the proportion of
spherical cells as a percentage of all the cells present for
columns 100 μm wide from the rostral pole of the AVCN to the
caudal region of the AVCN. (c) Histogram showing the packing
density (number of cells per μm) in columns 100 μm wide from
the rostral pole to the caudal region of the AVCN for all the
cells and for the spherical cells alone. (d) Histogram showing
the number of spherical cells in bin widths of 20 μm^2 from the
smallest cell area measured (190-209 μm^2) to the largest
(390-409 μm^2).

It was possible that the procedure used might be averaging out a
group of large spherical cells as a result of a two dimensional analysis
of what is essentially a curved three-dimensional structure, although
there was no obvious sign of such a group of cells in horizontal or
frontal sections either. A histogram of the numbers of spherical cells
with different cell areas was therefore plotted to look for a bimodal
distribution of the population into large and small cells (Fig. 1d).
Whatever binwidth was selected there was a unimodal distribution of cell
sizes. This suggests that there are not two distinct populations of
spherical cells but one, which varies from cells with sectional areas of
190 μm^2 to 410 μm^2.

Thus a division between the large and small spherical cell regions does not seem to occur in the guinea pig, at least on the basis of cell size and so the guinea pig AVCN differs somewhat from that of the cat. The large spherical cells of cat appear to have areas of about 700 um^2 and the small spherical cells of about 400 um^2. It is possible that the guinea pig spherical cells are the equivalent to the small spherical cells in cat and that large spherical cells do not exist in this species. Alternatively, there may also be a continuum in cat from large spherical cells to small spherical cells which is more pronounced than in the guinea pig, and the cell sizes may simply be greater on average. Material from cat is currently being examined using the same techniques to resolve these questions.

GLOBULAR CELLS

The guinea pig also has a rather different distribution of globular cells compared to that of the cat. They occur in great abundance caudal to the nerve root with fewer rostral to it, and the PVCN bulges according-ly. In Golgi material, cells which correspond to spherical and globular cells have a 'bushy' appearance (Cant and Morest, 1985) - and in guinea pig Golgi sections, bushy cells are seen throughout the VCN.

Ten cells are shown here (Fig. 2) which have been drawn along an axis running roughly caudal to rostral in a horizontal section. Cells 1-5 are in positions caudal to the nerve root and 6-10 are from positions rostral to it. Cells 1-5, which are from regions where globular cells are found in Nissl sections, have a more widely spaced dendritic bush compared to the spikier, denser pattern of cells 6-10 which are from regions where spherical cells are found in Nissl.

When viewed in horizontal sections there also appears to be a difference in the orientation of the dendritic fields of the two cell types. Spherical bushy cells have their initial dendritic trunk pointing in any direction whilst the main dendritic trunk of the globular bushy cells tends more often to be pointing from the cell body in a medial direction.

It seems therefore that it may be possible in Golgi stained material (and hence presumably in cells stained intracellularly with the horse radish peroxidase technique), at least in the guinea pig, to distinguish between spherical bushy cells and globular bushy cells not only on the basis of their position in the VCN, but also from their dendritic morphology.

Spherical bushy cells are known to receive large end bulbs of Held and have a so-called 'primary-like' physiological response pattern. The globular bushy cells appear to be contacted by modified end-bulbs of Held. Their physiological responses appear to be more controversial - whether they have primary-like, modified primary-like or non-primary-like responses (for instance, a chopper pattern) is still being investigated (see Young, 1985, for review). Using field responses, Evans and Nelson (1973) suggested that in cat there is a distinction between the rostral-most pole of the AVCN where little is seen in the way of inhibition whereas back towards the nerve root inhibitory side bands begin to appear. Whether differing physiological patterns can be found which correspond to the morphological patterns of the two types of bushy cells seen in guinea pig is currently under investigation at Keele.

ACKNOWLEDGEMENTS

This work has been supported by the Wellcome Trust and the MRC.

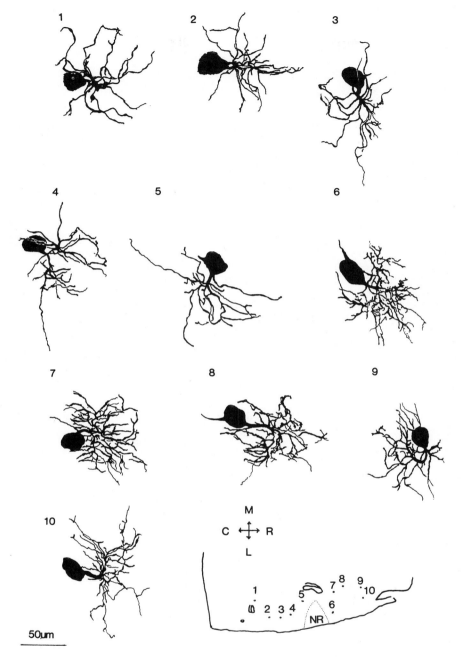

Fig. 2. A series of bushy cells drawn from a horizontal section of the
dorsal region of the guinea pig VCN. (The DCN has been trimmed
straight - see inset.) Cells 1-5 are located caudal to the nerve
root (NR) and are from positions which correspond to regions
containing globular cells in Nissl stained material, whilst cells
6-10 are from positions rostral to the nerve root which correspond
to regions containing spherical cells in Nissl stained material.
Note that cells 1-5 have more widely spaced, less densely branched
dendritic bushes than cells 6-10. (D=dorsal, V=ventral, M=medial,
L=lateral.)

Thanks are due to Mr. S. Murray for technical assistance, and to Dr. K. K. Osen for all her help and encouragement.

REFERENCES

Cant, N. B. and Casseday, J. H., 1986, Projections from the anteroventral cochlear nucleus to the lateral and medial superior olivary nuclei, J. Comp. Neurol. 247:457-476.

Cant, N. B. and Morest, D. K., 1985, The structural basis for stimulus coding in the cochlear nucleus of the cat, in: "Hearing Science", Berlin, C. ed., Taylor and Francis, London.

Evans, E. F. and Nelson, P. G., 1973, The responses of single neurons in the cochlear nucleus of the cat as a function of their location and anaesthetic state, Exp. Brain Res., 17:402-427.

Hackney, C. M. and Osen, K. K., 1985, The cochlear nucleus of the guinea pig - some light microscopic observations, Neuroscience Letts. Suppl., 25, S15.

Hackney, C. M. and Pick, G. F., 1986, The distribution of spherical cells in the anteroventral cochlear nucleus of the guinea pig, Brit. J. Audiol. 20:215-220.

Noda, Y. and Pirsig, W., 1974, Anatomical projection of the cochlea to the cochlear nuclei of the guinea pig, Arch. Oto.-Rhino.-Laryngol., 208:107-120.

Osen, K. K., 1969a, Cytoarchitecture of the cochlear nuclei in cat, J. Comp. Neurol., 136:453-484.

Osen, K. K., 1969b, The intrinsic organization of the cochlear nuclei in the cat, Acta-Oto-Laryngol.,67:352-359.

Young, E. D., 1985, Response characteristics of neurons of the cochlear nuclei, in: "Hearing Science", Berlin, C., ed., Taylor and Francis, London.

CELLULAR CONNECTIONS REVEALED BY TRANSNEURONAL TRANSPORT OF HRP

IN THE GUINEA PIG COCHLEAR NUCLEUS

D. J. Parker, E. F. Evans and C. M. Hackney

Department of Communication and Neuroscience
University of Keele
Keele, Staffs., ST5 5BG, UK

INTRODUCTION

The technique of intracellular injection of horseradish peroxidase (HRP) has been successfully used by a number of workers in order to label physiologically characterised cells in the cochlear nucleus (e.g. Rhode et al., 1983a, b; Rouiller and Ryugo, 1984, in the cat). We have extended it to the guinea pig cochlear nucleus (Parker et al., 1987). In addition to permitting a correlation between cell morphology and physiological response characteristics, this technique has conveniently resulted in transneuronal transport of HRP. Transneuronal transport of HRP, well described in other systems, is the process by which HRP in the injected cell gains access to other neurones, whether or not their synaptic contacts are involved. The process is facilitated by a number of factors; e.g. a large injection of HRP into the cell, adequate physiological condition of the animal (and, therefore, of the donor and recipient cells), a long post-injection survival time (Nassel, 1981), and increased activity of the neurones involved by appropriate stimulation (Harrison et al., 1984). This communication describes the physiological characterisation and HRP-labelling of a single giant (large multipolar) cell, the soma of which was situated in the guinea pig anteroventral cochlear nucleus (AVCN), and the transneuronal transport of HRP from this cell into a variety of different cell types in the cochlear nucleus. A preliminary communication of these data has been made (Parker et al., 1986).

METHODS

The pigmented guinea pig was anaesthetised using neuroleptanaesthesia (Evans, 1979a) and surgically prepared according to Evans (1979b). The cochlear nucleus was approached from the opened posterior fossa as in Evans and Nelson (1973). Intracellular and quasi-intracellular recordings of spikes were made using a micropipette filled with 4% HRP (Sigma type VI) in 0.5M KCl and having an impedance after bevelling of 80 megohms. The receptive field properties of the cell were mapped by the method of Evans (1979b), whereby 50ms tone-bursts were presented according to a pseudo-random frequency-intensity sequence, during which spikes were counted. A peri-stimulus-time (PST) histogram (using 500 μs bin widths) was also recorded in response to 256 tone-bursts (50ms duration) at the cell's characteristic frequency (CF) and at 96 dB SPL. Data were collected from quasi-intracellular responses with 4mV monophasic positive spikes and a d.c. potential of -20mV. Following intracellular penetration (probably intradendritic), the spike voltage increased to 6mV and the d.c. potential stabilised at -55mV. HRP was then administered via a single iontophoretic injection of 15 nA.min. The animal was in good physiological condition (as determined by the threshold of the cochlear action potential) for 5.3 hours post-injection survival, during which the auditory system was acoustically stimulated. The animal was then perfused with vascular flush and fixative and the cochlear nucleus sectioned (50-60μm) in the parasagittal plane. Subsequent histological processing was carried out according to Brown and Fyffe (1984), using the Hanker-Yates method (Hanker et al., 1977), with cobalt intensifications (Adams, 1977).

RESULTS

The receptive field of the cell is presented in figure 1 (A). It had a solely excitatory response area larger than those of cochlear nerve fibres with similar CF (14.8kHz); i.e. could be classified as type 1 or 2 according to Evans and Nelson (1973). The rate-level function at CF in figure 1 (B) shows minimal spontaneous activity and monotonic rate-level characteristics over a 50dB range. The PST histogram in response to 50ms tone-bursts at 96 dB SPL, presented in figure 1 (C), was of the transient/onset chopper type (Pfeiffer, 1966; Bourk, 1976; Rhode et al., 1983a).

Microscopical analysis of the histological material revealed the dark brown HRP reaction product within the soma, axon, dendrites, and dendritic terminal swellings of the injected cell. Fig. 2 shows a high resolution camera lucida reconstruction of this cell. Its soma and dendritic tree were situated posterodorsally in the AVCN. It had extensive dendritic arborisation which spanned nearly 2mm from the AVCN to the PVCN. Its axon could be traced as far as the region of origin of the intermediate and dorsal acoustic striae. On the basis of its size (38μm across the widest point of the soma) and shape it was classified as a giant (or large multipolar) cell.

Associated with the dendritic terminal swellings were thirteen patches of light brown HRP reaction product (stippled areas A to M in Fig. 2). These patches were assumed to be transneuronally filled cell somas. This assumption was later confirmed by Nissl counterstaining and, subsequently, an attempt was made to classify them into specific cell types. In addition to using soma size and shape, this was achieved by characterising the Nissl patterns (Osen, 1969; Hackney and Osen, 1985). It was tentatively concluded that: A, B, D, E and L were multipolar cells, C and G were spherical cells, J and K were small cells, H and M were giant cells and there was one globular cell (I). K could not be securely identified.

An important finding is that the dendritic tree of the giant cell cuts across many iso-frequency planes. In addition, although it is not evident from Figure 2, the transneuronally filled cells were distributed widely in the AVCN from the most lateral (A) to the most medial (M).

CONCLUSIONS AND DISCUSSION

The combined physiological and anatomical evidence suggests that the injected giant (large multipolar) cell was probably involved in information gathering over a large area of the ventral cochlear nucleus. In support of this, its dendritic processes cut through many iso-frequency planes, in a manner consistent with its receptive field being wider than that expected from primary afferent fibres. Recent evidence by Wenthold (1987) and Osen (1987) indicates that this class of cell (giant) may be

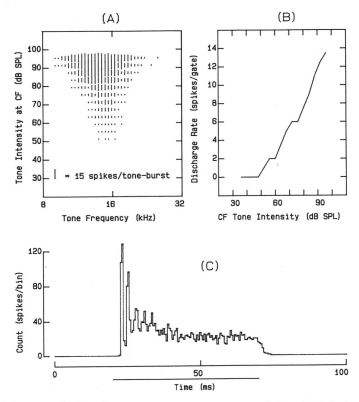

Fig. 1. (A) Map of the frequency-response area of the HRP-injected cell. The length of each bar represents the number of spikes counted in response to each 50 ms tone bursts, randomized in frequency-intensity space. The response area is wider than those found for primary afferent fibres with similar CF (14.8kHz); (B) rate-level function at CF showing that there was minimal spontaneous activity and monotonic discharge rate-level characteristics; (C) PST histogram constructed in response to 50 ms tone-bursts (indicated by the bar below the abscissa) at CF and 96 dB SPL, showing an onset/transient chopper response type.

glycinergic and has its axonal projection to the contralateral cochlear nucleus. The influence of the injected cell on its postsynaptic target cells may, therefore, have been inhibitory. However, it was not possible to trace the axon of the injected cell to the contralateral cochlear nucleus, although, it did travel dorsomedially, eventually sending collaterals both ventrally and caudally before expiry of the HRP label.

The observed transneuronal transport of HRP provided a unique opportunity to identify cell types in contact with the injected cell. Despite the possibility of some misclassification, it is evident that the injected cell had contacts with a variety of different cell types. The exact nature of the contacts is not known, but they could be synaptic rather than simply representing close proximity of the dendrites of the injected cell to the somas of the transneuronally filled cells. Without electron microscopic evidence, however, it is not possible to distinguish between the two possibilities. If the transneuronal transport was across synaptically related contacts, it leads to the conclusion that the giant cell received input from a wide variety of other cells in the ventral cochlear nucleus.

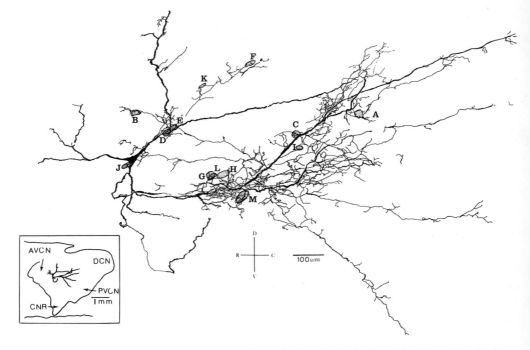

Fig. 2.   Camera lucida reconstruction of the injected cell. The stippled
          areas A to M are faintly labelled cell somas, associated with
          dendritic terminal swellings of the injected cell. The inset
          indicates the outlines of the anteroventral (AVCN), posteroventral
          (PVCN), dorsal (DCN) and cochlear nerve root (CNR) subdivisions
          of the cochlear nucleus, and shows the location of the injected
          cell. The transneuronally filled cells were tentatively
          identified as multipolar cells (A, B, D, E and L), spherical
          cells (C and G), small cells (J and K), giant cells (H and M)
          and a globular cell (I). K could not be identified.

ACKNOWLEDGMENTS

     The authors are indebted to Mr. R. Brunt, Mr. R. Winyard and Miss
J. Smithers for technical assistance, and to Mr. S. Murray for photo-
graphic assistance. This work was supported by the Medical Research
Council and the Wellcome Trust.

REFERENCES

Adams, J., 1977, Technical considerations on the use of horseradish
     peroxidase as a neuronal marker, Neurosci., 2:141-145.
Bourke, T. R., 1976, Electrical responses of neural units in the
     anteroventral cochlear nucleus of the cat, Ph. D. thesis, M. I. T.
     Press.
Brown, A. G. and Fyffe, R. E. W., 1984, Intracellular staining of
     mammalian neurones. Academic Press.
Evans, E. F., 1979a, Neuroleptanaesthesia for the guinea pig, Arch.
     Otolaryngol., 105:185-186.
Evans, E. F., 1979b, in: "Auditory Investigations: the Scientific and
     Technological Basis", Beagley, ed., Oxford University Press, 324-367.

Evans, E. F. and Nelson, P. G., 1973, Responses of single neurons in the cochlear nucleus of the cat as a function of their location and anaesthetic state, Exp. Brain. Res., 17:402-427.

Hackney, C. M. and Osen, K. K., 1985, The cochlear nucleus of the guinea pig: some light microscopic observations, Neurosci. Letters, Suppl 21, S15.

Hanker, J. S., Yates, R. E., Metz, C. B. and Rustioni, A., 1977, A new, specific, sensitive and non-carcinogenic reagent for the demonstration of horseradish peroxidase, Histochem. J., 9, 789-792.

Harrison, P. J., Hultborn, H., Jankowska, E., Katz, R., Storai, B. and Zytnicki, D., 1984, Labelling of interneurones by retrograde transsynaptic transport of horseradish peroxidase from motoneurones in rats and cats, Neurosci. Lett., 45:15-19.

Nassel, D. R., 1981, Transneuronal labelling with horseradish peroxidase in the visual system of the house fly, Brain Res., 206:431-438.

Osen, K. K., 1969, Cytoarchitecture of the cochlear nuclei in the cat, J. Comp. Neurol., 136:453-489.

Osen, K. K., 1987, Anatomy of the mammalian cochlear nuclei; a review. This volume.

Parker, D. J., Evans, E. F. and Hackney, C. M., 1986, Correlations between single neurone receptive fields and morphology in the guinea pig cochlear nucleus, Brit. J. Audiol., Proceedings of Meeting on Experimental Studies of Hearing and Deafness, Brighton 1986.

Parker, D. J., Evans, E. F. and Hackney, C. M., 1987, Receptive field properties and temporal discharge patterns related to neuronal morphology in the guinea pig cochlear nucleus, Brit. J. Audiol., Proceedings of Meeting on Experimental Studies of Hearing and Deafness, Oxford 1987.

Pfeiffer, R. R., 1966, Classification of response patterns of spike discharge for units in the cochlear nucleus: tone burst stimulation, Exp. Brain. Res., 1:220-235.

Rhode, W. S., Oertel, D., and Smith, P. H., 1983a, Physiological response properties of cells labelled intracellularly with horseradish peroxidase in cat ventral cochlear nucleus, J. Comp. Neurol., 213:448-463.

Rhode, W. S., Smith, P. H. and Oertel, D., 1983b, Physiological response properties of cells labelled intracellularly with horseradish peroxidase in cat dorsal cochlear nucleus, J. Comp. Neurol., 213: 426-447.

Rouiller, E. M. and Ryugo, D. K., 1984, Intracellular marking of physiologically characterised cells in the ventral cochlear nucleus of the cat, J. Comp. Neurol., 225:167-186.

Wenthold, R. J., 1987, Glycine immunoreactivity in the cochlear nucleus of the guinea pig: evidence for a glycinergic pathway connecting the two cochlear nuclei. A. R. O. abstracts, 212-213.

MORPHOMETRIC AND CYTOARCHITECTURAL STUDY OF THE DIFFERENT NEURONAL

TYPES IN THE VCN OF THE RAT

E. Saldaña, J. Carro*, M. Merchan, and F. Collia

Chair of Biology, Faculty of Medicine
University of Salamanca
37007, Salamanca, Spain
*Chair of Psychology, Faculty of Psychology
University of Salamanca, Spain

INTRODUCTION

The study and typification of the cytoarchitecture of the cochlear nuclei have been addressed from several points of view. The aim of the present study was to attempt to classify morphometrically the different neuronal types of the cochlear nuclei by correlating the Nissl and Golgi techniques.

MATERIAL AND METHODS

Seventy five Wistar rats were used for this study; 15 animals weighing between 225 and 300 grams were processed by the Nissl method (cresyl violet) and 6, from three to six weeks old, were processed by the variation of the Golgi method proposed by Sotelo and Palay (1968).

Keeping in mind six morphological criteria (shape of neuronal soma, characteristics of the neuronal soma outlines, dendrite characteristics, axon beginnings, size and topography of the neuronal soma), the neurons impregnated by the Golgi method were classified in five main groups: bushy, stellate, large multipolar, octopus and cochlear root globular neurons.

Neurons previously identified in parasagittal sections were measured with the help of a Kontron MOP AM/03 digitial board. The area, perimeter, diameter and shape factor of the neuronal soma were determined for each neuron (Fig. 2).

The data obtained were processed statistically. A descriptive analysis of the different neuronal types was carried out with the SPSS statistical package (Nie et al., 1979). Using the BMDP (Dixon, 1985) statistical package, by means of analysis of variance it was observed that the five groups under study were different. The BMDP package also permitted a differential functions (one for each neuron type) which in turn allowed us to assign any problem cell to the group to which it had the greatest probablity of belonging. To do so, it was only necessary to measure the problem cell and substitute the figures for its area, perimeter, etc., in the five classification functions. The function giving the highest

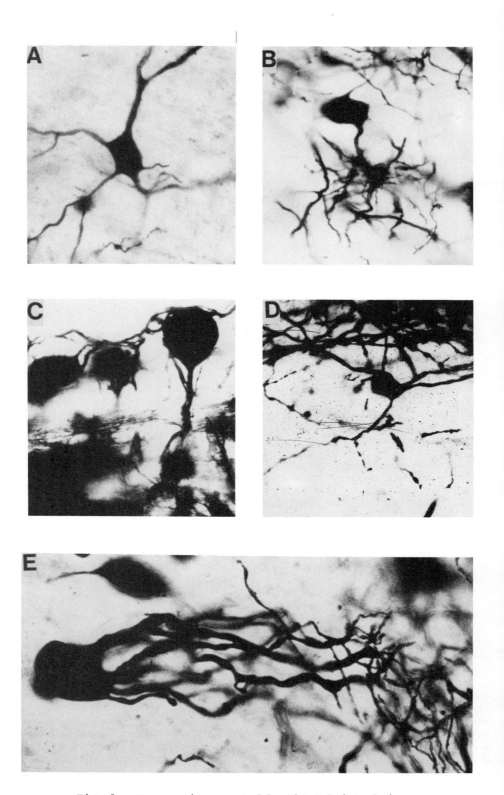

Fig. 1. Neurons impregnated by the Golgi technique.

Fig. 1. (cont'd)
A. Multipolar neuron. Note the large polygonal soma and the long scantily branched dendrites. x 1625.
B. Bushy neurons. Note the rounded soma, the single dendrite stem with a profusely divided hillock and the axon beginning opposite to the main dendrite. x 1625.
C. Cochlear root globular neuron. Chain of globular neurons stained with the Golgi method. The rounded soma and thick dendrite stem (arrow) perpendicular to the fibers of the cochlear nerve stand out clearly. x 1625.
D. Superficial stellate neuron. Note its greater axis parallel to the dendritic network underlying the surface of the cochlear nucleus. x 2250.
E. Octopus neurons. The marked polarity of their dendrites is noteworthy.

value indicated which type the cell belonged to.

Following this, we quantified a large number of neurons stained by the Nissl method, subjecting them to the differential functions. The latter had been previously corrected in order to adjust to the size difference among the neurons using the different methods (Golgi and Nissl). To calculate the actual difference in cell size (not in tissue), we measured 304 globular neurons stained by the Nissl method and 159 globular neurons stained using the Golgi method since globular neurons are clearly identifiable by both staining methods. With the Golgi method the cells were seen to be 25.92% larger.

Having identified the neurons stained with cresyl violet, the cytological characteristics of each neuron type can by observed with this technique.

Cytoarchitectural maps can be constructed assigning the space coordinates to each neuron stained by the Nissl method.

RESULTS

The neurons impregnated by the Golgi method were classified (according to the above-mentioned morphological criteria) in five main groups:bushy, superficial stellate, large multipolar, octopus and cochlear root globular neurons.

The principal morphological attributes observed were:
- Bushy neurons: rounded or oval soma, with a single deformation in its outline. Powdery Nissl grains. Single nucleolus with a variable position.
- Superficial stellate neurons: flat or triangular soma., scanty cytoplasm, flat nucleus. High karyoplasmatic ratio.
- Large multipolar neurons:triangular or polygonal outline. Abundant cytoplasm with some Nissl bodies, almost always peripheral. Large and easily identifiable nucleolus.
- Octopus neurons:long or triangular soma. Granular Nissl bodies. Nucleus consistently displaced to the convex zone of the cell, which usually has a cleft in the face directed towards the concavity.
- Root neurons: oval or pyriform soma, prominent Nissl bodies, with a large perinuclear Nissl body. Eccentric nucleus with a prominent nucleolus.

DISCUSSION

From the study of the arithmetic means, processed with the SPSS and

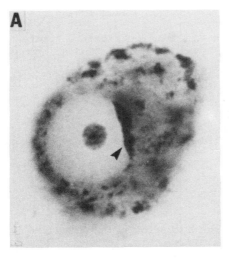
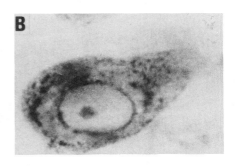
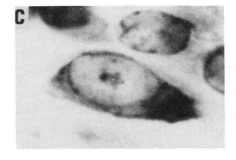
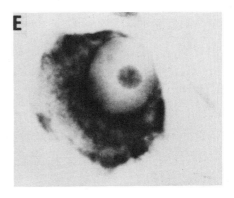
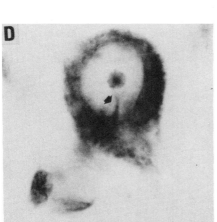

Fig. 2. A. Globular neuron of the cochlear root stained by the Nissl
method. Oval, voluminous soma with easily visible Nissl bodies;
large perinuclear Nissl body (arrow). Eccentric position of the
nucleus, large nucleolus inside. x 1625.
B. Bushy neuron stained by the Nissl method. Its outline shows
a marked deformation suggestive of the beginning of a thick
dendritic stem. x 1625.
C. Superficial stellate neuron stained by the Nissl method.
Fusiform polygonal soma with intensely stained scanty cytoplasm.
x 1625.
D. Octopus neuron stained by the Nissl method. Cytoplasm occupied
by Nissl substance with a granular appearance. The nucleus always
occupies the posterior part of the cell, the point opposite the
beginning of the main dendrites (arrow); Note presence of a
marked indentation on the anterior pole of the nucleus. x 1625.
E. Large polygonal multipolar neuron, Nissl bodies basically in
a peripheral and perinuclear position. Note large nucleus. x 1625.

BMDP packages, the following was deduced:

- The neurons with the greatest diameter and perimeter are those of the root and the smallest ones the superficial stellate neurons.
- The neurons with the greatest maximum diameter are those of the root and those with the smallest diameter the bushy and superficial stellate neurons.
- The roundest neurons are those of the root and the bushy type while the most irregular type are the octopus neurons.

From the study of the coefficients of variability (standard deviation/ arithmetic mean) the most homogeneous type of neurons are seen to be the bushy neurons and the most heterogeneous the octopus type.

The discriminant analysis performed in order to determine whether the cells measured can be identified from the objective parameters obtained and to observe the effectiveness of these parameters, permit the following conclusions:

- The most useful parameter when distinguishing neuronal types is the perimeter, with 45% efficiency.
- The maximum efficiency in the classification of neuronal types is obtained with simultaneous use of several parameters: area, perimeter, and shape factor, with an efficiency greater than 70%.

CONCLUSIONS

From the foregoing, it is inferred that the morphometric study of a population of neurons impregnated with the Golgi method permits statistical distinction of different types with a high discrimination capacity.

As a result of the discriminant capacity, different discriminating functions are obtained, one for each neuronal type, whose unknowns are the values of the parameters used.

To identify a problem cell, its values are substituted in all the discriminating functions and the cell belongs to the group from that proving to have the highest discriminating value.

By applying this method it is possible to identify neurons stained by the Nissl technique, establishing an equivalence pattern between the neurons observed with the Golgi method and the morphological image of the neurons obtained with the Nissl technique. This multitechnique analysis, Golgi method + morphometry + statistical analysis + Nissl method, permits one to gain knowledge of the exact cellular distribution of a particular zone of the nervous system.

REFERENCES

Dixon, W. J., 1985, Biomedial Computer Program (BMDP). University of California Press. Berkeley, Los Angeles.
Nie, N. H. et al., 1979, Statistical package for the Social Sciences, McGRaw-Hill, New York.
Sotelo, C. and Palay, S., 1968, The fine structure of the lateral vestibular nucleus of the rat, J. Cell Biol., 26:151-179.

RELATIONSHIP BETWEEN THE LEVEL OF ORIGIN OF PRIMARY FIBERS IN THE RAT COCHLEA AND THEIR SPATIAL DISTRIBUTION IN THE RAT COCHLEAR NUCLEI

M. Merchan, D. E. Lopez, E. Saldaña, and F. Collia

Chair of Biology, Faculty of Medicine
University of Salamanca
C/Fonseca, 37007, Salamanca, Spain

INTRODUCTION

The distribution pattern of the primary afferents to the cochlear nuclei was proposed by Cajal (1909) and Lorente de Nó (1933). Later it was analyzed by different authors in several different species (for a review, see Amesen et al., 1978). These authors concluded that the primary fibers branch in a "V" shape and that they have a geometric distribution that in all cases is related to a frequency; that is, to a particular origin in the cochlea. Our group has analyzed the problem with transganglionic HRP transport and silver staining methods (Merchan, 1985 a, b) and has concluded that apart from the tonotopic "V" shapes reported by Lorente de Nó (1933) there are nerve fibers that give rise to two primary plexuses (anterior and posterior, Fig. 1) that have a different distribution.

The primary anterior plexus is mainly distributed throughout the more lateral regions of the AVCN (zone with bushy and stellate neurons). The posterior primary plexus is distributed in the posterior territory of the PVCN (zone with octopus neurons) and has two terminal forms, either on the dorsal nucleus (the deepest zone) or on the boundary between the DCN and the VCN (the grain cell zone) and the surface of the AVCN.

MATERIAL AND METHODS

Thirty Wistar rats weighing 200 g were used. The animals were anaesthetized with equithesin and the left cochlea was exposed, making a small orifice in the apical spiral (10), another at the beginning of the medial spiral (5) and another at the end of the medial spiral (5). Orifices were also made at the beginning of the basal spiral (5) and another at the end of this spiral (5). A small metal electrode was inserted into the orifice; this was connected to an electrocauterization device, thereby producing small lesions in the organ of Corti.

After a degeneration period of 4 days, the rats were perfused with 15% saline formalin and the cochlea were decalcified, embedded in paraffin and serially sectioned to contrast the area and characteristics of the lesion.

The cochlear nuclei were sectioned serially along the parasagittal plane at 25 μm and observed according to the technique of Fink-Heimer to observe the extension and topographic distribution of the degenerated fibers.

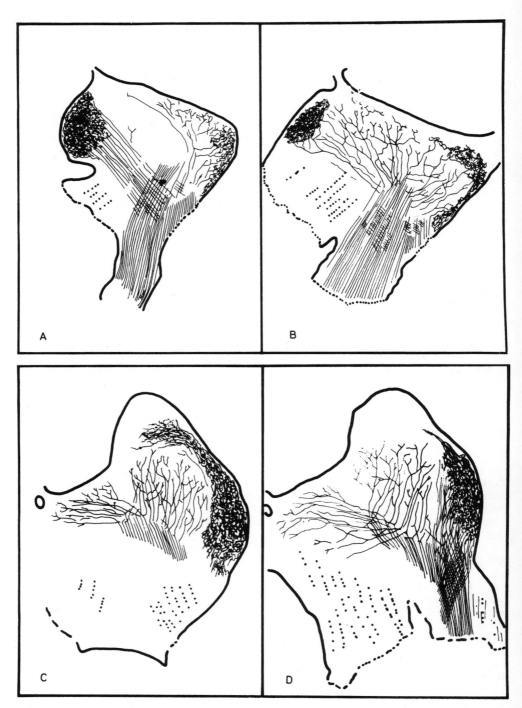

Fig. 1. Camera lucida drawing of parasagittal sections from lateral to
medial (A-D) representing the primary cochlear fibers in the
cochlear nuclei after transganglionic tracing with HRP injected
into the cochlea. Figure A shows the existence of a dense plexus
in the anterior region (asterisk). Figure B illustrates the
bunch-like disposition of the primary afferents. Figures C and

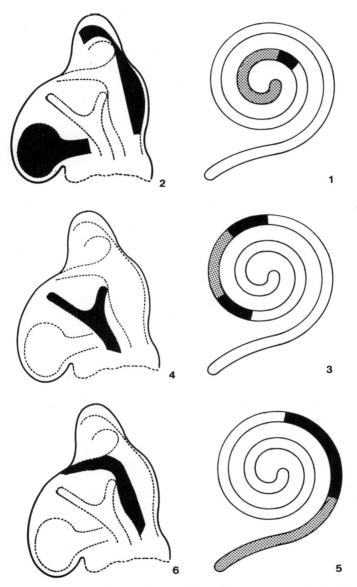

Fig. 2. Schematic drawing of the cochlear lesion at apical (1), medial (3) and basal (5) level.In the three figures the black area represents the zone of most intense lesion (neuronal loss greater than 85%). The gray area represents the zone of least neuronal loss (neuronal loss greater than 15%). (2) Pattern resulting from apical lesions. The degenerated territories correspond to the posterior-most

Fig. 2. (cont'd) region of the posterior plexus and to the anterior plexus. (4) Pattern resulting from medial lesions. The degenerated territories correspond to the central zone of the cochlear nuclei, in particular to the territory (without plexuses) corresponding to the distribution of the tonotopic "V"s. (6) Pattern resulting from basal lessions. An area of degeneration can be seen corresponding to the anterior-most region of the posterior plexus. As a result and according to distribution, most of the fibers of this localization on the boundary between the dorsal and ventral nuclei and the superficial region of the anterior nucleus also appear in this territory.

The results were drawn with a camera lucida and these pictures were used to construct schema accounting for all the affected fibers.

RESULTS

After focal lesion of the cochlea at different levels, three patterns of degeneration of the primary fibers in the cochlear nuclei are obtained.

The lesion in the apical spiral (Fig. 2-1) produces a fibrillar degeneration in the cochlear nuclei that in parasagittal sections covers the anterior and ventral territory of the anterior division of the ventral nucleus and a zone occupying the posterior zone of the posterior division of the ventral nucleus which curves inwards and penetrates the deep portion of the dorsal cochlear nucleus (Fig. 2-2).

The lesion in the medial spiral (Fig. 2-3) yields a "V"-shaped image of degeneration that occupies the boundary between the anterior and posterior division of the ventral cochlear nucleus and later extends, bifurcating in the anterior and posterior direction (Fig.2-4). The zone of degeneration does not reach the surface of the ventral nucleus. The shorter posterior zone of degeneration does not penetrate the territory of the posterior plexus (Fig. 1-1, 2-4).

Finally, the lesion of the basal spiral (Fig. 2-5) produces a zone of degeneration that extends linearly in the ventral dorsal direction along the anterior part of the posterior division of the ventral cochlear nucleus and on reaching the boundary between the ventral and dorsal cochlear nuclei curves in the anterior direction until it reaches the surface of the antero-ventral cochlear nucleus (Fig. 2-6).

SUMMARY

The anterior-most region of the posterior plexus and the anterior plexus have a common origin, either because they correspond to fibers originating at the same cochlear level that uncoil in different directions or because they are branches of a common fiber whose bifurcation is not possible to observe with the HRP technique (Fig. 1).

The classic pattern of Lorente de Nó coincides only with the degeneration patterns corresponding to the lesions of the medial levels of the cochlea (Fig. 2-2).

The degeneration pattern of the more basal level corresponds to the territory of the distribution of dendrites of the octopus neurons (anterior zone of the posterior plexus) and to the area of distribution of the third plexus (boundary of the VCN and DCN and surface of the AVCN) which means that it is located in the zone of the grain cells and superficial stellate neurons.

The findings of the present work allow us to conclude that the classic tonal scheme should be revised, at least in the case of the rat.

REFERENCES

Arnesen, A. R., Osen, K. K. and Mugnaini, E., 1978, Temporal and spatial sequence of anterograde degeneration in the cochlear nerve fibers of the cat, J. Comp. Neurol., 178:679-696.
Lorente de Nö, 1933, Anatomy of the eighth nerve. The central projection of the nerve endings of the internal ear, Laryngoscope, 43:1-38.
Merchan, M. et al., 1985a, Distribution of the primary afferent fibers in the cochlear nuclei. A silver and horseradish peroxidase (HRP) study, J. Anat., 141:121-130.
Merchan, M. et al., 1985b, Distribution of primary cochlear afferents in the bulbar nuclei of the rat: a horseradish peroxidase (HRP) study in parasagittal sections, J. Anat., 144:71-80.
Ramon y Cajal, 1909, Histologie du Systeme Nerveux de L'Homme et des Vertebres, Paris:Maloine.

THE CENTRAL PROJECTION OF INTRACELLULARLY LABELED AUDITORY NERVE FIBERS:

MORPHOMETRIC RELATIONSHIPS BETWEEN STRUCTURAL AND PHYSIOLOGICAL PROPERTIES

E. M. Rouiller and D. K. Ryugo

Institut de Physiologie
Rue du Bugnon 7, 1005 Lausanne, Switzerland
Center for Hearing Science
John Hopkins Univ. School of Medicine
Baltimore, MD, USA

INTRODUCTION

The acoustic receptor cells in the cochlea contact the peripheral processes of primary auditory neurons, called spiral ganglion (SG) cells. The central processes of SG cells bundle together to form the auditory nerve (AN) and project into the cochlear nucleus (CN). Through this pathway, transduced acoustic information is conveyed directly as input to the brain. Essentially all of our knowledge concerning the nature of this input via the AN is derived from the myelinated axon of type-I SG cells. Type I neurons innervate inner hair cells exclusively (Kiang et al., 1982) and represent 90-95% of the SG population (Spoendlin, 1971). Virtually nothing is known about the functional properties of the type-II SG cells which innervate outer hair cells and comprise the remainder of the SG population.

Most aspects of primary auditory neuron's responses to simple acoustic stimuli can be predicted once the tuning curve and spontaneous discharge rate (SR) are known (Kiang, 1984). The SR is related to the threshold where units having low or medium SRs ($\ll 18$ spike/s) have relatively high thresholds, whereas units with high SRs (> 18 spike/s) have low thresholds. Single type I SG neurons of the different SR classes differ significantly in their response properties (Liberman, 1978), and in the caliber and position of their synapse on the inner hair cells (Liberman, 1982). Given such differences, we analyzed the pattern of the central arborizations among type I SG cells in order to describe the morphological characteristics of type-I axons with respect to their physiological properties.

METHODS

For this study, the activity of SG neurons was recorded intra-axonally in the ANs of 15 cats using beveled glass micropipettes. Following the determination of a tuning curve and SR, the corresponding single AN fiber was injected with the tracer horseradish peroxidase (HRP, 10% solution in 0.05 M Tris buffer, pH=7.3, containing 0.15 M KCl) through the recording pipette. Postinjection tuning curves and SRs plus a continuous negative resting potential were used to ensure that the pipette remained inside the same axon. Approximately 24 hours after the injection, the animal was perfused, its brain was dissected and 40-60 microns thick Vibratome sections of the AN and CN were serially collected and reacted with diaminobenzidine. Details of the methods and criteria for the recovery of the labeled axons have been previously described (Fekete et al., 1984; Rouiller et al., 1986).

RESULTS

Successful experiments produced darkly labeled axons of type-I SG cells which could be completely reconstructed from serial sections with the aid of a light microscope and drawing tube. The arborization of 7 physiologically defined AN fibers is illustrated in Fig. 1. Each AN fiber bifurcates to give rise to an ascending branch going anteriorly in the anteroventral cochlear nucleus (AVCN) and a descending branch innervating posteriorly the posteroventral cochlear nucleus (PVCN) and usually (85% of the cases) the dorsal cochlear nucleus (DCN). The fibers having the lowest CFs (0.17 and 0.18 kHz) bifurcated immediately upon entering the CN and projected in the ventro-lateral region of the CN. Fibers having higher CFs bifurcated progressively higher in the nerve root region of the CN and projected more dorso-medially. These observations from adult cats reveal the cochleotopic projections on a fiber-by-fiber basis from inner ear to brain, and are consistent with the tonotopic organization of the CN (Rose et al., 1959; Bourk et al., 1981) and general descriptions reported for immature animals using Golgi methods (Lorente de No, 1981).

The terminal branches of the fibers were drawn at high magnification (1200x) where distinct swellings (endings) were morphometrically analyzed. These terminal structures varied considerably in shape and size as illustrated in Fig. 2, and were organized as small endings (simple boutons, string endings and small complex endings) or large complex endings (modified endbulbs and endbulbs of Held), according to criteria previously defined (Rouiller et al., 1986). Electron microscopic analysis of approximately 20 such structures was consistent with their being synaptic in function. In addition, distinct swellings were also observed along the length of many terminal branches. When the diameter of the fiber expanded at least 3-fold over a distance of less than 2 microns and returned to its original diameter in an equally abrupt fashion, the expansion was called "en passant swelling" (Fig. 2A). Four unmyelinated en passant swellings were studied with the electron microscope, all of which contained round vesicules and 2 of which exhibited membrane specializations typical of synapses.

Morphometric measurements were performed on 27 AN fibers whose anatomical and physiological identities were unambiguous. Twelve had SRs ≤18 spikes/s with high or intermediate thresholds to sound and 15 had high SRs (>18 spikes/s) with low thresholds. For each fiber, all terminal and en passant swellings were categorized, counted and their size estimated by silhouette planimetry. These data revealed a systematic correspondence between the morphological characteristics of the central axons of type I SG neurons and SR grouping. That is, the number of branches was greater for low and medium SR fibers than for high SR fibers (Fekete et al., 1984). The present analysis demonstrates that the number and size of small endings as well as the number of en passant swellings differed significantly across SR groups (Table I, part A). Low-medium SR fibers had, on average, twice as many small endings and en passant swellings as did high SR fibers, but their small endings were smaller in cross-sectional area.

On the other hand, modified endbulbs and endbulbs of Held did not exhibit number or size differences related to SR, but they did manifest

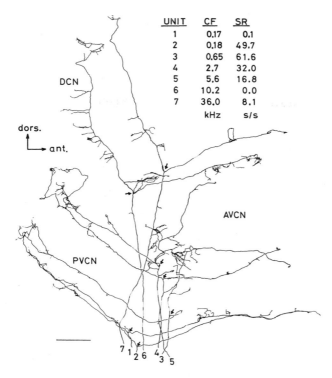

UNIT	CF	SR
1	0.17	0.1
2	0.18	49.7
3	0.65	61.6
4	2.7	32.0
5	5.6	16.8
6	10.2	0.0
7	36.0	8.1
	kHz	s/s

Fig. 1.   Camera lucida reconstruction of the central projection in the CN of 7 SG neurons intra-axonally recorded and labeled in the same AN. The scale bar is 500 microns. The 2 fibers having the lowest CFs (1 and 2) could not safely be identified individually and were therefore not considered in the morphometric analysis. Small arrows point the bifurcation for each fiber.

other quantifiable variations (Benson et al., 1986). Finally, counts were performed separatly with respect to ascending or descending branches in order to determine whether the number differences in small endings and en passant swellings related to SR were homogeneously distributed in the CN. The data revealed that the fiber differences in the number of small endings and en passant swellings were accounted for by the ascending branches innervating AVCN, whereas the descending branches confined to PVCN and DCN showed less variability related to SR (Table I, part B). We did not find morphological variations in endings related to fiber CF.

DISCUSSION

The present data extend previous observations that type-I SG neurons exhibit reliable structure-function relationships. CF is related to the longitudinal location of the terminal along the cochlear duct, and is topologically conferred from inner ear to CN. SR is related to fiber threshold, the caliber of the peripheral process, the position of the terminal upon the inner hair cell innervated, and is highly correlated to the morphology of the central arborization. The different SR types, which can be seen at all CF values, may play fundamentally different roles in the overall process of auditory perception. Low-medium SR fibers have relatively high thresholds and wider dynamic ranges than do high SR fibers (Evans and Palmer, 1980). Threshold and dynamic range are 2 important parameters relevant for "place" coding of complex sounds in

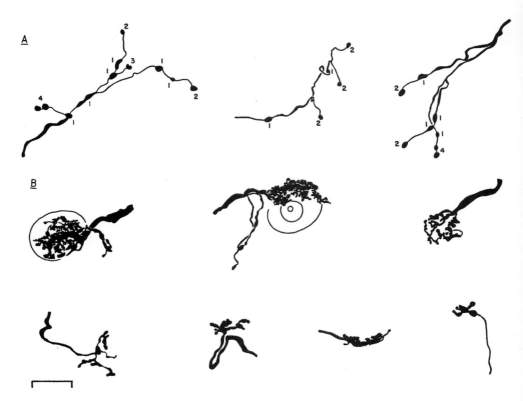

Fig. 2. Camera lucida drawings at high magnification of terminal branches of the axon of SG neurons illustrating the various types of endings. A: en passant swellings (1) and small size endings (2=simple terminal bouton, 3=small complex ending, 4=string of endings). B: large size terminal endings (endbulbs of Held on top and modified endbulbs on bottom). Scale bar is 20 microns.

Table I Part A:individual auditory nerve fiber characteristics with respect to SR:average number and size of endings with statistical comparisons across SR groups.

	High SR	Low-med. SR	p value
nb. of fibers	15	12	--
total nb. of endings	58.5 + 5.2	98.6 + 44.5	p < 0.01
nb. of small endings	53.8 + 19.5	94.2 + 41.3	p < 0.01
nb. of mod. endbulbs	2.9 + 0.5	2.8 + 1.0	ns
nb. of endbulbs of Held	1.7 + 0.2	1.6 + 0.2	ns
nb. of en passant swel.	34.9 + 18.8	74.1 + 42.9	p < 0.01
area of small endings*	6.6 + 1.6	4.8 + 3.4	p < 0.01
area of mod. endbulbs	39.5 + 17.2	23.0 + 18.2	ns
area of endbulbs of H.	225.8 + 85.7	238.2 +129.8	ns
area of en passant swel.	4.2 + 3.0	4.1 + 3.3	ns
* silhouette area measurements are in microns square.			

Table I

Part B: average number of small endings and en passant swellings computed
seperatly for terminal branches emitted by the ascending (AB) or
descending (DB) branches.

1) high SR fibers	AB	DB
nb. of small endings	21.7 ± 11.7	29.4 ± 11.4
nb. of en passant swel.	15.9 ± 10.7	17.2 ± 10.2
2) low-med SR fibers	AB	DB
nb. of small endings	50.6 ± 30.9	37.2 ± 15.0
nb. of en passant swel.	41.0 ± 32.9	28.4 ± 12.6

term of discharge rate profiles in the AN. At moderate and high intensity
levels, this representation for complex sounds is lost in the group of
high SR fibers because their discharge rate saturates or even decreases,
whereas low-medium SR fibers are still within their operational range
at such intensities (Sachs and Young, 1979; Miller and Sachs, 1983).
In addition, the spatial coding of pure tones within the population of
auditory nerve fibers was significantly less disrupted by the simultaneous
presentation of background noise in low-medium SR fibers than in high SR
ones (Costalupes, 1985). These properties of low-medium SR fibers may
therefore be essential for discriminating sounds at higher intensity levels
in view of the limited dynamic range and saturating discharge rate of
most AN fibers.

The significance of the morphometric differences of small endings and
en passant swellings related to SR is not well understood, but may also
be relevant to the processing of loud sounds. The increased number of
endings was confined to AVCN, and it may be that low-medium SR fibers are
especially involved in elucidating the middle ear muscles acoustic reflex
because of their elevated thresholds and because the reflex pathway
requires neurons of AVCN but not PVCN or DCN (Borg, 1973). The widespread
distribution of low-medium SR endings in AVCN may be related to the notion
that the number of active neurons would be proportional to the perceived
loudness of a sound stimulus (Stevens and Davis, 1938). Activation of
low-medium SR fibers by intense sounds would tend to produce a spread of
activity to additional neurons in the CN.

REFERENCES

Benson, T. E., Sento, S., and Ryugo, D. K., 1986, Endbulbs of Held and
sperical cells:activity dependent variations in morphology, Soc.
Neurosci. Abst., 12:1266.
Borg, E., 1973, On the neuronal organization of the acoustic middle
ear reflex:a physiological and anatomical study, Brain Res.,
49:101-123.
Bourk, T. R., Mielcarz, J. P., and Norris, B. E., 1981, Tonotopic
organization of the anteroventral cochlear nucleus of the cat,
Hearing Res., 4:215-241.
Costalupes, J. A., 1985, Representation of tones in noise in the
responses of auditory nerve fibers in cats:I Comparison with
detection thresholds, J. Neuroscience, 5:3261-3269.
Evans, E. F. and Palmer, A. R., 1980, Relationship between the dynamic
range of cochlear nerve fibers and their spontaneous activity,
Exp. Brain Res., 40:115-118.

Fekete, D. M., Rouiller, E. M., Liberman, M. C. and Ryugo, D. K., 1984,
    The central projections of intracellularly labeled auditory nerve
    fibers in cats, J. Comp. Neurol., 229:432-450.
Kiang, N. Y. S., 1984, Peripheral neural processing of auditory informa-
    tion, in: "Handbook of Physiology, Section I, Vol. III, Part 2",
    Darian-Smith, I., ed., American Physiological Society, Bethesda,
    639-674.
Kiang, N. Y. S., Rho, J. M., Northrop, C. C., Liberman, M. C. and
    Ryugo, D. K., 1982, Hair cell innervation by spiral ganglion cells
    in adult cat, Science, 217:175-177.
Liberman, M. C., 1978, Auditory nerve response from cat raised in a low-
    noise chamber, J. Acoust. Soc. Am., 63:442-455.
Liberman, M. C., 1982, Single neuron labelling in the cat auditory nerve,
    Science, 216:1239-1241.
Lorente de No, R., 1981, The primary acoustic nuclei, Raven Press,
    New-York, NY.
Miller, M. I. and Sach, M. B., 1983, Representation of stop consonants
    in the discharge pattern of auditory-nerve fibers, J. Acoust. Soc.
    Am., 74:502-517.
Rose, J. E., Galambos, R. and Hughes, J. R., 1959, Microelectrode studies
    of the cochlear nuclei of the cat, Bull. Johns Hopk. Hos., 104:
    211-251.
Rouiller, E. M., Cronin-Schreiber, R., Fekete, D. M. and Ryugo, D. K.,
    1986, The central projection of intracellularly labeled auditory
    nerve fibers in cats" An analysis of terminal morphology, J. Comp.
    Neurol., 249:261-278.
Sachs, M. B. and Young, E. D., 1979, Encoding of steady-state vowels in
    the auditory nerve: representation in terms of discharge rate,
    J. Acoust. Soc. Am., 66:470-479.
Spoendlin, H., 1971, Degeneration behavior of the cochlear nerve,
    Arch. Klin., Exp. Ohren-Nasen-Kehlkopf., 200:275-291.
Stevens, S. S. and Davis, H., 1938, Hearing:its psychology and physiology.
    John Wiley and Sons, NY, NY, 405-407.

NEUROTRANSMITTER MICROCHEMISTRY OF THE COCHLEAR NUCLEUS AND SUPERIOR OLIVARY COMPLEX

D. A. Godfrey, J. A. Parli, J. D. Dunn*, and C. D. Ross

Departments of Physiology and Anatomy*
Oral Roberts University
Tulsa, Oklahoma 74171, USA

The complex function of the nervous system is made possible by its underlying chemistry. One aspect of this chemistry involves the primary mechanism by which neurons communicate with each other, chemical neurotransmission. It is likely that the neurons of the auditory system are organized chemically as well as structurally and functionally. We may therefore look for populations of neurons employing particular neurotransmitters. These neuronal populations may serve specific functions in the auditory system, just as specific functions are served by cholinergic spinal motoneurons or dopaminergic nigrostriatal neurons.

Of the large number of transmitter candidates, the best established are smaller molecules including acetylcholine, norepinephrine, the excitatory amino acids aspartate and glutamate, and the inhibitory amino acids glycine and gamma-aminobutyrate (GABA) (Cooper et al., 1986). For these molecules, most or all of the generally accepted criteria for establishing a transmitter (Werman, 1966) have been satisfied at one or more types of synapses. These criteria include presence of the proposed transmitter substance and synthetic enzyme(s) in the presynaptic neuron, particularly at its terminals, evidence for release from the terminals upon stimulation of the neuron, presence of a receptor on the postsynaptic neuron, a means for terminating the action of the proposed transmitter (usually enzymatic degradation or uptake), a response to the substance by the postsynaptic neuron which resembles its response to natural stimulation, and synaptic effects of agonist and antagonist drugs which resemble their effects at synapses where the substance is well established as a transmitter.

Our approach to the study of neurotransmitter chemistry focuses upon only a few of these criteria, namely the presence of the proposed transmitter and enzymes of synthesis or degradation. Correlation of the results with those from other groups studying different types of evidence may then enable conclusions about the transmitter status of particular substances. In practice, because of the legacy of previous work on some transmitter candidates, their presence and/or that of their synthetic enzymes constitutes strong evidence for their involvement in neurotransmission. Examples of such substances are acetylcholine and norepinephrine. For other substances, such as aspartate and glutamate, their involvement in aspects of cellular metabolism unrelated to neurotransmission makes demonstration of their mere presence virtually irrelevant. They are present throughout the brain. More relevant information is the demonstration that a particular neuronal population is especially enriched in its content of these substances. For any substance, a necessary early step in evaluating the extent of its neurotransmitter role in a particular region is a quantitative assessment of its prevalence there, obtained by comparing measurements to those for other regions, especially regions where its transmitter function is considered very likely or unlikely. Another useful overall comparison is with the average concentration in the brain, obtained by measurements on homogenized whole brain.

The methods used in our studies, based on those of Lowry and Passonneau (1972), are designed to provide quantitative estimates of chemical composition on a microscopic scale, thus enabling correlation of measurements with anatomy and physiology. The basic methods (Fig. 1) involve removal and trimming of brain tissue, rapid freezing, sectioning, freeze-drying to remove water and thereby halt enzymatic activity, microdissection of samples from the freeze-dried sections, weighing of samples on specially constructed microbalances, loading into tubes, and chemical assay. In addition, by use of a drawing tube attachment on the dissecting microscope, the locations of the microdissected samples are mapped so that the chemical measurements may be correlated precisely with tissue structure (Godfrey and Matschinsky, 1976).

In associating proposed transmitters with particular neural pathways, a lesioning approach has been used in conjunction with the microchemical

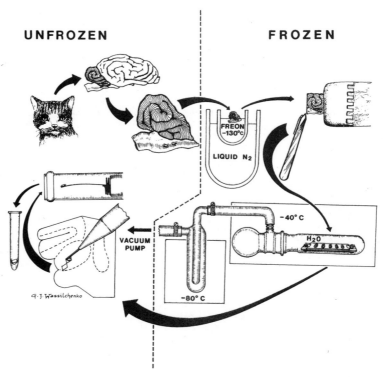

Fig. 1. Overview of the basic microchemical, or quantitative histochemical, methodology. Once the water has been removed from the frozen sections by the freeze-drying procedure, the microdissection, weighing and loading of tissue samples can be carried out at room temperature with relative humidity maintained below 50%. Drawing by George J. Wassilchenko.

Fig. 2.   "Life-size" drawing of a neuron with soma about 2 meters diame-
          ter, dendrites extending upward to about 10 meters, and axon
          projecting for many kilometers across the Vltava River and past
          Prague Castle. Neuron drawn by George J. Wassilchenko.

methods. Since the enzymes involved in transmitter metabolism are made in
the soma of a neuron and transported along its relatively long axon (Fig.
2) to reach the nerve terminals, where the supply of transmitter is
needed, transection of the axon should result in depletion of both enzymes
and transmitter from the region containing the distal portion of the axon
and its terminal field. Unfortunately, the interpretations of the experiments are complicated by possible retrograde effects in the soma of a
neuron after its axon is out, transneuronal effects upon neurons which
have lost some of their innervation, and nonspecific effects from inter-
ruption of blood supply or glial proliferation in the denervated region
(Cowan, 1970;Orkand, 1977). These complications can generally be addressed by careful and detailed analysis of the results if the substances
being measured are localized to specific neuronal populations (Godfrey et
al., 1983, 1987a,b). However, they present more difficulty for substances
as ubiquitously distributed as most amino acids.

     Our microchemical studies have been done primarily on rats and cats.
This presentation will emphasize results for cats. The surgical lesions
employed for the studies (Fig. 3) include cuts of pathways connecting the
cochlear nucleus and superior olivary complex and cuts within the cochlear
nucleus to transect pathways connecting its rostral and caudal parts
(Lorente de Nó, 1981). In sham cats, the cochlear nucleus was exposed,
but no cut was made. Postoperative survival time was 7-10 days.

Acetylcholine

     Much microchemical work has been devoted to the cholinergic system,
in which acetylcholine is the neurotransmitter. The enzyme of synthesis
for acetylcholine, choline acetyltransferase (ChAT), being in high concentration throughout the somata and axons of cholinergic neurons, but virtually absent from noncholinergic cells, is a definitive marker, more useful
than the enzyme of hydrolysis, acetylcholinesterase, or even acetylcholine
itself (Godfrey et al., 1985b).

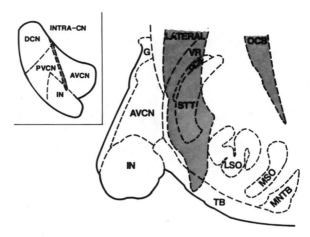

Fig. 3. Locations of surgical lesions, shown schematically on transverse section through cat brain stem. The large "lateral" cut
transected almost all centrifugal pathways to the cochlear
nucleus; the more superficial "OCB" cut transected only the
more dorsal pathways:dorsal and intermediate acoustic striae
and OCB;the intra-cochlear nucleus (intra-CN) cut transected
intrinsic rostral-caudal connections. In 2 cats, a cut aimed as
the "OCB" cut extended deeper and penetrated through most of the
MNTB.

Table 1.   Abbreviations

Asp:	aspartate
AVCN:	anteroventral cochlear nucleus
ChAT:	choline acetyltransferase
CN:	cochlear nucleus
DCN:	dorsal cochlear nucleus, including layers:
	m: molecular layer
	f: fusiform soma, or pyramidal cell, layer
	d: deep layer
G or Gr:	granular region
GABA:	gamma-aminobutyrate
GAD:	glutamate decarboxylase
Glu:	glutamate
IN:	interstitial nucleus (auditory nerve root)
LC:	locus coeruleus
LNTB:	lateral nucleus of the trapezoid body
LSO or L:	lateral superior olivary nucleus
MNTB or N:	medial nucleus of the trapezoid body
MSO or M:	medial superior olivary nucleus
OCB:	olivocochlear bundle
Pyr:	pyramid
PVCN:	posteroventral cochlear nucleus
STT:	spinal trigeminal tract
TB or T:	trapezoid body
VCN:	ventral cochlear nucleus

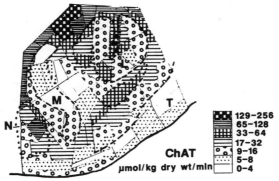

Fig. 4.   Distribution of ChAT activity in a section through the superior
          olivary complex of a cat. The section is adjacent to one which
          displayed prominent staining for acetylcholinesterase activity
          in the dorsal hilus of the LSO. Data are coded in the same way
          as done previously for the cochlear nucleus (Godfrey et al.,
          1977b). Thin lines are boundaries of cut samples. Thick lines
          are regional boundaries, many seen within the freeze-dried
          section itself (solid and dashed lines), others traced from
          adjacent stained sections (dotted lines). The unlabeled region
          outlined ventral to the LSO is the LNTB (Warr, 1975).

     The average ChAT activity for the cat cochlear nucleus is fairly low,
being less than 15% of that for cat brain. Within the cochlear nucleus,
the distribution of ChAT activity includes steep gradients, with highest
activities in parts of the granular region, especially most rostrally.
These granular region activities often exceed average brain values and are
roughly 10% of the values for the facial motor root, a virtually pure
population of cholinergic fibers, and the facial nucleus, a region popu-
lated by cholinergic somata (Godfrey et al., 1977b).

     In the superior olivary complex (Fig. 4), the distribution of ChAT
activity includes a range of values comparable to those in the cochlear
nucleus. Extremely low activities are found in and near the medial superior olivary nucleus (MSO), while highest values are dorsally located, including the dorsal hilus region of the lateral superior olivary nucleus
(LSO). At least to a first approximation, the higher ChAT activities correlate in location with the somata of olivocochlear neurons (Warr, 1975).

     The large lateral cut (Fig. 3) produced a profound (about 70%) reduc-
tion of ChAT activity throughout the lesion-side cochlear nucleus. The
olivocochlear bundle (OCB) cut had comparably large effects in the rostralmost part of the cochlear nucleus, but more moderate effects elsewhere
(Godfrey et al., 1985a):in many regions, the lesion-side ChAT activity
was reduced by only about 20% relative to the control-side activity. In
rat, the lesion-side reduction following OCB transections was about 20%
in all cochlear nucleus regions (Godfrey et al., 1987a).

     Based on our results combined with those of many others (reviewed in
Godfrey et al., 1985b), cholinergic terminals in the cochlear nucleus may
derive partly from interneurons, partly from dorsal centrifugal projections, including collaterals from the OCB (Adams, 1982; White and Warr,
1983;Liberman and Brown, 1986) and partly from ventral centrifugal projections probably located in or near the trapezoid body (Fig. 5). The 3
pathways may contribute roughly equally in the cat, whereas the trapezoid
body route predominates in rat. The centrifugal projections appear to
originate in or pass through the superior olivary complex.

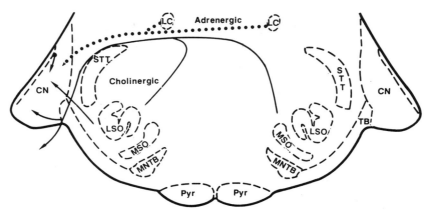

Fig. 5. Possible cholinergic and adrenergic pathways to the cochlear
nucleus represented on a schematic transverse section through
the brain stem.

Norepinephrine

Although we have not studied the adrenergic system, previous work
has indicated that the vast preponderance of norepinephrine in the rat
cochlear nucleus is associated with a bilateral projection from the locus
coeruleus (Kromer and Moore, 1980) (Fig. 5).

Amino acids

To study the distributions of amino acids in the cochlear nucleus,
high performance liquid chromatography (HPLC), using an assay based on
previous work (Hill et al., 1979), has been combined with the microdissection procedure. An advantage of this approach is that the concentrations
of many amino acids are simultaneously measured for each tissue sample.

Measurements have been made on the sets of sections from the same
lesioned cats used to study the cholinergic system (Fig. 3). The rostral
part of the anteroventral cochlear nucleus (AVCN) (region AA of Brawer
et al., 1974) has been studied in most of the cats (Table 2). GABA
concentration was reduced to about half on the lesion side of cats with
the complete lateral cut, but was not affected when the cut extended
just through the OCB. Aspartate was 20-30% lower on the lesion side
following both the lateral and OCB cuts. Glutamate and taurine were not
greatly affected by any of the cuts. Glutamine was elevated on the lesion

Table 2. Effects of Lesions on Amino Acid Concentrations in Rostral AVCN
Lesion-side value as % of control-side value
Data for 2 cats with each of 3 lesion types

	Lateral cut		OCB cut		Intra-CN cut	
Aspartate	76	63	79	82	93	113
Glutamate	91	83	84	102	96	100
Glutamine	165	191	133	150	131	204
Glycine	125	178	87	67	88	127
Taurine	98	86	93	93	--	76
Alanine	138	112	62	82	--	109
GABA	52	45	96	100	93	90

Table 3. Average Amino Acid Concentrations in Superior Olive Regions
Data from a cat with a sham lesion on the contralateral side
mmol/kg dry wt, mean ± S. E.

	No. samples	GABA	Glycine	Glutamate	Aspartate
LSO	13	2 ± 0.2	24 ± 1	25 ± 0.4	14 ± 0.4
MSO	7	2 ± 0.2	15 ± 1	28 ± 1	14 ± 1
MNTB	9	3 ± 1	16 ± 1	21 ± 1	14 ± 1
LNTB	4	4 ± 0.5	14 ± 0.4	21 ± 1	13 ± 1
Periolivary	19	2 ± 0.2	12 ± 0.5	20 ± 1	10 ± 0.4
TB	7	1 ± 0.1	9 ± 0.5	17 ± 1	10 ± 1

side in all cases, presumably related to its relatively high concentration in glia (Patel and Hunt, 1985), which proliferate in the region of
terminal degeneration (Orkand, 1977). The results for glycine and alanine,
with lesion-side concentration lower than control-side following OCB cut,
but higher following the more complete lateral cut, can not be interpreted
without further study.

Additional results will be presented for the 4 amino acids which are
major transmitter candidates - GABA, glycine, glutamate and aspartate.

GABA

The GABA concentrations in the cochlear nucleus tend to be low
compared to other brain regions (Fisher and Davies, 1976; Godfrey et al.,
1977a). The average for the rat cochlear nucleus is about a third of the
value for whole brain (Parli et al., 1987). GABA concentrations in cat
range from very low in the auditory nerve root and much of the ventral
cochlear nucleus (VCN) to highest values in the superficial layers of the
dorsal cochlear nucleus (DCN) comparable to those in the cerebellar cortex
(Godfrey et al., 1977a), which contains some of the best established GABA-
ergic neurons (Cooper et al. 1986). Both GABA and its enzyme of synthesis, glutamate decarboxylase (GAD), seem especially concentrated in the
DCN in all species examined (Tachibana and Kuriyama, 1974;Fisher and
Davies, 1976;Wenthold and Morest, 1976;Godfrey et al., 1978), and both
terminals and somata, not including those of fusiform cells, in the superficial layers of the DCN have been found to immunoreact for GABA and GAD
(Moore and Moore, 1984;Mugnaini, 1985;Thompson et al., 1985; Schwartz
and Yu, 1986; Wenthold et al., 1986b; Peyret et al., 1986; Roberts and
Ribak, 1987).

All superior olivary complex regions so far examined have been found
to contain rather low GABA concentrations (Table 3), resembling some of
the low regional averages in ventral parts of the cochlear nucleus.

Consistent with its effect in rostral AVCN, the large lateral cut
resulted in decreased GABA concentrations throughout the lesion-side
cochlear nucleus (Fig. 6). The decreases of 35-43%, in the DCN layers and
35% in caudal posteroventral cochlear nucleus (PVCN) are close to the
reported decreases in GABA uptake and release following similar lesions
in guinea pigs:30-40% and 20-25%, respectively (Potashner et al., 1985a).

The available data support GABA association with interneurons in the
cochlear nucleus as well as innervation from centrifugal fibers (Fig. 9).
The somata of the interneurons seem especially prominent in the DCN,
based on retrograde transport of GABA (Ostapoff et al., 1985) and the
immunohistochemical labeling studies, but some labeled somata have also

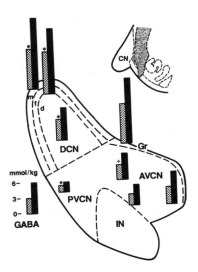

Fig. 6. Average concentrations of GABA in lesion-side (speckled bar)
and control-side (solid bar) cochlear nucleus subregions of
a cat with a large lateral lesion (as shown in diagram above),
shown on a schematic side view of the cochlear nucleus. Numbers
of samples contributing to averages range from 2 to 13. Location of base of bar pair gives approximate location of region
represented. Statistical significance of difference between
lesion and control sides indicated by + for p < 0.05 and *
for p < 0.01.

been reported in the VCN (Adams and Mugnaini, 1984;Wenthold et al.,
1986b;Roberts and Ribak, 1987). The routes of all the GABA centrifugal
fibers to the cochlear nucleus are not fully established. Based on our
data, those to the rostral AVCN follow a ventral course, probably in or
near the trapezoid body. This agrees with immunohistochemical labeling
results of Osen (presented at this symposium). It also fits with the
results of Potashner et al. (1985a), since their cuts missed the rostral
part of the trapezoid body and did not affect GABA uptake or release in
the AVCN. The results of Fisher and Davies (1986) do not contradict this
conclusion since their trapezoid body cuts, which did not affect GAD
activity in the ipsilateral cochlear nucleus as compared to the contralateral side, were placed medial to the superior olivary complex and
would thus be expected to affect only its crossed projections, and to
the same extent on both sides. We have not yet studied the effect of the
OCB cuts on GABA concentrations more caudally in the cochlear nucleus,
but results of Davies (1977) suggest that some of the GAD activity in the
cochlear nucleus derives from fibers entering dorsally in the acoustic
striae.

     Despite the low GABA concentrations in the superior olivary complex,
several studies suggest that GABA pathways to the cochlear nucleus derive
from the lateral (LNTB) and ventral nuclei of the trapezoid body and include crossed as well as ipsilateral connections (Ostapoff et al., 1985;
Saint Marie et al., 1986;Adams and Wenthold, 1987).

Glycine

     In contrast to GABA, glycine concentrations in the cochlear nucleus
are very high overall, as high as those in the gray matter of the spinal
cord, where evidence is strong for a transmitter role (Godfrey et al.,
1977a;Cooper et al., 1986). The average for rat cochlear nucleus is more

than twice that for whole brain (Parli et al., 1987). Within the cat cochlear nucleus, glycine concentrations are high throughout, with highest values in the DCN (Godfrey et al., 1977a), apparently contained in both somata and terminals (Schwartz, 1985; Altschuler et al., 1986; Frostholm and Rotter, 1986; Baker et al., 1986; Wenthold et al., 1986a).

Glycine concentrations are also high in the superior olivary complex, especially in the LSO (Table 3), where, according to increasing evidence, there are terminals of a glycinergic pathway from the ipsilateral MNTB (Moore and Caspary, 1983; Sanes et al., 1985; Schwartz, 1985; Baker et al., 1986; Wenthold et al., 1986a; Bledsoe et al., 1987).

From our lesion studies and results of others (Wu and Oertel, 1986; Adams and Wenthold, 1987), it would seem that much glycine in the cochlear nucleus is associated with interneurons (Fig. 9). However, there is also evidence for glycine pathways entering the cochlear nucleus (Staatz-Benson and Potashner, 1987). One such pathway may derive from the contra-lateral cochlear nucleus (Wenthold, 1987), following a dorsal route (Cant and Gaston, 1982), and there is recent evidence for projections from the LNTB (Adams and Wenthold, 1987) (not shown in Fig. 9). Perhaps our confusing results for glycine, with apparent reductions throughout the control-side cochlear nucleus following the large lateral cuts, relate to some peculiarity of the glycine pathways.

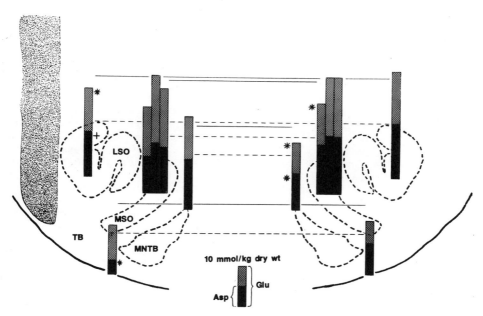

Fig. 7. Average concentrations of glutamate and aspartate bilaterally in superior olivary nuclei and trapezoid body of 2 cats with lateral brain stem cuts. Approximate lesion location is shown as stippled area. Numbers of samples contributing to averages range from 4 to 20. Statistically significant differences between the two sides of the brain, or between lateral and medial sides of the MSO, are given by + for $p < 0.05$ and * for $p < 0.01$. Horizontal lines (solid for glutamate, dashed for aspartate) give average data for the control side of a cat with a sham lesion (Table 3); each line corresponds to the bars at which it ends.

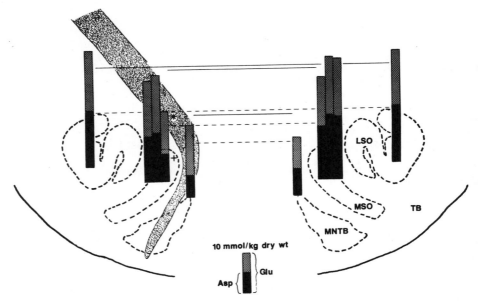

Fig. 8. Average concentrations of glutamate and aspartate bilaterally in superior olivary nuclei of a cat with a brain stem cut. Numbers of samples for averages range from 3 to 10. Other details as in Fig. 7.

Glutamate

Glutamate concentrations in the cochlear nucleus are similar to those in most brain regions (Godfrey et al., 1977a). In the rat, the average concentration in the cochlear nucleus resembles that for whole brain (Parli et al., 1987), but is lower than that for retina or auditory cortex. The distribution of glutamate concentrations in the cochlear nucleus is rather uniform (Godfrey et al., 1977a), and the same appears to be true in the superior olivary complex. Nevertheless, the use of lesions has provided suggestive evidence for specific association of glutamate with ascending auditory pathways. In the cochlear nucleus, initial cochlear ablation results (Wenthold and Morest, 1976; Wenthold and Gulley, 1977) have been supplemented by a large body of other types of evidence supporting a transmitter role for glutamate in the auditory nerve (reviewed in Wenthold, 1985; Potashner et al., 1985b; Martin, 1985; Caspary et al., 1985). Further, the uniformity of the glutamate distribution in the cochlear nucleus can be explained if glutamate is also specifically associated with granule cells (Potashner et al., 1985b; Schwartz, 1985), since the somata and axons of these neurons occupy the regions relatively deficient in auditory nerve innervation.

Lesions have also been found to affect glutamate concentrations in the superior olive (Figs. 7 and 8). The lateral cut of almost all projections from one cochlear nucleus resulted in an approximately 20% decrease in average glutamate concentration in the ipsilateral LSO and on the side of each MSO facing the lesion, and a 27% decrease in the contralateral medial nucleus of the trapezoid body (MNTB). These would be the locations where degeneration of terminals from cochlear nucleus neurons should occur subsequent to the lesion (Warr, 1982). When the lesion was medially located (Fig. 8), it directly damaged one MNTB, but otherwise cut only decussating fibers. In this case, there was no decrease in glutamate concentration in the LSO, a decrease medially in each MSO, and a decrease bilaterally in the MNTB, again correlating with the distribution of transected cochlear nucleus projections. Although some-what preliminary, there results suggest that the cochlear nucleus projec-tions to the superior olivary complex are enriched in glutamate. This could support a transmitter role, consistent with ultrastructural simi-larities between terminals in the LSO and auditory nerve terminals (Cant, 1984). Also like auditory nerve terminals, those in the LSO and MSO have been reported not to demonstrate uptake of glutamate or aspartate (Schwartz, 2985).

Aspartate

As for glutamate, aspartate concentrations in the cochlear nucleus are similar to those in most other brain regions sampled (Godfrey et al., 1977a). The average for rat cochlear nucleus resembles the value for whole brain (Parli et al., 1987). Within the cat cochlear nucleus, however, the distribution of aspartate differs from that of glutamate, glycine, or GABA in that concentrations are less in granular regions and DCN molecular layer than in the deeper regions. This correlation with the innervation pattern of the auditory nerve led to the initial suggestion that aspartate's relation to auditory nerve fibers should receive further study (Godfrey et al., 1977a). Subsequent work (reviewed in Wenthold, 1985; Potashner et al., 1985b; Martin, 1985; Caspary et al., 1985), has established aspartate as the most promising candidate for an auditory nerve transmitter. Still, the evidence remains incomplete and has been challenged, along with that for glutamate (Oertel, 1984; Schwartz, 1985).

Aspartate concentrations in the superior olivary complex are compar-able to those in the cochlear nucleus, and tend to be higher in the major nuclei than in the periolivary regions (Table 3). The lesions affecting glutamate concentrations in the superior olivary complex tended to have similar effects on aspartate (Figs. 7 and 8). Although the percentage decreases for both amino acids following the lesions were rather small, generally 20-30%, they were not much less than the 30-40% decreases reported for VCN regions following cochlear ablation (Wenthold, 1978).

The large lateral lesion resulted in 15-35% decreases in aspartate concentrations in the cochlear nucleus, in all except granular regions, in addition to its effect in the superior olive, possibly suggesting

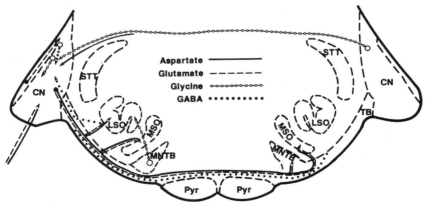

Fig. 9.   Possible auditory pathways utilizing amino acid transmitters, displayed on a schematic transverse section through the brain stem.

a centrifugal aspartate-rich pathway to the cochlear nucleus (Fig. 9).
In view of the potential complications to interpreting the lesion
results for amino acids, however, and the preliminary nature of the
evidence, this pathway can be only very tentatively suggested.

CONCLUSIONS

Pathways connecting to the cochlear nucleus and superior olivary
complex which may use particular transmitters are shown schematically in
Figs. 5 and 9. These pathways should be considered only as preliminary
suggestions based on the available incomplete evidence. Glutamate and
aspartate seem to have some specific association with ascending auditory
pathways, whether or not as transmitters. Acetylcholine and norepinephrine, on the other hand, seem to be preferentially associated with
descending pathways and may provide feedback modulation of auditory
processing. Glycine and GABA seem associated with a variety of intrinsic
and descending pathways, with glycine having a more prominent role in
general than GABA. Both appear to be involved in introducing inhibition
into the response patterns of cochlear nucleus neurons (Caspary et al.,
1979), but their functional role remains unclear (Martin, 1985). Obviously, much more work will be needed to sort out the functions of these and
other transmitter candidates in the auditory system. Microchemical
studies will constitute one important part of this work by providing
comprehensive quantitative information on the relative prominence of
particular transmitter systems among particular groups of neurons.

ACKNOWLEDGMENTS

The authors are grateful to John Schemenaur for technical assistance,
Janet Doty-Potts for graphics and Tracy Hartness for photography work.
Supported by NIH grant no. NS 17176 and ORU intramural funds.

REFERENCES

Adams, J. C., 1982, Collaterals of labyrinthine efferent axons, Soc.
    Neurosci. Abstr., 8:149.
Adams, J. C. and Mugnaini, E., 1984, GAD-like immunoreactivity in the
    ventral cochlear nucleus, Soc. Neurosci. Abstr., 10:393.
Adams, J. C. and Wenthold, R. J., 1987, Immunostaining of GABA-ergic and
    glycinergic inputs to the anteroventral cochlear nucleus, Soc.
    Neurosci. Abstr.,13:1259.
Altschuler, R. A., Betz, H., Parakkal, M. H., Reeks, K. A. and Wenthold,
    R. A., 1986, Identification of glycinergic synapses in the cochlear
    nucleus through immunocytochemical localization of the postsynaptic
    receptor, Brain Res., 369:316-320.
Baker, B. N., Glendenning, K. K. and Hodges, P., 1986, Acoustic chiasm:
    distribution of glycine and GABA receptors in the brainstem auditory
    nuclei of cat, Abstr. 9th Midwinter Res. Mtng., ARO:7.
Bledsoe, S. C. Jr., Altschuler, R. A., Wenthold, R. J. and Prasad, V.,
    1987, Immunocytochemical localization of glycine in the guinea pig
    superior olivary complex:lesion studies, Abstr. 10th Midwinter Res.
    Mtng.,ARO:157.
Brawer, J. R., Morest, D. K. and Kane, E. C., 1974, The neuronal architec-
    ture of the cochlear nucleus of the cat, J. Comp. Neurol., 155:251-
    300.
Cant, N. B., 1984, The fine structure of the lateral superior olivary
    nucleus of the cat, J. Comp. Neurol., 227:63-77.
Cant, N. B. and Gaston, K. C., 1982, Pathways connecting the right and
    left cochlear nuclei, J. Comp. Neurol., 212:313-326.

Caspary, D. M., Havey, D. C. and Faingold, C. L., 1979, Effects of micro-
    iontophoretically applied glycine and GABA on neuronal response
    patterns in the cochlear nuclei, Brain Res., 172:179-185.
Caspary, D. M., Rybak, L. P. and Faingold, C. L., 1985, The effects of
    inhibitory and excitatory amino-acid neurotransmitters on the
    response properties of brainstem auditory neurons, in: "Auditory
    Biochemistry", Drescher, D. G., ed., Thomas, Springfield, 198-226.
Cooper, J. R., Bloom, F. E. and Roth, R. H., 1986, "The Biochemical Basis
    of Neuropharmacology", Oxford Univ., New York.
Cowan, W. M., 1970, Anterograde and retrograde transneuronal degeneration
    in the central and peripheral nervous system, in: "Contemporary
    Research Methods in Neuroanatomy", Nauta, W. J. H. and Ebbesson,
    S. O. E.,eds., Springer-Verlag, New York, 217-251.
Davies, W. E., 1977, GABAergic innervation of the mammalian cochlear
    nucleus, in: "Inner Ear Biology", Portmann, M. and Aran, J.-M., eds.,
    INSERM, Paris, 68:155-164.
Fisher, S. K. and Davies, W. E., 1976, GABA and its related enzymes in
    the lower auditory system of the guinea pig, J. Neurochem., 27:1145-
    1155.
Frostholm, A. and Rotter, A., 1986, Autoradiographic localization of
    receptors in the cochlear nucleus of the mouse, Brain Res. Bull.,
    16:189-203.
Godfrey, D. A. and Matschinsky, F. M., 1976, Approach to three-dimensional
    mapping of quantitative histochemical measurements applied to studies
    of the cochlear nucleus, J. Histochem. Cytochem., 24:697-712.
Godfrey, D. A., Carter, J. A., Berger, S. J., Lowry, O. H. and Matschinsky,
    F. M., 1977a, Quantitative histochemical mapping of candidate
    transmitter amino acids in the cat cochlear nucleus, J. Histochem.
    Cytochem., 25:417-431.
Godfrey, D. A., Williams, A. D. and Matschinsky, F. M., 1977b, Quantita-
    tive histochemical mapping of enzymes of the cholinergic system in
    cat cochlear nucleus, J. Histochem. Cytochem., 25:397-416.
Godfrey, D. A., Carter, J. A., Lowry, O. H. and Matschinsky, F. M.,
    1978, Distribution of gamma-aminobutyric acid, glycine, glutamate
    and aspartate in the cochlear nucleus of the rat, J. Histochem.
    Cytochem., 26:118-126.
Godfrey, D. A., Park, J. L., Rabe, J. R., Dunn, J. D. and Ross, C. D.,
    1983, Effects of large brain stem lesions on the cholinergic system
    in the rat cochlear nucleus, Hearing Res., 11:133-156.
Godfrey, D. A., Beranek, K. L., Carlson, L., Dunn, J. D. and Ross, C. D.,
    1985a, Centrifugal cholinergic projections to subregions of cat
    cochlear nucleus, Soc. Neurosci. Abstr., 11:1052.
Godfrey, D. A., Park, J. L., Dunn, J. D. and Ross, C. D., 1985b, Cholin-
    ergic neurotransmission in the cochlear nucleus, in: "Auditory
    Biochemistry", Drescher, D. G., ed., Thomas, Springfield, 163-183.
Godfrey, D. A., Park-Hellendall, J. L., Dunn, J. D. and Ross, C. D.,
    1987a, Effect of olivocochlear bundle transection on choline acetyl-
    transferase activity in the rat cochlear nucleus, Hearing Res.,
    28:237-251.
Godfrey, D. A., Park-Hellendall, J. L., Dunn, J. D. and Ross, C. D.,
    1987b, Effects of trapezoid body and superior olive lesions on
    choline acetyltransferase activity in the rat cochlear nucleus,
    Hearing Res., 28:253-270.
Hill, D. W., Walters, F. H., Wilson, T. D. and Stuart, J. D., 1979, High
    performance liquid chromatographic determination of amino acids in
    the picomole range, Analyt. Chem., 51:1338-1341.
Kromer, L. F. and Moore, R. Y., 1980, Norepinephrine innervation of the
    cochlear nuclei by locus coeruleus neurons in the rat, Anat. Embryol.,
    158:227-244.
Liberman, M. C. and Brown, M. C., 1986, Physiology and anatomy of single
    olivocochlear neurons in the cat, Hearing Res., 24:17-36.

Lorente de Nó, R., 1981, "The Primary Acoustic Nuclei", Raven, New York.

Lowry, O. H. and Passonneau, J. V., 1972, "A Flexible System of Enzymatic Analysis", Academic, New York.

Martin, M. R., 1985, The pharmacology of amino acid receptors and synaptic transmission in the cochlear nucleus, in: "Auditory Biochemistry", Drescher, D. G., ed., Thomas, Springfield, 184-197.

Moore, J. K. and Moore, R. Y., 1984, GAD-like immunoreactivity in the cochlear nuclei and superior olivary complex, Soc. Neurosci. Abstr., 10:843.

Moore, M. J. and Caspary, D. M., 1983, Strychnine blocks binaural inhibition in lateral superior olivary neurons, J. Neurosci., 3:237-242.

Mugnaini, E., 1985, GABA neurons in the superficial layers of the rat dorsal cochlear nucleus:light and electron microscopic immunocytochemistry, J. Comp. Neurol., 235:61-81.

Oertel, D., 1984, Cells in the anteroventral cochlear nucleus are insensitive to L-glutamate and L-aspartate; excitatory synaptic responses are not blocked by D-alpha-aminoadipate, Brain Res., 302:213-220.

Orkand, R. K., 1977, Glial cells, in: "Handbook of Physiology. Section 1. The Nervous System", Brookhart, J. M. and Mountcastle, V. B., eds., Am. Physiol. Soc., Bethesda, 855-875.

Ostapoff, E. M., Morest, D. K. and Potashner, S. J., 1985, Retrograde transport of ^{3}H-GABA from the cochlear nucleus to the superior olive in guinea pig, Soc. Neurosci. Abstr., 11:1051.

Parli, J. A., Schemenaur, J. E., Godfrey, D. A. and Ross, C. D., 1987, Amino acid concentrations in rat auditory, olfactory and visual structures, Soc. Neurosci. Abstr., 13:1288.

Patel, A. J. and Hunt, A., 1985, Concentration of free amino acids in primary cultures of neurons and astrocytes, J. Neurochem., 44:1816-1821.

Peyret, D., Geffard, M. and Aran, J.-M., 1986, GABA immunoreactivity in the primary nuclei of the auditory central nervous system, Hearing Res., 23:115-121.

Potashner, S. J., Lindberg, N. and Morest, D. K., 1985a, Uptake and release of gamma-aminobutyric acid in the guinea pig cochlear nucleus after axotomy of cochlear and centrifugal fibers, J. Neurochem., 45:1558-1566.

Potashner, S. J., Morest, D. K., Oliver, D. L. and Jones, D. R., 1985b, Identification of glutamatergic and aspartatergic pathways in the auditory system, in:"Auditory Biochemistry", Drescher, D. G., ed., Thomas, Springfield, 141-162.

Roberts, R. C. and Ribak, C. E., 1987, GABAergic neurons and axon terminals in the brainstem auditory nuclei of the gerbil, J. Comp. Neurol., 258:267-280.

Saint Marie, R. L., Ostapoff, E. M. and Morest, D. K., 1986, Co-localization of ^{3}H GABA and GABA-like immunoreactivity in superior olivary neurons retrogradely labeled from guinea pig cochlear nucleus, Soc. Neurosci. Abstr., 12:1269.

Sanes, D. H., Geary, W. A. II. and Wooten, G. F., 1985, The quantitative distribution of ^{3}H-strychnine binding in the lateral superior olivary nucleus of the gerbil, Soc. Neurosci. Abstr., 11:1051.

Schwartz, I. R., 1985, Autoradiographic studies of amino acid labeling of neural elements in the auditory brainstem, in: "Auditory Biochemistry", Drescher, D. G., ed., Thomas, Springfield, 258-277.

Schwartz, I. R. and Yu, S.-M., 1986, An anti-GABA antibody labels subpopulations of axonal terminals and neurons in the gerbil dorsal cochlear nucleus and superior olivary complex, Soc. Neurosci. Abstr., 12:128-137.

Staatz-Benson, C. and Potashner, S. J., 1987, Uptake and release of glycine in the guinea pig cochlear nucleus, J. Neurochem., 49:128-137.

Tachibana, M. and Kuriyama, K., 1974, Gamma-aminobutyric acid in the lower auditory pathway of the guinea pig, Brain Res., 69:370-374.

Thompson, G. C., Cortez, A. M. and Lam, D. M. K., 1985, Localization of GABA immunoreactivity in the auditory brainstem of guinea pigs., Brain Res., 339:119-122.

Warr, W. B., 1975, Olivocochlear and vestibular efferent neurons of the feline brain stem: their location, morphology and number determined by retrograde axonal transport and acetylcholinesterase histochemistry, J. Comp. Neurol., 161:159-182.

Warr, W. B., 1982, Parallel ascending pathways from the cochlear nucleus: neuroanatomical evidence of functional specialization, Contrib. Sens. Physiol., 7:1-38.

Wenthold, R. J., 1978, Glutamic acid and aspartic acid in subdivisions of the cochlear nucleus after auditory nerve lesion, Brain Res., 143: 544-548.

Wenthold, R. J., 1985, Glutamate and aspartate as neurotransmitters of the auditory nerve, in: "Auditory Biochemistry", Drescher, D. G., ed., Thomas, Springfield, 125-140.

Wenthold, R. J., 1987, Evidence for a glycinergic pathway connecting the two cochlear nuclei:an immunocytochemical and retrograde transport study, Brain Res., 415:183-187.

Wenthold, R. J. and Gulley, R. L., 1977, Aspartic acid and glutamic acid levels in the cochlear nucleus after auditory nerve lesion, Brain Res., 138:111-123.

Wenthold, R. J. and Morest, D. K., 1976, Transmitter related enzymes in the guinea pig cochlear nucleus, Soc. Neurosci. Abstr., 2:28.

Wenthold, R. J., Altschuler, R. A., Huie, D., Parakkal, M. H. and Reeks, K. A., 1986a, Immunocytochemical localization of glycine in the cochlear nucleus and superior olivary complex of the guinea pig, Soc. Neurosci. Abstr., 12:1265.

Wenthold, R. J., Zemple, J. M., Parakkal, M. H., Reeks, K. A. and Altschuler, R. A., 1986b, Immunocytochemical localization of GABA in the cochlear nucleus of the guinea pig, Brain Res., 380:7-18.

Werman, R., 1966, Criteria for identification of a central nervous system transmitter, Comp. Biochem. Physiol., 18:745-766.

White, J. S. and Warr, W. B., 1983, The dual origins of olivocochlear neurons in the albino rat, J. Comp. Neurol., 219:203-214.

Wu, S. H. and Oertel, D., 1986, Inhibitory circuitry in the ventral cochlear nucleus is probably mediated by glycine, J. Neurosci., 6:2691-2706.

CHOLINERGIC, GABA-ERGIC, AND NORADRENERGIC INPUT TO COCHLEAR GRANULE CELLS IN THE GUINEA PIG AND MONKEY

Jean K. Moore

Deparment of Anatomical Sciences
SUNY at Stony Brook
Stony Brook, NY, 11794

In most mammals, the granule cell zone of the cochlear nuclei forms a peripheral capsule surrounding both the dorsal and ventral nuclei. Part of the capsule is a cellular layer composed of granule cells and "star" stellate cells (Moore, 1986). The heaviest accumulation of these small neurons occurs on the lateral free surface of the nuclei, but thinner layers of cells extend between the nuclei and around the medial edge of the cochlear complex (Fig. 1, guinea pig). The granule cell area also includes the dorsal nucleus molecular layer, a lamina formed by axons of granule cells in all parts of the capsular zone (Fig. 1, Mugnaini et al., 1980). Much of the afferent input to the granule cell area appears to arise either from intrinsic cochlear axons or from descending projection systems originating more centrally in the nervous system. Some of these projections have been demonstrated by histological methods for localizing neurotransmitters and transmitter-related enzymes. In particular, a concentration of cholinergic, GABA-ergic, and noradrenergic terminals has been demonstrated in the peripheral granule cell area of mammals such as the cat, rat, chinchilla, and guinea pig. In the case of the cholinergic and GABA-ergic projections systems, the input to the cochlear granule cell zone may arise from side branches of systems of axons projecting to the cochlea.

It has been noted previously that the granule cell capsule is greatly reduced in higher primates (Moore, 1980). In monkeys such as the macaque, all that remains of the granule cell region is a rather restricted lateral external granular layer which does not extend over the rostral and medial surfaces of the complex or between the dorsal and ventral nuclei (Fig. 1, macaque). The reduction of the granule cell area in higher primates raises several questions in regard to the systems which provide its afferent input. Are projections into the peripheral zone of the primate cochlear nuclei reduced proportional to the loss of granule cells? And if so, is there a concomitant reduction in related efferent system which innervate the inner ear? To date, there have been no investigations of these intrinsic and descending projections systems in primates from the standpoint of their relationship to the cochlear nuclei. Because the distinctly primate features of cyto- and myeloarchitecture seen in monkeys also characterize the human cochlear complex (Moore and Osen, 1979; Moore, 1980), information about the organization of these systems in primates is relevant to understanding the process of efferent control of auditory input in man. For this reason, a series of comparative histochemical and immunohistochemical investigations have been carried out in this laboratory. The experimental animals used were one nonprimate species, the guinea pig (Cavia porcellus) and one primate, the macaque monkey (Macaca nemistrina or Macaca fascicularis). Guinea pig brain stems are those of animals perfused in this laboratory. Perfused brain stems of macaques were obtained from Dr. Anita Hendrickson, University of Washington.

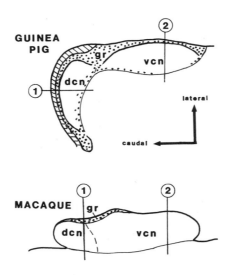

Fig. 1.   Horizontal sections through the cochlear nuclei of the guinea pig and macaque. In the guinea pig, the heaviest concentration of granule cells is on the lateral surface of the nuclei (stippled area), with their axons forming a superficial molecular layer (shaded area). In the macaque, the external granular layer covers only a portion of the lateral side of the nuclei (stippled area). Abbreviations: dcn, dorsal cochlear nucleus; gr, granule cell layer; vcn, ventral cochlear nucleus.

Cholinergic Input to the Granule Cell Area

Cholinergic projections to the cochlear nuclei have been investigated by the Koelle histochemical method for the demonstration of acetyl-cholinesterase (AChE; Osen and Roth, 1969). The results of these studies are shown schematically in Figure 2. In the guinea pig, AChE staining in the main body of both the dorsal and ventral nuclei consists only of scattered precipitate and light intracytoplasmic staining of the larger neurons. However, staining for cholinesterase is distinctly heavier in the outer layers of the nuclei. In the dorsal nucleus (Fig. 2, guinea pig DCN), there is moderately dense staining in the molecular layer, grading into somewhat lighter staining in the granule-fusiform cell layer. In this cellular layer, there is a periodicity to the staining in that there are alternate dense and light bands of precipitate oriented orthogonal to the nuclear surface. It has not been possible to determine if these bands coincide with clusters of fusiform cells. In the ventral nucleus (Fig. 2, guinea pig VCN), there is a dense, homogeneous, and sharply bounded outer layer of staining which corresponds to the superficial granule cell layer as seen in adjacent Nissl sections. These observations in the guinea pig are similar to the results of histochemical studies done in the cat, mouse, rat, and chinchilla (Osen and Roth, 1969; Martin, 1981; Osen et al., 1984). Though there are some intraspecific differences, especially in regard to degree of staining of the molecular layer, all of these studies demonstrated

a concentration of AChE + axons and terminals in the superficial layers of
the nuclei. The histochemical studies are in accord with biochemical assays
done in the cat, showing that the highest levels of acetylcholine and
choline acetyltransferase are found in the superficial layers of the nuclei
(Godfrey et al., 1977a). There is also good concordance with the results of
receptor studies in the mouse (Frostholm and Rotter, 1986) which showed
that the ligand for muscarinic adrenergic receptor sites (^{3}-H-quinuclidynil
benzilate) is concentrated in the granular and molecular layers.

AChE

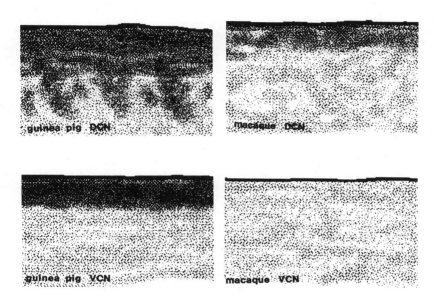

Fig. 2.   Schematic representation of the distribution and density of
staining in cochlear nuclei processed for histochemistry/
immunohistochemistry. Illustrations done by computer graphics
(MacIntosh MacPaint). DCN indicates a section through the nuclei
at level 1 in Figure 1; VCN indicates a section through level 2.
Staining for acetylcholinesterase by the Koelle histochemical
method.

    AChE histochemistry in the macaque cochlear nuclei reveals a rather
different picture. There is a superficial zone of AChE reactivity which
is narrower and less intense that the areas of staining seen in the
guinea pig (Fig. 2, macaque DCN). This band coincides with the width and
extent of the external granular layer over the dorsal and posteroventral
nuclei. There is no deeper extension of the staining within the dorsal
nucleus and no banding, in accord with the fact that there is no well
defined internal granule-fusiform cell layer in monkeys (Moore, 1980).
Anterior to the external granule layer, there is no zone of AChE positivity
along the lateral surface of the nucleus (Fig. 2, macaque VCN).

    In both the guinea pig and the monkey, there is AChE staining in
structures located more centrally in the brainstem. These structures
include neuronal somata in the periolivary region and fascicles of axons
at the periphery of the brainstem. A number of fascicles of axons run in
the dorsal acoustic stria and enter the caudal end of the cochlear complex.
Other groups of axons curve between the spinal trigeminal nucleus and
the cerebellar peduncles. These AChE-positive olivary neurons have been

shown to be the source of efferent projections to the cochlea in both
nonprimates (Warr, 1975) and primates (Thompson and Thompson, 1986;
Carpenter et al., 1987). Osen and coworkers noted that the AChE+ axons
entering the cochlear nuclei appeared to arise as side branches of the
olivocochlear system. However, even very large lesions transecting the
efferent olivocochlear pathways do not completely abolish cholinesterase
reactivity in the granule cell layer (Osen et al., 1984). Similar lesions
isolating the cochlear nuclei lower the levels of ChAT in the granule cell
area by 65-75% and of AChE by less than 50% (Godfrey et al., 1983). These
results seem to show that the origin of the cholinergic input to the
cochlear nuclei is partly extrinsic, from the olivocochlear system, and
partly from intrinsic cochlear sources. Certainly the eighth nerve does
not seem likely to be the source of this input, as assays have shown very
low levels of cholinergic enzymes here (Fex and Wenthold, 1976; Godfrey
et al., 1977a).

GABA-ergic Input to the Granule Cell Area

GABA-ergic input has been investigated by immunohistochemistry
employing an antiserum to glutamic acid decarboxylase (GAD), the synthesizing enzyme for GABA (Oertel et al., 1981). The protocol used for GAD
immunohistochemistry in this laboratory produced primarily staining of
immunoreactive puncta which are interpreted as axon terminals. In the
dorsal nucleus of the guinea pig, the molecular and granular layers
contain a moderately dense mixture of medium-sized and fine terminals
(Fig. 3, guinea pig DCN). More rostrally, GAD immunoreactive terminals
are densely packed, with a somewhat patchy distribution, in the granule
cell layer over the free surface of the nucleus (Fig. 3, guinea pig VCN).
In both nuclei, the dense accumulation of terminals stands in contrast
to the more scattered array of GAD-immunoreactive puncta in the deeper
parts of the nuclei. This pattern of staining seen in the guinea pig,
with a significantly greater density of GAD+ terminals in the granular
and molecular layers, is very similar to results obtained by GAD and
GABA immunohistochemistry by other investigators in the rat (Mugnaini,
1985; Moore and Moore, 1987), gerbil (Roberts and Ribak, 1987), and
guinea pig (Thompson et al., 1987). Very high levels of GABA have also
been demonstrated by biochemical assay in the peripheral granule cell
and molecular layers (Godfrey et al., 1977b, 1978). These studies are
in good agreement with receptor binding studies showing that density of
uptake of 3-H-muscimol is greatest in these layers (Frosthom and Rotter,
1986).

GAD immunostaining in the macaque cochlear nuclei gives results
which are rather different from what is seen in the guinea pig and
other nonprimate mammals, but quite similar in distribution to the pattern
of staining seen in this monkey with AChE. Over the center of the somewhat increased density of terminals (Fig. 3, macaque DCN). Rostral to
this granular layer, at the surface of the anterior ventral nucleus, there
is no region of increased terminals staining (Fig. 3, macaque VCN). In
all deeper regions of the nuclei, the pattern of terminal distribution
of GAD-immunoreactive terminals is identical to that in the guinea pig
(Moore, 1988).

The immunohistochemical protocol used for GAD in the guinea pig and
macaque was not optimal for staining of cell somata. However, other
studies employing antibodies to GAD and GABA (Thompson et al., 1985;
Wenthold et al., 1986; Peyret et al., 1986; Moore and Moore, 1987; Roberts
and Ribak, 1987) have shown a number of immunoreactive small neurons in
the granular and molecular layers. The size and distribution of these
small neurons has suggested to most investigators that they are stellate

GAD

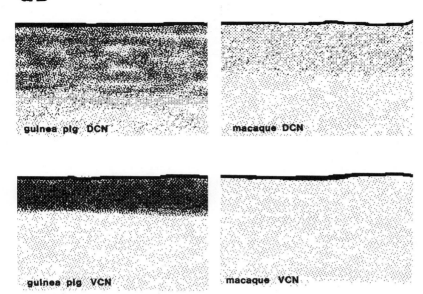

Fig. 3. Schematic representation of the distribution and density of
staining in cochlear nuclei processed for histochemistry/immuno-
histochemistry. Illustrations done by computer graphics
(MacIntosh MacPaint). DCN indicates a section through the nuclei
at level 1 in Figure 1; VCN indicates a section through level 2.
Immunoreactivity to an antiserum to glutamic acid decarboxylase,
demonstrating GABA-ergic terminals.

and cartwheel cells, and this assumption has been confirmed by EM immunohistochemistry for cells of this type within the dorsal nucleus (Murgnaini,
1985). It seems likely that the stellate cells in other parts of the
cochlear complex are also GABA-positive and may be an intrinsic source of
terminals in the granule cell layer. In addition, the possibility of
extrinsic sources of GABA must also be considered. In the guinea pig and
monkey, as in the rat (Moore and Moore, 1987), small fascicles of GAD+
axons run in the acoustic stria and subpeduncular route, along the medial
side of the cochlear nuclei, and through the vestibular nerve to its distal
end. Sectioning one segment of this pathway, the dorsal stria, reduces
GAD in the dorsal nucleus by 30% (Davies, 1977) thus demonstrating the
efferent nature of the fibers. These efferent fibers may arise from GABA-
ergic periolivary neurons which have been demonstrated by the previously
mentioned immunohistochemical studies. The neurons in the olivary complex
are strikingly similar in their distribution and neuronal morphology to
AChE-positive cells, and uptake by these periolivary neurons of [3]-H-GABA
injected into cochlear nuclei indicates that these cells do project into
the cochlear complex (St. Marie et al., 1986). Some of the fascicles of
GAD+ axons continue to the distal end of the vestibular nerve, and thus
may be the source of GAD+ and GABA+ terminals observed in the organ of
Corti in the rat, guinea pig, and squirrel monkey (Fex and Altschuler,
1984; Thompson et al., 1986). These fascicles of GAD-immunoreactive axons
in the vestibular nerve appear more prominent in the macaque than in the
guinea pig.

Noradrenergic Input to the Granule Cell Area

   Noradrenergic projections to the cochlear nuclei have been studied by use of an antibody to the synthesizing enzyme for noradrenalin, dopamine -β-hydroxylase (Eugene Tech). In the guinea pig, a moderate number of fine varicose axons are seen within the main body of the cochlear nuclei. Because the immunoreaction takes place only at the surface of the tissue sections, the plexus formed by these immunoreactive fibers appears in the immunohistochemical material as disconnected axon fragments. Many of these axons run horizontally and branch into a moderately dense plexus in the outer layers of the dorsal nucleus (Fig. 4, guinea pig DCN). A somewhat larger number of fibers are seen to form a branching plexus in the granule cell layer over the ventral nucleus (Fig. 4, guinea pig VCN). The only other animal in which the distribution of noradrenergic axons in the cochlear nuclei has been described in detail is the rat (Kromer and Moore, 1977; Levitt and Moore, 1979). In this species, in which there is a very weak development of the cochlear granule cell layer, the investigators did not note a concentration of noradrenergic axons at the periphery of the nuclei, but the differential density is clear in hyperinnervated brain stem material.

DBH

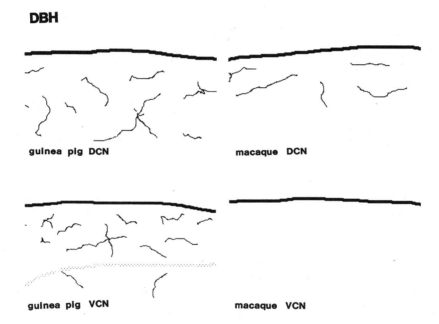

guinea pig DCN   macaque DCN

guinea pig VCN   macaque VCN

Fig. 4.   Schematic representation of the distribution and density of
          staining in cochlear nuclei processed for histochemistry/immuno-
          histochemistry. Illustrations done by computer graphics
          (MacIntosh MacPaint). DCN indicates a section through the nuclei
          at level 1 in Figure 1; VCN indicates a section through level 2.
          Immunoreactivity to an antibody to dopamine-β-hydroxylase,
          showing nonadrenergic axons.

In sections through the macaque cochlear nuclei processed for DBH immunohistochemistry, there is a slightly greater concentration of the immunoreactive axons at the outer surface of the nuclei in the region of the external granular layer (Fig. 4, macaque DCN), but no axons are present at the rostral end of the complex where no granular layer is present (Fig. 4, macaque VCN).

In the brainstem of both the guinea pig and macaque, there is clear DBH immunoreactivity in the noradrenergic neurons of the locus coeruleus and lateral tegmental cell groups. In addition, there are fascicles of extremely thin varicose axons in the acoustic stria and subpeduncular route. All of these axons appear to enter the cochlear complex, as noted in the rat (Kromer and Moore, 1979). In terms of their morphology, the axons innervating the cochlear nuclei are fine, regularly varicose axons of the type innervating all brain stem sensory nuclei and the cerebellum (Levitt and Moore, 1979). They differ from the thicker, more irregular axons arising in the lateral tegmental noradrenergic nuclei. Injections of HRP into the cochlear nuclei and dorsal stria backfill locus coeruleus neurons (Adams and Warr, 1976), and HRP injections into the locus anterogradely label axons in the cochlear nuclei (Kromer and Moore, 1980). Further, bilateral lesions of the locus cause a disappearance of the catecholaminergic axons in all brain stem sensory nuclei, including the cochlear nuclei (Levitt and Moore, 1980). It thus seems conclusive that the locus coeruleus is the sole source of the plexus innervating the cochlear nuclei. There is no convincing evidence in either guinea pig or monkey material of axons running to the distal end of the vestibular nerve.

Conclusion

The guinea pig resembles other nonprimate mammals in having a greater concentration of cholinergic, GABA-ergic, and noradrenergic terminals in the superficial granular and molecular layers of the cochlear nuclei. In the cholinergic and GABA-ergic systems, there is some evidence suggesting that there are intrinsic sources of this input within the cochlear nuclei. In addition, there are efferent axons of both types entering the nuclei from the brain stem. In both systems, neurons in the olivary complex are a potential source of these extrinsic fibers. Noradrenergic input arises only in the brain stem, from the locus coeruleus, and the efferent fibers follow the same routes as the cholinergic and adrenergic efferents to reach the cochlear nuclei.

In the monkey, there is a clear reduction in extent and density of the terminal zone of all three systems, corresponding to the reduction of the granule cell layer. Similar observations on the cholinergic input to the cochlear nuclei were made in the reeler mouse in which there is a marked reduction of the granule cell layer (Martin, 1981). There is not, however, any obvious reduction in the system of cholinergic and GABA-ergic axons running in the vestibular nerve to form the efferent projection to the inner ear. This is also the case in the cholinergic efferent system in the human brainstem, in which the olivocochlear bundle shows normal staining but there is no terminal zone within the cochlear complex (Moore and Osen, 1979). It is still unclear whether the efferent projections to the cochlear nuclei and the cochlea are separate parallel systems or a single system of dichotomizing axons, but it appears that the two types of efferents can indeed vary independently.

ACKNOWLEDGEMENTS

GAD antiserum was donated by I. Kopin, NIH. This work was supported by the Deafness Research Foundation.

REFERENCES

Adams, J. C. and Mugnaini, E., 1987, Patterns of glutamate decarboxylase immunostaining in the feline cochlear nuclei complex using silver enhancement and electron microscopy, J. Comp. Neurol., 262:315-401.
Adams, J. C. and Warr, W. B., 1976, Origins of axons in the cat's acoustic striae determined by injection of horseradish peroxidase into severed tracts, J. Comp. Neurol., 170:107-122.

Carpenter, M. B., Chang, L., Pereira, A. B., Hirsh, L. B., Bruce, G. and Wu, J.-Y., 1987, Vestibular and cochlear efferent neurons in the monkey identified by immunohistochemical methods, Brain Res., 408: 275-280.

Davies, W. E., 1977, GABA-ergic innervation of the mammalian cochlear nucleus, Coll. Inst. Nat. Sante. Rech. Med. 68:155-164.

Fex, J. and Altschuler, R. A., 1984, Glutamic acid decarboxylase immuno-reactivity of olivocochlear neurons in the organ of Corti of guinea pig and rat, Hearing Res. 15:123-131.

Fex, J. and Wenthold, R. J., 1978, Choline acetyltransferase, glutamate decarboxylase, and tyrosine hydroxylase in the cochlea and cochlear nucleus of the guinea pig, Brain Res., 172:179-185.

Frostholm, A. and Rotter, A., 1986, Autoradiographic localization of receptors in the cochlear nuclei of the mouse, Brain Res. Bull., 16: 189-203.

Godfrey, D. A., Carter, J. C., Berger, S. J., Lowry, O. H. and Matchinsky, F. M., 1977b, Quantitative histochemical mapping of candidate transmitter amino acids in cat cochlear nucleus, J. Histochem. Cytochem., 26:118-126.

Godfrey, D. A., Carter, J. C., Lowry, O. H. and Matchinsky, F. M. 1978, Distribution of gamma aminobutyric acid, glycine, glutamate, and aspartate in the cochlear nucleus of the rat, J. Histochem. Cytochem., 26:118-126.

Godfrey, D. A., Park, J. L., Rahe, J. R., Dunn, J. D. and Ross, C. D., 1983, Effects of lesions of the olivocochlear bundle on levels of cholinergic enzymes in the cochlear nuclei, Hearing Res., 11:133-156.

Godfrey, D. A., Williams, A. D. and Matchinsky, F. M., 1977a, Quantitative histochemical mapping of enzymes of the cholinergic system in cat cochlear nucleus, J. Histochem. Cytochem., 25:397-416.

Kromer, L. F. and Moore, R. Y., 1977, Cochlear nucleus innervation by central norepinephrine neurons in the rat, Brain Res., 118:227-537.

Kromer, L. F. and Moore, R. Y., 1980, Norepinephrine innervation of the cochlear nuclei by locus coeruleus neurons on the rat, Anat. Embryol., 158:227-244.

Levitt, P. and Moore, R. Y., 1979, Origin and organization of brain stem catecholamine innervation in the rat, J. Comp. Neurol., 186:505-528.

Levitt, P. and Moore, R. Y., 1980, Organization of brain stem noradrenalin hyperinervation following neonatal 6-hydroxydopamine treatment in rat, Anat. Embryol., 158:133-150.

Martin, M., 1981, Acetylcholinesterase-positive fibers and cell bodies in the cochlear nuclei of normal and reeler mutant mice, J. Comp. Neurol., 197:153-167.

Moore, J. K., 1980, The primate cochlear nuclei: Loss of lamination as a phylogenetic process, J. Comp. Neurol., 193:609-629.

Moore, J. K., 1986, The cochlear nuclei:Relationship to the auditory nerve, in: "The Cochlea: Neurobiology of Hearing", R. Altschuler, D. Hoffman and R. Bobbin, eds, Academic Press, New York, 283-301.

Moore, J. K. and Moore, R. Y., 1987, Glutamic acid decarboxylase-like immunoreactivity in the brain stem auditory nuclei of the rat, J. Comp. Neurol., 260:157-174.

Moore, J. K. and Osen, K. K., 1979, The cochlear nuclei in man, Am. J. Anat., 154:393-418.

Mugnaini, E., 1985, GABA neurons in the superficial layers of the rat dorsal cochlear nucleus. Light and electron microscopic immuno-histochemistry, J. Comp. Neurol., 325:61-81.

Mugnaini, E., Warr, W. B. and Osen, K. K., 1980, Distribution and light microscopic features of granule cells in the cochlear nuclei of the cat, rat and mouse, J. Comp. Neurol., 191:581-606.

Oertel, W. H., Schmechel, E. E., Mugnaini, E., Tappaz, M. L. and Kopin, J., 1981, Immunocytochemical localization of glutamate decarboxylase in rat cerebellum with a new antiserum, Neurosci., 6:2715-2735.

130

Osen, K. K., Mugnaini, E., Dahl, A.-L. and Christiansen, A. H., 1984, Histochemical localization of acetylcholinesterase in the cochlear and superior olivary nuclei: A reappraisal with emphasis on the cochlear granule cell system, Arch ital. Biol., 12:169-212.

Osen, K. K. and Roth, K., 1969, Histochemical localization of cholinesterases in the cochlear nuclei of the cat, with notes on the origin of acetylcholinesterase-positive afferents and the superior olive, Brain Res., 16:165-185.

Peyret, D., Gifford, M. and Aran, J.-M., 1986 GABA immunoreactivity in the primary nuclei of the auditory central nervous system. Hearing Res. 23:115-121.

Roberts, R. and Ribak, C. E., 1987, GABA-ergic neurons and axon terminals in the brain stem auditory nuclei of the gerbil, J. Comp. Neurol., 258:267-280.

St. Marie, R. L., Ostapoff, E. M. and Mores, D. K., 1986, Colocalization of tritiated GABA and GABA-like immunoreactivity in superior olivary neurons retrogradely labeled from guinea pig cochlear nucleus, Soc. Neurosci. Abs., 12:1269.

Thompson, G. C., Cortez, A. M. and Lam, S. M.-K., 1985, Localization of GABA immunoreactivity in the auditory brainstem of guinea pigs, Brain Res., 339:119-122.

Thompson, G. C., Cortez, A. M. and Igarashi, M., 1986, GABA-like immunoreactivity in the squirrel monkey organ of Corti, Brain Res., 372: 72-79.

Thompson, G. C. and Thompson, A. M., 1986, Olivocochlear neurons in the squirrel monkey brainstem, J. Comp. Neurol., 254:246-258.

Warr, W. B., 1975, Olivocochlear and vestibular efferent neurons of the feline brain stem. Their location, number and morphology determined by retrograde axonal transport and acetylcholinesterase histochemistry, J. Comp. Neurol., 161:159-182.

Wenthold, R. J., Zempel, J. M., Parakkal, M. H., Reeks, K. A. and Altschuler, R. A., 1986, Immunocytochemical localization of GABA in the cochlear nucleus of the guinea pig, Brain Res. 380:7-18.

GLUTAMATE DECARBOXYLASE IMMUNOSTAINING IN THE HUMAN COCHLEAR NUCLEUS

Joe C. Adams

Medical University of South Carolina
Department of Otolaryngology and Communicative Sciences
Charleston, South Carolina, 29425 USA

Knowledge of the structure and function of individual neurons of the human auditory system must remain indirect because of the lack of non-invasive means of studying single cells in living tissue. Considerable progress can be made, however, by establishing structure-function relations of cells in animals and then inferring functions of morphological cell types in humans. Some progress has been made in this approach to the study of the human cochlear nucleus (see Adams, 1986) but most of the needed work clearly lies ahead. An extremely powerful tool for use in the study brain structure and function is immunocytochemistry. This technique permits the visualization chemically distinct cells or portions of cells by the use of antibodies directed against specific molecular sequences. The technique is particularly useful when antibodies permit visualization of antigens whose function is known to play a key role in synaptic transmission. This is true in the case of antisera against glutamate decarboxylase (GAD), a specific synthetic enzyme essential for the production of the inhibitory neurotransmitter gamma amino butyric acid (GABA). The present report is concerned with the form and distribution of nerve cells and processes in the human cochlear nucleus that immunostain with a well characterized GAD antibody.

It is apparent from evidence obtained by investigators using a variety of methods that GABA plays a large role in the functioning of all central auditory structures. To date the nucleus that has been most intensively studied using immunocytochemistry is the cochlear nucleus (See Adams and Mugnaini, 1987 for references). There appear to be virtually no cells within the cochlear nucleus that are not either GABA-ergic or that are not contacted by GABA-ergic terminals. Clearly, GABA plays a major role in neural processing in the cochlear nucleus. An immunocytochemical study of GAD in the human cochlear nucleus will give some indication of the extent to which GABA is utilized there and will permit comparison of human and animal tissue with respect to the distribution of GABA-ergic processes.

METHODS

The results reported here are from 9 cases that ranged in age from 31 weeks gestation to 56 years. None were known to have neurological disorders or head trauma. At the termination of the autopsies (4 to 12 hours postmortem) the brainstem was immersed in 10% formal saline contain-

133

ing 0.5% zinc salicylate (pH 6.0, 4°C) (Mugnaini and Dahl, 1983). After approximately one hour in this fixative the tissue was trimmed of unwanted structures and cut into blocks (usually in the frontal plane) approximately 1 cm in thickness. These blocks were placed in a similar fixative

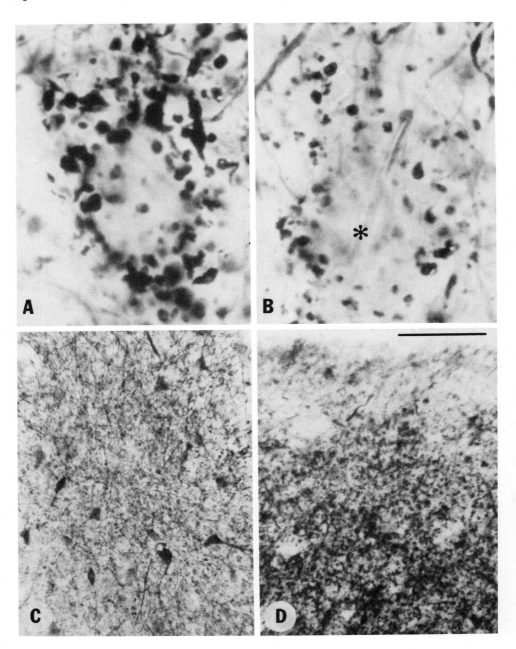

Fig. 1.   A. GAD immunostained terminals enveloping a bushy cell. B. GAD immunostained terminals apposed to an octopus cell. The asterisk is over the soma. C. GAD immunostained cells in the dorsal cochlear nucleus. D. Immunostaining of the dorsal cochlear nucleus. Note the sparsity of staining near the surface (top). Calibration bar in D represents 20 micra for A and B, 90 micra for C, and 50 micra for D.

modified by substituting zinc dichromate for zinc salicylate and with the pH adjusted to 5.0 (Mugnaini and Oertel, 1985). The blocks were left in this fixative overnight stirring on a stirplate that was kept in a refrigerator. On the following day, infiltration of the tissue in an increasing series of sucrose concentrations was begun. After 3 or more days of refrigerated stirring the tissue was considered sufficiently infiltrated with 25% sucrose for cryoprotection and was cut on a freezing stage microtome, usually at 40 micra. Selected sections were incubated in sheep anti-GAD antiserum overnight or longer and the immunoreactive sites visualized using the ABC method as previously described (Adams and Mugnaini, 1987).

RESULTS

In the ventral cochlear nucleus there are two clear staining patterns of terminals on principal cells. The most striking pattern of immuno-stained terminals consists of coarse processes that surround the somata of medium size, generally round cells with roughly 18-24 micra average diameter. The coarse terminals predominate but a range of bouton sizes down to less than one micron are found mixed with the coarse terminals. The density of terminals on cells receiving this pattern is usually much greater than the density found on cells receiving only smaller terminals. This type ending is illustrated in Fig. 1A. In cat tissue endings of this sort have been associated with bushy cells and have been shown to contain pleomorphic synaptic vesicles (Adams and Mugnaini, 1987). Primary terminals on bushy cells, which these immunostained terminals resemble somewhat when viewed with the light microscope, were not found to be immunostained in cat or rat material when examined with the electron microscope (ibid.). The similarity of the GAD immunostained terminals in cat and human material is striking. The size, shape and distribution of cells in the human ventral cochlear nucleus that receive these terminals is a good match for the size, shape, and distribution of bushy cells seen in human Golgi impregnations (Adams, 1986). All the evidence indicates that these coarse GAD immunostained terminals in human material are enveloping bushy cells.

The second prominent pattern of immunostained terminals is shown in Fig. 1B. This pattern is characterized by finer, more sparse processes that loosely envelop large (25-30 micra average diameter) cells which have large primary dendrites. Frequently the immunostained terminals appose the dendrites of these cells. This pattern of terminals closely matches that previously described on octopus cells in cat material (Adams and Mugnaini, 1987). The size, shape, and distribution of cells receiving GAD-immunostaining terminals of this pattern in human tissue supports the interpretation that it is octopus cells in the human cochlear nucleus that receive terminals of this type.

Another pattern of immunostaining terminals in the human ventral cochlear nucleus is that of relatively fine boutons which, instead of conspicuosly covering the soma of a particular cell type, are found in the neuropil surrounding accumulations of smaller (10-15 micra average diameter) multipolar cells. These accumulations of cells occur to a degree along the medial and lateral borders of ventral portions of the cochlear nucleus, are more pronounced at the dorsal cap of the anterior portion of the nucleus, and are most pronounced caudally, in the region where the rostral portion of the dorsal cochlear nucleus lies above the posterior portion of the ventral nucleus. The cells situated within the immunostained neuropil seldom have substantial accumulations of immuno-stained boutons apposed to their somata.

Cells associated with one of these three staining patterns are

usually found in clusters of a common cell type or even in large fields
of cells of a common type. As one might expect, in rostral portions of
the nucleus a large proportion of cells are covered with terminals that
appear to be the bushy cell type pattern. However, as is the case in the
cat, cells with the bushy cell pattern are found throughout the ventral
nucleus, including the most caudal part. In some cases a few of these
cells are found within deep layers of the dorsal cochlear nucleus.
Although cells receiving such terminals are found throughout the rostro-
caudal extent of the ventral nucleus, they are somewhat sparse in dorsal
portions, where there is an abundance of fine immunostained processes in
the neuropil.

There is a good correlation between the locations of higher density
of neuropil immunostaining and the locations of immunostained cell bodies.
Within the ventral portion of the nucleus staining of the neuropil is
most conspicuous in the dorsal quarter in the region caudal to the nerve
root. In this same rostro-caudal portion of the nucleus there are also
narrow bands of neuropil immunostaining along the medial border of the
nucleus. Similar bands are even more prominent along the lateral margin.
It is within these regions that immunostained cell bodies are most often
found. Most of these cells are small (5-10 micra in minimum diameter).
They are both multipolar and fusiform in shape. These cells appear to
correspond to the small cells described in Golgi material (Adams, 1986).
Only occasionally are larger immunoreactive cells seen. These too are
most often associated with regions of neuropil immunostaining.

The density of immunostaining in the dorsal nucleus is far greater
than in the ventral nucleus. The neuropil of the dorsal nucleus is very
heavily stained except for the most superficial layer (Fig. 1D). Large
numbers of cells within the dorsal nucleus are covered with immunostained
terminals. There are numerous immunostained cells with the dorsal nucleus
(Fig. 1C). As in the ventral nucleus most immunostained cells are small
and can be either multipolar or fusiform in shape.

It seems likely that many of the immunostained terminals that are
present within the cochlear nucleus arise from intrinsic neurons. The
corresponding locations of immunostained cells and small immunostained
boutons in the neuropil suggests that the former may be the origins of
the latter. Another likely source of immunostained terminals is the
superior olivary complex. There are immunostained cells in periolivary
cell groups and there are coarse immunostained axons within the
trapezoid body that appear to enter the ventral nucleus and then disperse.
Much the same results have been reported in the cat (Adams and Mugnaini,
1987) where it has been demonstrated that most periolivary cells project
to the cochlear nucleus (Adams, 1983).

DISCUSSION

The widespread distribution of GAD immunostaining terminals in the
human cochlear nucleus indicates that in humans, as in animals, GABA
plays a large role neural processing of auditory information. Furthermore,
in the ventral cochlear nucleus it appears that the organization of GABA-
ergic terminals with respect to specific cell classes is much the same
as was previously described in the cat. Cells that receive secure
excitatory inputs on their somata, bushy cells, have abundant coarse
GABA-ergic terminals on their somata. Cells that have smaller primary
terminals distributed over both somata and dendrites, octopus cells, have
GABA-ergic terminals with similar sizes and distributions. Cells that may
be presumed to have few primary terminals on their somata, multipolar
cells, have few GABA-ergic somatic terminals, but rather seem to be
located preferentially in regions where fine GABA-ergic boutons fill the

neuropil and presumably terminate on multipolar cell dendrites. Thus, it appears that there is a rough equivalence of the size and distribution of GABA-ergic inputs with the size and distribution primary inputs to principal ventral cochlear nucleus cell types in both animals and humans.

The apparent equivalence of GABA-ergic and primary inputs to principal cells of the ventral cochlear nucleus suggests that there is a need for the possibility of suppression of peripheral excitation within the ventral cochlear nucleus. There are well known systems for decreasing the cochlea's response to sound, the middle ear muscles and the medial olivo-cochlear system. It seems likely that a general suppression of peripheral input is not the function of the GABA-ergic endings in the cochlear nucleus. The fact that the size and distribution of GABA-ergic terminals matches those of primary terminals for different cell types indicates that the quality and quantity of inhibition is taylored for different cell types. If morphological cell types are thought of as representing different modes or channels of processing peripheral input, it appears that the different channels have their individual systems for suppressing peripheral input. It may be that different channels are suppressed on different occasions, depending upon the nature of the processing occurring at different times. Such possibilities present compelling subjects for investigation in future work.

The characteristic immunostaining of small boutons in the small cell cap and along the medial and lateral borders of the ventral nucleus draws attention to these regions as having a common GABA-ergic innervation pattern and suggests a common function may also be involved. Others (Moore and Osen, 1979; Heiman-Patterson and Strominger, 1985) have noted that the small cell cap portion of the ventral cochlear nucleus in humans is better developed than in cat. It appears that the small cell cap in humans contains a substantial proportion of multipolar cells. Immunostain-ing with the GAD antibody demonstrated that the dorsal, small cell region has a characteristic staining of its neuropil and that this characteristic staining is also present where there are other collections of multipolar and small cells. Similar immunostaining in the small cell cap and border regions of the cat ventral cochlear nucleus has been described (Adams and Mugnaini, 1987). A characteristic neuropil along the borders of the anteroventral cochlear nucleus has previously been noted and a correla-tion of this neuropil with a high density of cells that project to the inferior colliculus has been described in the cat (Adams, 1977, 1979). A concentration of small multipolar (stellate) cells along the borders of the cat anterior cochlear nucleus has been reported (Cant and Morest, 1979). The similarities in cats and humans of the morpology, distribu-tion and characteristic GAD immunostaining of neuropil associated with multipolar cells suggest that these cells have similar functions and connections in the two species. This notion needs confirmation by other evidence. For the present, it has the appeal of fitting with a general trend that suggests that the organization of ventral cochlear nucleus cell types and their GABA-ergic inputs are quite similar in humans and cats.

There are two obvious features of GAD immunostaining of the cochlear nucleus that are different between human tissue and that of rodents and cats. The first is that in animals the neuropil of the granule cell domain stands out because of its intense staining (Mugnaini, 1985; Adams and Mugnaini, 1987; Moore and Moore, 1986). The present results add to the evidence that the physiology of the human dorsal cochlear nucleus is surely different from that in lower animals because its contents and organization are so different between species.

On the other hand, the similarities present in the GAD immunostaining within the ventral cochlear nucleus of humans and animals are gratifying. These similarities substantiate previous work showing morphological similarities of cell types in human and cat material (Adams, 1986). The present findings that the morphology and immunostaining properties of terminals associated with various cell types is similar in humans and animal tissue helps confirm the assumption that the functions of the cell types in the ventral cochlear nucleus are similar in humans and animals. As a consequence one is encouraged to pursue more intense investigations of the physiology, morphology, cytochemistry and other characteristics of the cat ventral cochlear nucleus with reasonable expectations that the findings will have relevance to human hearing.

ACKNOWLEDGEMENTS

The antiserum used in this work (1440) was developed by Drs. Wolfgang Oertel, Donald Schmechel and Marcel Tappaz in the laboratory of Dr. Irwin Kopin at the National Institute of Mental Health. Development of its use for immunocytochemistry was done by Dr. Enrico Mugnaini. This work was supported by NIH grant NS 21307.

REFERENCES

Adams, J. C., 1977, Organization of the margins of the anteroventral cochlear nucleus, Anat. Rec., 187:520.

Adams, J. C., 1979, Ascending projections to the inferior colliculus. J. Comp. Neurol., 183:519-538.

Adams, J. C., 1983, Cytology of periolivary cells and the organization of their projections in the cat. J. Comp. Neurol., 215:275-289.

Adams, J. C., 1986, Neuronal morphology in human cochlear nucleus, Arch. Otolaryngol., 112:1253-1261.

Adams, J. C. and Mugnaini, E., 1987, Patterns of glutamate decarboxylase immunostaining in the feline cochlear nuclear complex studied with silver enhancement and electron microscopy. J. Comp. Neurol., 262:375-401.

Cant, N. B. and Morest, D. K., 1979, Organization of the neurons in the anterior division of the anteroventral cochlear nucleus of the cat. Light-microscopic observations. Neuroscience 4:1909-1923.

Heiman-Patterson, T. D. and Strominger, N. L., 1985, Morphological changes in the cochlear nuclear complex in primate phylogeny and development. J. Morphol., 186:289-306.

Moore, J. K. and Moore, R. Y., 1987, Glutamic acid decarboxylase-like immunoreactivity in brainstem auditory nuclei of the rat. J. Comp. Neurol., 206:157-174.

Moore, J. K. and Osen, K. K., 1979, The cochlear nuclei in man. Am. J. Anat., 154:393-418.

Mugnaini, E., Warr, W. B., and Osen, K. K., 1980a, Distribution and light microscopic features of granule cells in the cochlear nuclei of cat, rat, and mouse. J. Comp. Neurol., 191:581-606.

Mugnaini, E., Osen, K. K., Dahl, A. L., Friedrich, V. L., and Korte, G., 1980b, Fine structure of granule cells and related interneurons (termed Golgi cells) in the cochlear nuclear complex of cat, rat, and mouse. J. Neurocytol.,9:537-570.

Mugnaini, E. and Dahl, A., 1983, Zinc aldehyde fixation for light micro-
    scopic immunocytochemistry. J. Histochem. Cytochem., 31:1435-1438.
Mugnaini, E., 1985, GABA neurons in the superficial layers of the rat
    dorsal cochlear nucleus: light and electron microscopic immunocyto-
    chemistry. J. Comp. Neurol., 235:61-81.
Mugnaini, E. and Oertel, W. H., 1985, An atlas of the distribution of
    GABA ergic neurons and terminals in the rat CNS as revealed by GAD
    immunocytochemistry, in: "Handbook of Chemical Neuroanatomy. Vol. 4:
    GABA and Neuropeptides in the CNS. Part 1." Bjorkland, A., Hokfelt,
    T., eds. Amsterdam, Elsevier, 436-608.

FORWARD MASKING OF SINGLE NEURONS IN THE COCHLEAR NUCLEUS

F. A. Boettcher, R. J. Salvi* and S. S. Saunders*

University of Texas at Dallas, U.S.A.
*SUNY University at Buffalo, U.S.A.

INTRODUCTION

Temporal resolution describes the ability of the auditory system to analyze rapid fluctuations in sound intensity. A common psychophysical procedure for estimating temporal resolution is the forward masking paradigm. In the traditional forward masking study, a listener is asked to detect a brief probe stimulus which is preceded in time by a masking stimulus. When the time interval between masker offset and probe onset is greater than 200 ms, the threshold of the probe tone is typically unaffected by the masker. However, as the masker-probe interval decreases, the threshold of the probe increases. Furthermore, the threshold of the probe increases as the intensity and duration of the masker increase (Luscher and Zwislocki, 1949; Harris, 1950; Elliott, 1962; Wilson, 1970).

In recent years, physiologists have begun to explore the neural correlates of forward masking using both single unit and gross potential techniques (Smith, 1977; Harris, 1977; Kramer and Teas, 1982). Evoked potential studies carried out with humans indicate that certain peaks in the evoked response waveform are more resistant to forward masking than others. For example, the masker appears to cause a significantly greater reduction in the amplitude of wave I than wave V, but a greater increase in the latency of wave V than in wave I (Kramer and Teas, 1983; Thornton and Coleman, 1979; Kevanishvilli and Lagidze, 1979). These findings suggest that there is some form of temporal recoding between the neural generators for wave I and wave V. Surprisingly, it has also been reported that the amplitude of wave V may actually be enhanced in certain forward masking conditions (Ananthanarayan and Gerken, 1983). Thus, the neural correlates of forward masking appear to be more complex in the central nervous system than at the periphery.

In order to obtain a more detailed understanding of the neural events underlying forward masking, researchers have turned to single unit studies in order to investigate the process of short-term adaptation and the recovery from adaptation. Most of the data available to date have come from units in the auditory nerve (Smith, 1977; Harris, 1977; Harris and Dallos, 1979; Salvi et al., 1986). The response patterns of units in the auditory nerve are fairly homogeneous. In general, the firing rate to the probe increases as the time interval between masker and probe increases so that a unit is almost completely recovered at a masker-probe interval of 200-300 ms. The decrease in firing rate to the probe stimulus also increases with masker duration out to approximately 100 ms. In addition, the decrement in probe firing rate increases with masker intensity. Smith (1977) and Harris (1977) suggest that the decrement reaches an asymptote once the firing rate to the masker saturates.

All of the auditory nerve fibers exiting the cochlea synapse in the CN. However, the CN is not a simple relay nucleus, but instead is actively involved in recoding the pattern of neural activity flowing into the central auditory pathway. This is exemplified by the heterogeneous array of Post-stimulus time (PST) histograms obtained from CN units using tone bursts. Since the temporal response patterns to tone bursts are more complex, one would expect a greater variety of forward masking recovery functions in the CN compared to the VIII nerve. We are aware of only one report by Watanabe and Simada (1971) dealing with forward masking in the CN. Forward masking effects were seen in the few units they studied; however, there was no attempt to quantitatively characterize the results or to look for differences in forward masking response patterns across different classes of units in the CN. In a related study, Moller (1969) described differential responses of CN units when stimulus repetition rate was varied. "Transient" units, which responded with a single spike per tonal stimulus, responded preferentially to specific click rates. At click rates above and below the preferred rate, the response was inconsistent, but at the preferred rate the unit fired to each click. The range of preferred rates varied from 200 to 800 clicks per second. While this investigation did not study forward masking per se, it does suggest that the adaptation and recovery process may vary significantly across units in the CN.

The present study was undertaken to characterize forward masking patterns of units of the CN and to correlate such patterns with unit types based on PST histogram shape. As a basis for comparison, we recorded from VIII units using a protocol identical to that used in the CN. These results are potentially important for understanding the neural basis of forward masking at higher levels of the nervous system and for interpreting and understanding the results from evoked potential studies.

METHODS

Subjects and anesthesia. Healthy chinchillas aged 1 to 2 years were utilized as subjects. Animals used for CN recordings were anesthetized initially with a mixture of ketamine and xylazine and were given subsequent doses of ketamine alone. Animals used in VIII nerve experiments were anesthetized with Dial in urethane.

Surgical approach. Each subject was initially tracheotomized and the right pinna and bony ear canal were removed to place the sound source close to the tympanic membrane. A silver ball electrode was placed on the round window in order to monitor the whole-nerve action potential during the course of the experiment. A ventral-posterior section of the right skull was removed and the underlying cerebellum was aspirated to expose the right cochlear nucleus or VIII nerve.

Recording techniques. Micropipettes with resistances of 5-12 Mohms were utilized to monitor activity in the CN and electrodes with resistances of 20-50 Mohms were used to record from VIII nerve fibers. Spike discharges were amplified and spike arrival times from stimulus onset were recorded using a PDP 11/23 computer system and stored on disc for offline analysis.

Stimulus control. Once a unit was encountered, a measure of spontaneous rate was made over a 10 second interval. An automated tuning curve
program was then utilized to determine the characteristic frequency (CF)
and threshold of a unit. All stimuli used in the forward masking paradigm
were pure tones presented at the CF of a unit. The stimulus paradigm for
collecting forward masking data may be summarized as follows. Each 1000
ms stimulus cycle consisted of the following sequence:(a) 20 ms of silence
for determination of spontaneous rate, (b) a probe tone (control) 20 ms

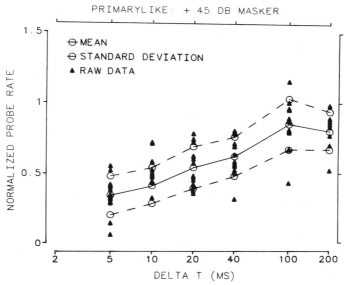

Fig. 1.   Mean forward masking recovery functions for 13 auditory
          ("VIII") nerve units are shown with +/- 1 standard deviation.
          Raw data from auditory nerve single units are presented
          as an overlying scatterplot. Masker level is 45 dB above
          threshold.

in duration presented at an intensity 15 dB above the threshold of
the unit, (c) 50 ms of silence for recovery from adaptation to the probe,
(d) a 100 ms masker presented in 15 dB steps from threshold to 75 dB
above threshold, (e) a silent period ("DT") which was varied from 5 to
200 ms, and (f) a masked probe, with characteristics identical to those
of the control probe.

    Data analysis. The results were analyzed using PST histograms (2ms
bin width) and spike counts were made for all signals. The ratio of the
firing rate for the masked probe to the firing rate for the control probe
was used to estimate the extent of masking (Harris, 1977; Salvi et al.,
1986).

RESULTS

    The preliminary results described here were recorded from 62 CN
units and 13 VIII nerve units. Units in the CN were categorized on the
basis of the shapes of their PST histograms obtained with tone bursts at
CF presented 30 to 50 dB above threshold (Pfeiffer, 1965; Evans and
Nelson, 1977; Godfrey et al., 1979). Most of the results are from units
classified as primarylike, primarylike-notch, and chopper units and
sufficient results are available to compare group mean data for these unit

types with those from the VIII nerve. In addition, interesting data from a few on and pausers units are presented.

Auditory Nerve. In order to put the results from the CN into proper perspective, it is useful to review the forward masking data from a sample of VIII nerve fibers. Mean forward masking recovery functions and raw data for 13 VIII nerve units are shown in Fig. 1. The abscissa shows the normalized probe rate as a function of the masker-probe interval (DT).

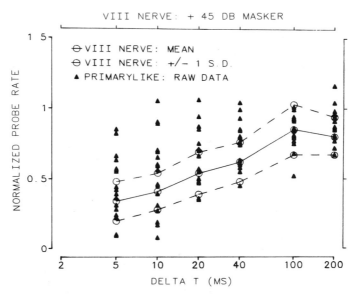

Fig. 2. Mean forward masking recovery functions for 13 auditory ("VIII") nerve units are shown with +/- 1 standard deviation. Raw data from 17 primarylike units of the cochlear nucleus are shown as an overlying scatterplot.

The normalized probe rate is lowest at a DT of 5 ms. As the masker-probe interval increases, the normalized probe rate increases so that by approximately 200 ms, the response is almost fully recovered. Increasing masker intensity, and therefore the firing rate, causes the normalized probe rate to decrease, particularly at short DTs. Because the response of VIII nerve units is fairly homogeneous, a mean forward masking function +/- 1 standard deviation was computed. This provides a metric for evaluating forward masking data from the CN.

Cochlear nucleus units. To facilitate comparisons between CN and VIII nerve data, raw data for primarylike units of the CN were overlaid on the VIII nerve group means +/- 1 standard deviation. Fig. 2 compares the results from a sample of primarylike units (N = 14) with those from the VIII nerve. It is apparent that the range of normalized probe rates is greater in primarylike units than in VIII nerve units. In addition, at DTs of 40 ms or less, 40-50% of the primarylike units show normalized probe rates greater than 1 standard deviation above the mean for VIII nerve units. Thus, a large proportion of primarylike units are less susceptible to forward masking than are VIII nerve units.

The range of normalized probe rates for primarylike-notch and chopper
units was similar to primarylike units. Thus, we will not present data
from primarylike-notch or chopper units graphically.

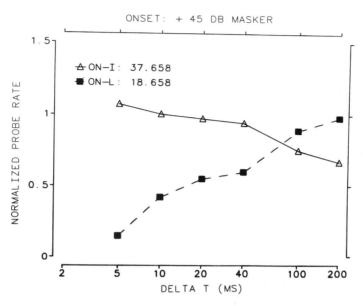

Fig. 3.   Forward masking recovery functions for 2 on units of the CN.
          Unit 37.658 is an On-I unit and unit 18.658 is an On-L unit.

Group comparison. Since a reasonable number of primarylike, primarylike-notch, and chopper units were studied, it was possible to compare
the mean data from these groups with those from the VIII nerve. At +30
and +45 dB masker levels, primarylike and chopper units are the most
resistant to masking at short DTs. The differences in the groups disappear
at longer DTs. Primarylike-notch units as a group appear to be similar to
VIII nerve units. In general, it may be summarized that primarylike,
chopper, and primarylike-notch units have forward masking functions
qualitatively similar to those of VIII nerve units. However, at short
DTs (40 ms or less), primarylike and chopper units are slightly less
susceptible to masking than are VIII nerve units.

Onset units. We do not yet have sufficient data to describe group
results for on units. Therefore, we will present interesting results from
two on units, an On-I unit and an On-L unit. Fig. 3 shows masking
functions for an On-I and an On-L unit. Unit 37.658, an On-I unit which
responds with only one spike at stimulus onset, shows little or no masking
even at the highest masker level. Unit 18.658, an On-L unit, was similar
to VIII units in susceptibility to forward masking.

Pauser units. The PST histograms of pauser units typically show a
strong onset peak, followed by a brief (4-8 ms) cessation of activity and
a resumption of activity which may (a) be stable over time ("pauser") or
(b) gradually build up ("pauser-buildup"). The units which we have
studied are interesting in that they have very non-monotonic forward
masking functions. The pauser units (N = 3) show much greater masking

at DTs of 20-100 ms than at shorter or longer DTs. The pauser-buildup units (N = 2) show enhanced response at DTs of 5-10 ms, significant masking at 20-100 ms DTs, and some recovery by 200 ms DT.

SUMMARY AND CONCLUSIONS

We have qualitatively described the forward masking behavior of units of the cochlear nucleus of the chinchilla. Whereas units in the VIII nerve recover from adaptation in a fairly homogeneous manner, a range of forward masking recovery patterns has been found in the CN. Preliminary data for primarylike and chopper units suggest that both groups are more resistant to forward masking than are VIII nerve units. Data on primary-like-notch units suggests that the mean recovery functions are similar to VIII nerve units.

Preliminary data from several on units suggest that On-I units may be extremely resistant to forward masking, while On-L units respond in a manner similar to VIII nerve units. Pauser units have unusual forward masking patterns, with enhanced probe response common at short DT and significant masking at moderate DTs. As was previously stated, these results come from a preliminary investigation of forward masking in the cochlear nucleus. Further goals of the project are to gather a large data base for each unit type in the CN in order to quantitatively compare groups. In addition, we are interested in the effects that inhibitory stimuli have on a forward masking patterns. Traditionally, recovery functions have been fit to exponential functions. However, due to the complex nature of the recovery patterns in some classes of CN units, a different approach to characterizing the recovery process appears to be necessary in the CN.

ACKNOWLEDGMENTS

This research was supported by the following grants: NIH IROINS16761 and NIOSH 5RO10H01152. We wish to thank K. Rockwood and G. Appleton for the generous donation of chinchillas used in this study.

REFERENCES

Ananthanarayan, A. K. and Gerken, G. M., 1983, Post-stimulatory effects on the auditory brainstem response: Partial masking and enhancement, Electroenecph. clin. Neurophysiol., 55:223-226.

Elliot, L. L., 1962, Backward and forward masking of probe tones of different frequencies, J. Acoust. soc. Amer., 34:1116-1117.

Harris, D. M., 1977, Forward masking and recovery from short-term adaptation in single auditory nerve fibers, Ph. D. dissertation, Northwestern University, Evanston, Illinois.

Harris, D. M. and Dallos, P., 1979, Forward-masking of auditory nerve fiber response, J. Neurophysiol., 42, 1083-1107.

Kevanishvilli, A. and Lagidze, Z., 1979, Recovery function of the human brainstem auditory evoked potential, Audiology, 18:472-484.

Salvi, R. J., Saunders, S. S., Ahroon, W. A., Shivapuja, B. G. and
    Arehole, S., 1986, Psychophysical and physiological aspects of
    temporal processing in listeners with noise-induced hearing loss,
    in: ""Basic and applied aspects of noise-induced hearing loss,
    R. J. Salvi, D. Henderson, R. P. Hamernik and V. Colletti, eds.,
    Plenum Press, New York.
Smith, R. L., 1977, Short-term adaptation in single auditory nerve
    fibers: Some post-stimulatory effects, J. Neurophysiol., 40:1098-
    1112.
Thornton, A. R. D. and Coleman, M. J., 1975, The adaptation of cochlear
    and brainstem auditory evoked potentials in humans, Electroenceph.
    clin. Neurophysiol., 39:399-406.
Watanabe, T. and Simada, Z., 1971, Auditory temporal masking:an electro-
    physiological study of single neurons in the cat's cochlear nucleus
    and inferior colliculus, Jap. J. Physiol., 21:537-549.

RESPONSES OF CAT VENTRAL COCHLEAR NUCLEUS NEURONES TO VARIATIONS IN

THE RATE AND INTENSITY OF ELECTRIC CURRENT PULSES

C. L. Maffi, Y. C. Tong and G. M. Clark

Department of Otolaryngology
University of Melbourne
Victoria 3052, Australia

INTRODUCTION

The speech processing strategies used with contemporary cochlear implant hearing prostheses have been designed primarily on the basis of psychophysical data (Millar et al., 1984). Further improvement of these strategies requires precise and predictable control of the neural population by the electric stimulus. To achieve this, the relationship between the response patterns of the neural elements and the electric stimulus needs detailed investigation.

There have been several studies of single auditory neurone responses to electric stimulation which involved sinusoidal and pulsatile current waveforms presented through intracochlear and extracochlear stimulating electrodes (Kiang and Moxon, 1972; Clopton and Glass, 1984; Hartmann et al., 1984; van den Honert and Stypulkowski, 1984; Javel et al., 1987). These studies all reported neural responses highly sunchronized with the cycles of electric stimuli. The synchronized activity produced by electric stimulation has been shown to occur in neurones over a broad spectrum of acoustic characteristic frequencies, and to persist to higher electric stimulus frequencies than for stimulation with acoustic tones. A wide range of response latencies has been recorded. Increasing stimulus intensity has been demonstrated to produce a greater degree of synchronization and a shortening of response latency. Stimulus-evoked discharge rates have been found to increase rapidly as a function of stimulus current level.

The aim of the present study was to determine some of the fundamental response properties of ventral cochlear nucleus neurones to variations in the rate and intensity of biphasic electric current pulses. In particular, this study focused on the responses of ventral cochlear nucleus units that exhibit a "primary-like" (PRI-like) response to acoustic stimulation. Ventral cochlear neurones associated with this response are thought to receive direct auditory nerve fibre input (Brawer and Morest, 1975; Rhode et al., 1983). Our major interest was in describing the temporal firing patterns of these auditory units in response to electric stimuli used in the Melbourne cochlear implant (Millar et al., 1984).

Observing the details of auditory neurone responses to electric stimulation may help in selecting electric stimuli for improved speech comprehension in cochlear implant patients.

METHODS

Normally-hearing adult cats with no signs of ear infection were used in this study. Animals were initially anaesthetized with an intraperitoneal injection of sodium pentobarbital (40 mg/kg) and thereafter maintained at a surgical level of anaesthesia with supplemental doses of barbiturate. A tracheal cannula was inserted, and the animal's body temperature maintained at 37.5 $^{\circ}$C by a regulated heating blanket.

The cat's left bulla was exposed and opened. A banded electrode array was threaded approximately 6 mm into the scala tympani via the round window, and was firmly tied in place with polyester mesh. The electrode array comprised three circumferential platinum bands, each of 0.3 mm width and 0.4 mm diameter, mounted approximately 1 mm apart on a silicone-rubber carrier. Two bands selected from the array functioned as a bipolar pair to deliver the stimulus current. Data were obtained using the pair of stimulating electrode bands for which threshold current level for eliciting neural activity was lowest. These were usually the most apical and basal bands.

Electric stimuli were biphasic charge-balanced constant-current pulses generated by an optically isolated current source. Pulse duration (set to 100 µsec/phase), amplitude and repetition rate were set under computer control. The pulses were delivered in trains of 50 msec duration, with 2.5 msec rise/fall times and delivered at a rate of 5 bursts per second.

The left cochlear nucleus was exposed surgically. Single unit activity in the ventral cochlear nucleus was recorded extracellularly with glass micropipettes. The microelectrodes were filled with 1.7 M KCl and had impedances of 8 - 12 M at 1 kHz.

Discharge spikes of units were obtained over 200 stimulus repetitions and summed onto histograms. Each unit's characteristic frequency (CF) was determined using acoustic tones, and the input-output relationships to both acoustic and electric stimulation were recorded.

RESULTS

Following implantation, acoustic and electric stimulus thresholds gradually increased throughout the experiment. Acoustic tuning curves and input-output functions could, however, be recorded from single ventral cochlear nucleus units up to 4 days postimplantation.

PRI-like ventral cochlear nucleus units were identified according to Pfeiffer's (1966) classification. Data were collected from 20 PRI-like ventral cochlear nucleus units, and the responses of representative examples are presented in this paper.

The time structure of a PRI-like ventral cochlear nucleus unit's response to a train of biphasic current pulses is illustrated in Fig. 1. Each row of histograms in Fig. 1 shows the response of the unit to a particular electric stimulus intensity: that which produced threshold firing rate in the uppermost row (0.80 mA), firing at an intermediate rate in the middle row (0.85 mA), and saturation firing rate in the lowermost row (1.10 mA).

Neural responses were always closely synchronized with the current pulses in the train, even at threshold firing rates. The electric stimuli

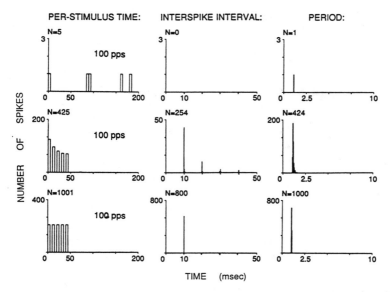

PER-STIMULUS TIME: INTERSPIKE INTERVAL: PERIOD:

Fig. 1.   Discharge spikes recorded from a ventral cochlear nucleus unit
          (CF 3.15 kHz). Stimuli: biphasic current pulses 100 μsec/phase,
          presented at 100 pps in trains of 50 msec duration at 5 trains/
          sec. Upper, middle and lower row: current level of 0.80 mA,
          0.85 mA and 1.10 mA respectively.

consisted of 5 current pulses, one occurring every 10 msec for 50 msec.
Accordingly, stimulus-evoked discharge spikes always occurred within the
first 50 msec of the per-stimulus time (PST) histograms and corresponded
to the 5 stimulus pulses. This is also seen in the interspike interval
(ISI) histograms: ISIs were equal to integral multiples of the stimulus
period (10 msec).

     The high synchronicity of firing to each current pulse is also
reflected in the period histograms in Fig. 1, which indicate that discharge spikes always occurred during a very narrow portion of the period.
The current pulse coincides with the beginning of the histogram time
axis. A delay of 1.0 - 1.3 msec was evident between the occurrence of
the current pulse and the resulting discharge spikes.

     The unit's firing pattern and rate changed with increasing stimulus
intensity. At threshold intensity (upper row of histograms in Fig. 1)
only the first current pulse in the train elicited a discharge spike.
Progressively higher stimulus intensities (subsequent rows) elicited
firing to an increasing number of pulses. At sufficiently high stimulus
intensity (lowest row), every biphasic pulse elicited a spike. This
resulted in a rectangular PST histogram and ISIs all equal to the
stimulus period, and corresponded to saturation firing rate.

     Firing rate in response to electric stimulation increased rapidly
and monotonically with stimulus intensity, as seen in Fig. 2A. Dynamic
ranges of PRI-like ventral cochlear nucleus units to electric stimulation
were typically 3 - 4 dB, and saturation firing rates, as noted previously,
were equal to the stimulus pulse rates.

     The temporal firing patterns and rate/intensity relationships
described above were consistent over the range of stimulus pulse rates
studied (100, 200, 400 and 800 pps). However, as shown in Fig. 2B, the

higher the pulse rate (or the shorter the pulse period), the higher the stimulus intensity required for every current pulse to elicit a discharge spike.

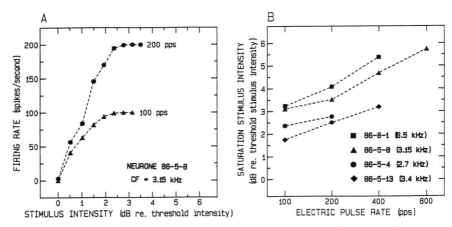

Fig. 2. A: firing rate/stimulus intensity relationship for unit in Fig. 1. Stimuli: as per Fig. 1, at 100 and 200 pps for lower and upper curve respectively. B: relationship between electric pulse rate and electric stimulus intensity required to elicit one discharge spike per current pulse, for ventral cochlear nucleus units of various CF (indicated in parentheses). Stimuli: as per Fig. 1, at 100, 200, 400 and 800 pps.

DISCUSSION

Having examined the relationship between the neural responses and the electric stimulus, it is possible to consider what modifications of the electrically-elicited responses might improve the communication performance of cochlear implant patients. One strategy may be to minimize the observed differences between the artificially- (electrically-) elicited responses and those elicited naturally (acoustically).

The acoustically-driven responses of PRI-like ventral cochlear nucleus units recorded in this study were consistent with those published in the literature (Rose et al., 1959; Pfeiffer, 1966; Rhode et al., 1983). In response to pure tone bursts at their CF, many PRI-like ventral cochlear nucleus units exhibited a monotonically increasing rate-intensity curve. Dynamic ranges were typically 30 - 40 dB. Thus electric stimuli elicited rate/intensity curves of similar shape to those elicited acoustically, with a much narrower dynamic range. To overcome this difference, in the cochlear implant acoustic signals are compressed to fit within the narrower dynamic range of responses to electric stimulation. Since the number of discernible intensity-difference steps is comparable for both stimuli (Hochmair-Desoyer et. al., 1981), the acoustic signal can be compressed and still be faithfully represented over the narrower dynamic range.

Within the dynamic range of the response, acoustic PST histograms were characterised by an initial peak in firing level and a subsequent plateau of firing at a lower level until the end of the tone burst, ISIs were distributed over a wide range and were most commonly below 10 msec, and finally discharge spikes could occur in any part of the acoustic tone period indicating probabilistic neural firing. This was in contrast with the highly deterministic nature of responses elicited from the same ventral cochlear nucleus units with electric stimulation, as described previously in this paper.

Furthermore, the PST firing patterns elicited by acoustic stimulation were consistently PRI-like over each ventral cochlear nucleus unit's dynamic range. In contrast, those elicited by electric stimulation changed throughout the unit's dynamic range, becoming progressively flatter and more unlike the acoustic PRI-like response with increasing stimulus intensity.

The discharge spikes recorded in the present study from PRI-like ventral cochlear nucleus units were very similar to those recorded from electrically-stimulated auditory nerve fibres by Javel et al. (1987). The similarity indicates that the PRI/PRI-like correspondence between the discharge patterns recorded from auditory nerve fibres and PRI-like ventral cochlear nucleus units respectively, occurred with both acoustic and electric stimulation of auditory nerve fibres. This in turn suggests that the ventral cochlear nucleus neurones were able to process the information from the electrically-stimulated auditory nerve fibres in a normal manner. However, the auditory nerve fibre input received by the ventral cochlear nucleus neurones had different temporal properties to the input elicited by acoustic stimulation. Consequently the ventral cochlear nucleus neurone output was likewise different from that elicited acoustically. It is likely that in the presence of electric stimulation, more central auditory neurones were not receiving the sort of input they might need in order to extract useful information from the stimulus.

In conclusion, it may be that electric stimuli which elicit more probabilistic neural discharges that are more akin to those elicited acoustically over their dynamic range, would produce more natural-sounding percepts in cochlear implant patients. Such stimuli may improve the communication performance of cochlear implant patients.

REFERENCES

Brawer, J. R. and Morest, D. K., 1975, Relations between auditory nerve ending and cell types in the cat's anteroventral cochlear nucleus seen with the Golgi Method and Nomarski Optics. J. Comp. Neurol., 160:491-506.
Clopton, B. M. and Glass, I., 1984, Unit responses at cochlear nucleus to electrical stimulation of the cochlea. Hearing Res., 14:1-11.
Hartmann, R., Topp, G. and Klinke, R., 1984, Discharge patterns of cat primary auditory fribres with electrical stimulation of the cochlea. Hearing Res., 13:47-62.
Hochmair-Desoyer, I. J., Hochmair, E. S., Burian, K. and Fischer, R. E., 1981, Four years of experience with cochlear prostheses. Med. Progr. Technol., 8:107-119.
van den Honert, C. and Stypulkowsky, P., 1984, Physiological properties of the electrically stimulated auditory nerve. II. Single fibre recordings. Hearing Res., 14:225-243.
Javel, E., Tong, Y. C., Shepherd, R. K. and Clark, G. M., Responses of cat auditory nerve fibres to biphasic electrical current pulses. Ann. Otol. Rhinol. Laryngol., 96 (Suppl. 128):26-30.
Kiang, N. Y. S. and Moxon, E. C., 1972, Physiological considerations in artificial stimulation of the inner ear. Ann. Otol., 81:714-730.
Millar, J. B., Tong, Y. C. and Clark, G. M., 1984, Speech processing for cochlear implant prostheses. J. Speech Hearing Res., 27:280-296.

Pfeiffer, R. R., 1966, Classification of response patterns of spike
    discharges for units in the cochlear nucleus: Tone-burst stimula-
    tion. Exp. Brain Res., 1:220-235.
Rhode, W. S., Oertel, D. and Smith, P. H., 1983, Physiological response
    properties of cells labelled intracellularly with horseradish
    peroxidase in cat ventral cochlear nucleus. J. Comp. Neurol., 213:
    448-463.
Rose, J. E., Galambos, R. and Hughes, J. R., 1959, Microelectrode studies
    of the cochlear nuclei of the cat. Bull. John Hopk. Hosp., 104:211-
    251.

COCHLEAR FREQUENCY SELECTIVITY AND BRAINSTEM EVOKED

RESPONSE (BER) DEPENDENCE ON CLICK POLARITY

R. Schoonhoven and V. F. Prijs

ENT Department, University Hospital

Leiden, the Netherlands

INTRODUCTION

The Brainstem Evoked Response (BER) is widely used in the study of peripheral auditory functioning, and has particular applications in audiology. Usually a wideband click is used as a stimulus. The response consists of a number of waves, the vertex positive peaks usually being indicated with Roman numerals I to V (VI, VII) according to the Jewett convention.

In this study we concentrate on the influence of click polarity on the recorded BER. Alternating (A) clicks are often used, but various investigators use rarefaction (R) or condensation (C) clicks separately. Significant differences between C and R responses have been reported (Ornitz and Walter, 1975; Kevanishvili and Aphonchenko, 1981; Moore, 1983), in particular in pathological hearing (e.g. Maurer, 1985). These differences concern the global appearance of the BER, peak amplitudes, absolute and inter-peak latencies and absence of waves in response to one stimulus which are clearly present in responses to others.

The aim of the present analysis is to investigate to what extent BER differences can be explained by different contributions to C and R response from different cochlear frequency bands. This question is studied with a high-pass noise masking paradigm and a subsequent narrow band analysis (cf Don and Eggermont, 1978; Parker and Thornton, 1978). Briefly, a high-pass noise masker is applied to the tested ear, the noise cut-off frequency being varied. This situation can be considered as a simulation of different degrees of high-frequency sensorineural hearing loss. Then, the responses corresponding to two subsequent noise cut-off frequencies are subtracted. The resulting narrow band BER is considered to represent the brainstem response to activation of the frequency band in the cochlea roughly between the two cut-off frequencies.

METHODS

Auditory Brainstem Evoked Responses were recorded upon click stimulation in twelve normally hearing human subjects (M/F) of ages 20-35 year. The click stimulus was a 100 μsec rectangular pulse applied to a TDH-39 earphone, presented at 70 dB nSL at a rate of 13/sec. The contralateral

ear was always masked with wideband noise. First the response to the pure
click was recorded. Then, the click was mixed with wideband noise at a

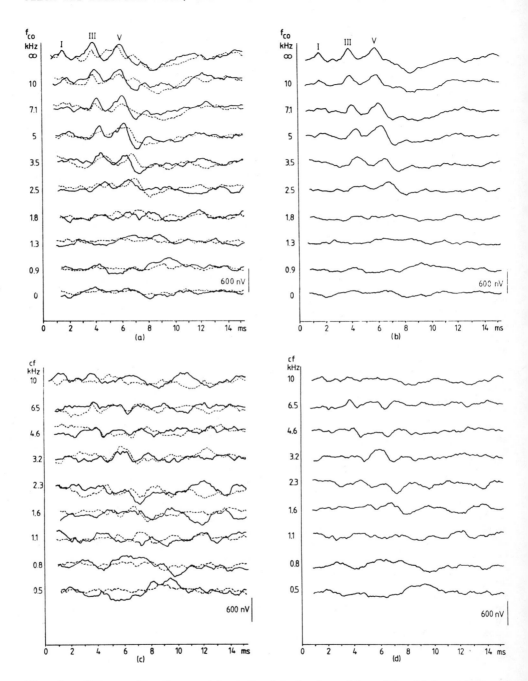

Fig. 1. BER results from subject A. (a) Condensation (C) click responses
(solid line), rarefaction (R) click responses (dashed line) and
(b) alternating (A) click responses during masking with high-
pass filtered noise (HPNBERs) at cut-off frequencies (f_{co}) as
indicated, and (c,d) narrow band responses (NBBERs) derived from
the curves in (a,b) corresponding to frequency bands with central
frequency (cf) as indicated.

level of 85 dB SPL, sufficient to almost completely mask the BER as well as the subjective click perception. Subsequently the noise was high-pass filtered before the mixing stage at cut-off frequencies of 10, 7.1, 5, 3.5, 2.5, 1.8, 1.3 and 0.9 kHz (half octave steps down from 10 kHz) and the BER was recorded for each filter setting. A cascaded pair of 48 dB/oct filters was used.

Recordings were made with active electrode on the forehead, reference on the ipsilateral mastoid and ground on the contralateral mastoid. The signal was filtered (30 Hz - 3 kHz), amplified and 3000 sweeps were averaged for each stimulus setting.

The above experimental paradigm was applied both for the rarefaction (R) and for the condensation (C) click. After the experiment, corresponding R and C responses were added numerically, the summation being considered as a fair approximation of the response to an alternating (A) click stimulus.

Finally, narrow band BERs were calculated by subtracting the BERs corresponding to each pair of subsequent noise cut-off frequencies. The slopes of the filters being finite, the central frequency of the narrow band is always slightly below the lower cut-off frequency.

RESULTS

In each of the subjects distinct differences between R and C responses were observed. However, a considerable qualitative and quantitative inter-subject variation was found. Therefore we present here the data of representative individual subjects rather than a group analysis. In Fig. 1 we show the BERs in the high-pass noise masking situation (HPNBER) and the corresponding narrow band responses (NBBER) for subject A, one of the subjects showing the more moderate differences between R and C responses. In Fig. 2 similar curves are given for subject B, one of the subjects showing quite pronounced differences between R and C responses.

The differences which were found in most subjects in HPNBER upon R and C stimulation are clearly illustrated in Figs 1a,b and 2a,b. These include
- differences in the global BER appearance causing unequivocal peak identification to become difficult (Fig. 2a)
- different absolute and inter-peak latencies of all peaks
- considerable differences in peak amplitudes, e.g. the relative IV-V amplitude (Fig. 2a)
- different amplitude behaviour of different peaks as a function of cut-off frequency
- running out of phase of waves III to V in R and C response, leading to (partial) cancellation of the A response, in particular in the low frequency range.

The NBBER curves (Figs 1c,d and 2c,d) show a number of distinct waves which, by relating them to the HPNBER curves, can be identified as Jewett peaks I-V generated by each particular cochlear frequency band. In order to further quantify the observations peaks I (where possible), III and V were identified in R and C HPNBERs and NBBERs. Their latency and amplitude (to succeeding trough) are shown in Fig. 3 for subject A.

DISCUSSION

One of the principal origins of different C and R BER responses lies at the level of basilar membrane motion which is assumed to be initially downward for C and upward for R clicks. The associated phase differences

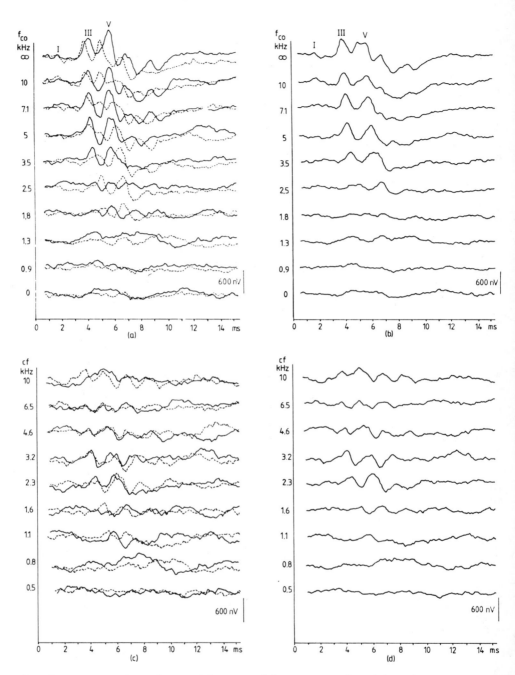

Fig. 2.   BER results from subject B. (a) Condensation (C) click responses
          (solid line), rarefaction (R) click responses (dashed line) and
          (b) alternating (A) click responses during masking with high-
          pass filtered noise (HPNBERs) at cut-off frequencies (f_{co}) as
          indicated, and (c,d) narrow band responses (NBBERs) derived
          from the curves in (a,b) corresponding to frequency bands with
          central frequency (cf) as indicated.

give rise to a temporal dispersion between VIIIth nerve C and R responses,

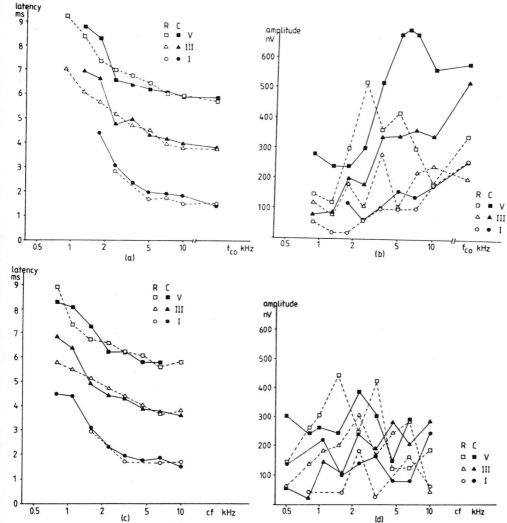

Fig. 3.   Results from subject A. Latency and amplitude of peak I, III
          and V in (a,b) for the high-pass noise masked click response
          (HPNBER) of Fig. la as a function of noise cut-off frequency
          (f_{co}), in (c,d) for the corresponding narrow band responses
          (NBBER) of Fig. lc as a function of central frequency (cf) of
          the band.

expected to be of the order of one half the period corresponding to each
particular frequency and therefore of increasing magnitude toward the
lower frequencies (e.g. Kiang, 1965).

     Associated effects are clearly found here in the HPNBER curves
(Figs. la, 2a). The temporal dispersion between C and R responses
increases towards lower noise cut-off frequencies, even until waves III
to V are getting virtually out of phase in the 1-2 kHz range. Here the A
response, due to partial cancellation of C and R responses, shows
considerably less pronounced peaks than either C or R response. These
phenomena are quite similar to observations in pathological hearing such
as high-frequency sensorineural hearing loss.

Apart from the global latency dispersion, however, it is observed that inter-peak latencies vary as a function of noise cut-off frequency as well (cf. Fig. 3). Also, waveshape variations are different: relative amplitudes of e.g. peaks III and V in C and R response show a considerable variability over the frequency range, and the global waveshapes may have such variable character that identification of peaks III, IV and V becomes quite arbitrary. Even the unmasked response may show large differences between C and R stimulation (Figs. 1a, 2a). Such responses are commonly thought to be dominated by highly synchronized contributions from the basal part of the cochlea, in which case such large differences would not be expected. These observations indicate that the later BER waves do not reflect a mere time delay with respect to the VIIIth nerve CAP, but that definite aspects of different C and R click representation in the brainstem nuclei are reflected in the BER as well.

The interpretation of the narrow band BERs (Figs. 1c,d and 2c,d) is less obvious. The small signal-to-noise ratio in particular for the early peaks and in the low frequency bands puts definite restrictions to a detailed interpretation of a second order effect as studied here. Identification of the early peaks becomes increasingly difficult for lower frequency bands but peaks III and in particular V can in most cases be traced to frequency bands below 1 kHz. This observation holds similarly for C, R and A responses. The latency curves of Fig. 3c show a picture quite consistent with observations in HPNBER (Fig. 3a). The amplitude behaviour does not allow definite conclusions neither from the response curves in Figs. 1c and 2c, nor from Fig. 3d.

As in other studies the inter-subject variability was found to be large, both qualitatively and quantitatively. For example, the question adressed by many authors whether it is the C or R click which yields the smallest latencies had to be answered differently in different subjects. This observation may partly be due to variability in middle ear transfer characteristics, which may considerably affect the phase behaviour of the stimulus at the level of the stapes, and thus the waveshape of the BER response (cf Gerull et al., 1985).

In conclusion, we showed that many of the observations concerning different C and R BERs, especially in audiological literature, could be simulated using the high-pass noise masking paradigm, which indicates that these phenomena can to a large extent be explained by cochlear mechanisms. In addition, manifestations of different C and R activation of the brainstem nuclei were identified in our analysis. The narrow band analysis supported these conclusions. Improvement of the signal-to-noise ratio and the development of a coherent interpretation of different narrow band contributions would be required to obtain further insight into the origins of C and R BER differences at the cochlear level.

REFERENCES

Don, M. and Eggermont, J. J., 1978, Analysis of click-evoked brainstem potentials in man using high-pass noise masking, J. Acoust. Soc. Am. 63:1084-1092.
Gerull, G., Mrowinski, D., Janssen, T. and Anft, D., 1985, Brainstem and cochlea potentials evoked by rarefaction and condensation single-slope stimuli, Scand. Audiol., 14:141-150.
Kevanishvili, Z. and Aphonchenko, V., 1981, Click polarity inversion effects upon the human brainstem auditory evoked potential, Scand. Audiol., 10:141-147.
Kiang, N. Y., 1965, Discharge patterns of single fibers in the cat's auditory nerve, MIT Press, Cambridge MA.
Maurer, K., 1985, Uncertainties of topodiagnosis of auditory nerve and

brainstem auditory evoked potentials due to rarefaction and condensation stimuli. Electroenceph. Clin. Neurophysiol., 62:135-140.

Moore, E. J. ed., 1983, Bases of auditory brainstem evoked responses, Grune and Stratton, New York.

Ornitz, E. M. and Walter, D. O., 1975, The effect of sound pressure waveform on human brainstem auditory evoked responses, Brain Res., 92:490-498.

Parker, D. J. and Thornton, A. R. D., 1978, Frequency specific components of the cochlear nerve and brainstem evoked responses of the human auditory system, Scand. Audiol., 7:53-60.

LATENCY "SHIFT" OF THE WAVE V IN BERA FOR LOWER FREQUENCIES?

K. Bareš and M. Navara

ENT Clinic, Charles University

Prague, Czechoslovakia

The normal intensity-latency curves of the BERA examination have a rather small dispersion. In the presence of high-pass noise masking, these curves are expected to be translated along both axes and to change their steepness. The interindividual dispersion becomes larger.

We analysed the influence of the cut-off frequency (2 kHz, 4 kHz) of the masker on the intensity -latency curves in 5 healthy women of 25 - 42, each separately. The normal hearing thresholds for pure tones and the normal hearing thresholds of clicks in the presence of masking by high-pass filtered white noise are overlapping for the frequencies under 2 kHz (Fig. 1, curve NHT). The BERA normal thresholds of the last mentioned stimulus are quite different. (Fig. 2, curve BERA).

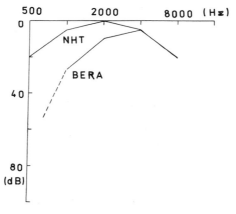

Fig. 1. Normal hearing and BERA thresholds. NHT=normal hearing thresholds of pure tone. BERA=normal BERA thresholds of click in the presence of masking by high-pass filtered white noise, for the frequency 8000 Hz threshold of tone burst of 1 ms duration. Averages of 5 probands.

Intensity-latency curves for different cut-off frequencies of the filter of the masker were compared. Obvious is the difference in (dB) among these curves - i. e. translation along the axis of intensities (Fig. 2). It was not possible to prove any differences in the axis of latencies or in the steepness among these curves, in the limits of our accuracy $1\sigma = 0.15$ ms.

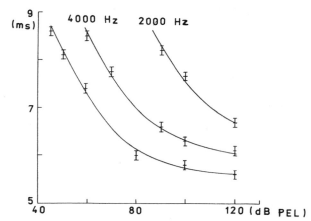

Fig. 2.   Intensity-latency curves for broad-band click in the presence of ipsilateral masking by white noise filtered by high-pass filter at 2000 Hz and at 4000 Hz. Average of 5 probands.

The latency shift for the responses of lower frequency BERA stimuli was possible to prove only as a consequence of the "threshold shift", i.e. higher thresholds of lower frequency stimuli.

Our results conform essentially to the data found by Thuemmler et al. (1984) and Kluba et al. (1985): they differ, however, from the views of some other authors (see Moore, 1983).

REFERENCES

Moore, E. J. 1983, Bases of Auditory Brain-Stem Evoked responses, Grune and Stratton, New York, 227-231.
Thuemmler, I. and Tietze, G., 1984, Derived acoustically evoked brainstem responses by means of narrow-band and notched-noise masking in normal-hearing subjects, Scand. Audiol. 13:129-137.
Kluba, J., Pethe, J., Aposchenko, V., Kwoles, R. and Specht, M., 1985, Zur Frequenzspaezifitaet von Tonbursts in BERA-Untersuchungen, H.N.O. Praxis, 10:279-284.

THE INFLUENCE OF RELAXATION, QUALITY AND GAIN OF SIGNALS

ON BERA LATENCY

Z. Novotný

ENT Department
University Hospital and Policlinic
Prague 2, Czechoslovakia

Achievement of true brain stem signals is the basic condition for a good BERA recording. Only signals without artifacts can be sufficiently enhanced and computer processed. In our study, the patients who were not sufficiently relaxed had prolonged wave V latency and flat waves. Prolonged latency was ascertained in 22 cases (35.5%) from 62 BERA tests. There was no significant difference as regards the hearing thresholds, sex or age.

The restoration of normal latency and waves was achieved after the patients' deeper relaxation with Haloperidol. Although sedatives themselves have no significant influence on BERA recordings, appropriate relaxation achieved with their help may be helpful for the test and should be made use of. Detailed findings of the investigated group of patients will be found in the following tables. Two short cases reports are joined.

Table 1. Age and BERA latency

| Age (years | Wave V latency | | |
	normal $\pm$ 2 s	shortened − 2 s	prolonged + 2 s
20 – 29	15	2	4
30 – 39	5	3	6
40 – 49	8	1	7
50 – 58	4	2	5

Case report 1: D. P., female, age 26 years, normal hearing. During the BERA test, low gain of signals (2 - 4), wave V latency prolonged, flat waves (Fig. 1).

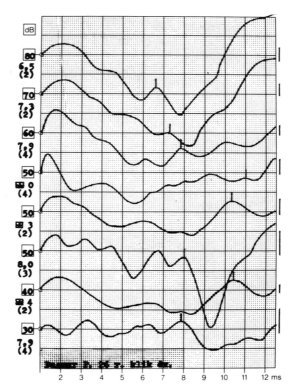

Fig. 1. Amplitude-intensity function of the brainstem response
in one patient (D. P.).

Case report 2: L. F., male, age 56 years, cochlear hearing loss of the
right ear. The first BERA test of the right ear proved low gain of
signals (4 - 5) and prolonged wave V latency (6.8 ms on 110 dB of click).

Table 2. Hearing threshold and BERA latency

Hearing threshold	Wave V latency		
	normal + − 2 s	shortened − 2 s	prolonged + 2 s
up to 20 dB	16	2	8
up to 40 dB	8	1	1
up to 60 dB	6	3	8
over 60 dB	2	2	5

The elevated hearing thresholds apply to patients with cochlear lesions.
"s" represents the standard deviation.

Repeated BERA test with a higher premedication proved a higher gain
of signals (6 - 9) and normal wave V latency (Fig. 2).

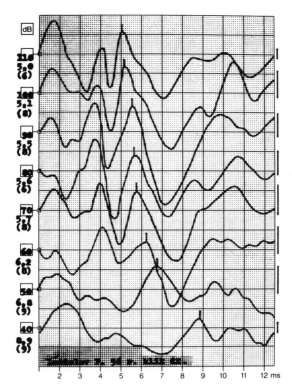

Fig. 2. Amplitude-intensity function of the brainstem response
in one patient (L. F.).

All the tests were performed on the BERA apparatus manufactured by
Hortmann GmbH. Signal gain degrees 1 - 4 are insufficient, sometimes
also 4 and 6, especially in lower click intensities.

REFERENCES

Adams, D. A., McClelland, R. J., Houston, H. G. and Gamble, W. G., 1985,
    The effects of diazepam on the auditory brain stem responses,
    Br. J. Audiol., 19:277 - 280.
Davis, H., 1976, Principles of electric response audiometry, Ann. Otol.
    85:Suppl. 28, 87.
Rosenhammer, H. J., Lindstrom, B. and Lundborg, T., 1980, On the use of
    click evoked electric brain stem responses in audiological diagnosis,
    Scand. Audiol., 9:93-100.

AUDITORY SUBCORTICAL NUCLEI

ORGANIZATION OF THE LATERAL LEMNISCAL FIBERS CONVERGING ONTO

THE INFERIOR COLLICULUS IN THE CAT: AN ANATOMICAL REVIEW

Motoi Kudo and Yasuhisa Nakamura

Department of Anatomy
School of Medicine, Kanazawa University
Kanazawa 920, Japan

REVIEW OF THE RESULTS OF ANTEROGRADE TRACING STUDIES

The ascending component of the lateral lemniscus, constituted by projection fibers originating in the cochlear nuclear complex, superior olivary complex and the nuclei of the lateral lemniscus, enters the inferior colliculus (IC) and terminates, as a whole, mainly in the ventrolateral division of the central nucleus of standard parcellation (i.e. the central nucleus of Morest and Oliver, 1984; Oliver and Morest, 1984, abbreviated as "ICC" in the present text), whereas the descending component of the corticofugal fibers from the auditory cortex and the commissural component from the contralateral IC terminate mainly in the dorsomedial division of the central nucleus and the pericentral nucleus of standard parcellation (i.e. the dorsal cortex of Morest and Oliver, abbreviated "ICD" in the present text). In the ICC, the ascending lateral lemniscal fibers, which have different origins in over 10 major auditory centers on bilateral sides, converge in a highly complicated manner as shown in Fig. 1.

In the present section, patterns of termination of projection fibers from each of the subcollicular auditory centers in the cat IC will be reviewed briefly. Most of the data have been obtained from autoradiographic experiments following injections of tritiated amino acid into a restricted small portion of each subcollicular nucleus.

Cochlear Nuclear Complex (CNC) (Fig. 2)

The projection from the CNC is largely contralateral (over 95%) and tonotopically organized. Among the great morphological variety of the neuronal population in the CNC, pyramidal neurons in the dorsal nucleus (DCN) and multipolar neurons in the antero- (AVCN) and postero- (PVCN) ventral nucleus (VCN) send direct branches to the IC. Most of the projection fibers from the contralateral DCN or VCN form a terminal focus along corresponding contour of the isofrequency laminae with some diffusely distributed throughout the ICC. While the fibers from the contralateral DCN (cDCN) tend to terminate caudo-ventro-medially in the ICC (Oliver, 1984), the fibers from the contralateral VCN (cVCN) centrally (Kudo unpublished observation), although the focus of termination can be shifted in accordance with high-to-low frequency organization within each nucleus.

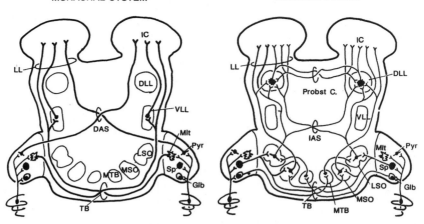

MONAURAL SYSTEM   BINAURAL SYSTEM

Fig. 1.   Diagrams of ascending auditory pathways in lower brain stem,
          where monaural and binaural systems are separately depicted.
          Note highly convergent projections onto inferior colliculus (IC)
          from variety of subcollicular nuclei on both sides. DAS, dorsal
          acoustic stria; Glb, globular neuron in VCN; IAS, intermediate
          acoustic stria; LL, lateral lemniscus; Mlt, multipolar neuron in
          VCN;Probst C., commissure of Probst; Pyr, pyramidal neuron in
          DCN;Sp, spherical neuron in VCN;TB, trapezoid body. See text for
          other abbreviations.

Superior Olivary Complex (SOC) (Fig. 3)

        The projection from the lateral superior olive (LSO) is bilateral,
while the projection from the medial superior olive (MSO) is ipsilateral.

        The projection fibers from the LSO terminate primarily in the ventro-
lateral part of the caudal half of the ICC, where they often form patches
of discontinuous terminal plexus with some terminals diffusely distributed
(Elverland, 1978; Glendenning et al., 1981; Glendenning and Masterton,
1983; Kudo unpublished observation). While the ipsilateral LSO (iLSO)
appears to contribute fibers to the low-to-middle frequency laminae of
the ICC (the dorsolateral sector), the contralateral LSO (cLSO) to the
middle-to-high frequency laminae (the ventromedial sector). However, a
precise topographical organization of the projection from the LSO onto the
ICC has not yet been fully clarified.

        The projection fibers from the ipsilateral MSO (iMSO) terminate
predominantly in the dorsolateral part of the ICC along the border of the
ICC and the external nucleus at the middle levels of the rostrocaudal
extent of the IC (Henkel and Spangler, 1983). This MSO recipient zone
corresponds with the low frequency sector of the ICC in accordance with
that low frequency (below 4 kHz) representation is disproportionately
large in the MSO, although the focus of termination can be shifted medio-
laterally within the dorsolateral low frequency sector of the ICC according to the high-to-low frequency organization within the MSO.

Nuclei of the Lateral Lemniscus (NLL) (Fig. 4)

        The projection from the dorsal nucleus (DLL) is bilateral with a

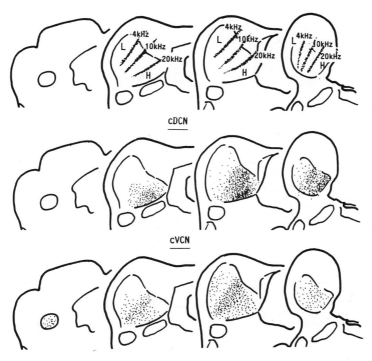

Fig. 2. Proposed schemata showing cochleotopical representation (upper, after Merzenich and Reid, 1974; Nudo and Masterton, 1984 and Servie're et al., 1984) and projection terminals from contralateral DCN (middle, after Oliver, 1984) and contralateral VCN (lower, Kudo unpublished data) in standardized frontal sections of inferior colliculus arranged rostrocaudally from left to right.

slightly contralateral predominance, while the projection from the ventral nucleus (VLL) is totally ipsilateral.

The projection fibers from the DLL terminate mainly in the dorsomedial half of the ICC in the proximity of ICD throughout the whole rostro-caudal extent of IC except its most caudal capsule (Kudo, 1981). In the dorsomedial half of the ICC, the fibers extending parallel with the fibro-dendritic laminae of the ICC branch away and form many fine patches of terminal plexus to show a comb-like appearance as a whole. A few fibers spill over into the dorsally adjacent ICD. Patterns of termination of the fibers from the contralateral DLL (cDLL) and those from ipsilateral DLL (iDLL) are identical but not symmetrical: notably the iDLL contributes more fibers to the rostral process of the IC but less to the caudal part of the ICC than the cDLL. Although the DLL has been said to have dorso-ventrally organized low-to-high frequency representation, topographical organization of this projection is uncertain.

The ipsilateral VLL (iVLL) projects strongly to the ICC (Kudo, 1981; Whitley and Henkel, 1984). However, the VLL is the only auditory nucleus whose projection to the ICC do not appear to be topographically organized, and the projection fibers terminate quite diffusely throughout the whole area of the ICC including the rostral process and the caudal cap of the

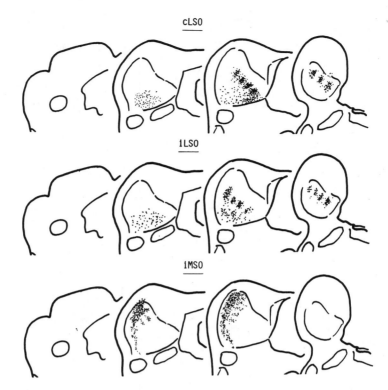

Fig. 3. Proposed schemata showing projection terminals from contralateral
LSO (upper) and ipsilateral LSO (middle) (after Elverland, 1978;
Glendenning et al., 1981 and Glendenning and Masterton, 1983)
and projection terminals from ipsilateral MSO (lower, Henkel and
Spangler, 1983) in standardized frontal sections of inferior
colliculus.

IC. Although the location of the autoradiographically labeled terminal
foci, if emerged, can vary among cases with different sites of tritiated
amino acid injections, no orderly topographical organization appears to
exist (Whitley and Henkel, 1984).

RETROGRADE TRACING STUDY

     The alternative anatomical approach is demonstrating subcollicular
auditory neurons sending their fibers to the ICC following retrograde
tracers injections into the ICC. This approach has two advantages: 1) All
neurons sending their fibers to the injection spot are consistently
labeled in a single case; 2) The obtained data, the number of retrogradely
labeled neurons for each subnucleus, are good for quantitative studies.

     In the present study, a special emphasis has been laid on the quanti-
tative analysis concerning the whole subcollicular neurons in order to
get the inter-relationships among the subnuclei as well as the relationship between the ICC and one of the subnuclei, although similar studies
have been done earlier in the cat (Roth et al., 1978; Adams, 1979; Brunso-
Bechtold et al., 1981).

174

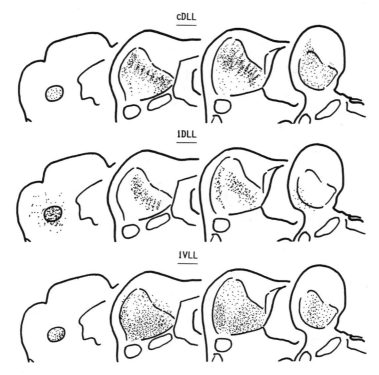

cDLL

lDLL

lVLL

Fig. 4.   Proposed schemata showing projection terminals from contralateral DLL (upper) and ipsilateral DLL (middle) (after Kudo,
1981) and projection terminals from ipsilateral VLL (lower,
after Kudo, 1981 and Whitley and Henkel, 1984) in standardized
frontal sections of inferior colliculus.

METHODS

Seventeen normal adult cats have been used in a series of our
experiments. During all surgical procedures the cats were deeply anesthetized with intraperitoneal injections of pentobarbiturate. Each cat
received a single injection of 2% WGA-HRP saline solution in a restricted
part of the ICC. The injection was made by air pressure through a glass
micropipette introduced stereotaxically through the cerebellum or under
direct vision after suction of the cerebellum. After a survival time of
24-48 hours, the cats were perfused with 0.9% saline followed by 8%
buffered formalin (pH 7.4) and then the brains were stored overnight
at 4°C in sucrose buffer (pH 7.4). Frozen sections were cut at 60 um in
the frontal plane, and every alternative section was collected serially
and incubated with tetramethylbenzidine (TMB) (Mesulam, 1978), immersed
in a saturated ammonium molybdate solution for 15 minutes to stabilize
the reaction product, mounted on slide glasses and counterstained with
neutral red. The number and distribution of the retrogradely labeled
neurons in the brain stem were recorded by examining all sections
processed on a brigt-or dark-field microscope.

QUANTITATIVE STUDY

Ten selected cases with various sites of injections were used for
the present quantitative study. As shown in Fig. 5, the injection spots

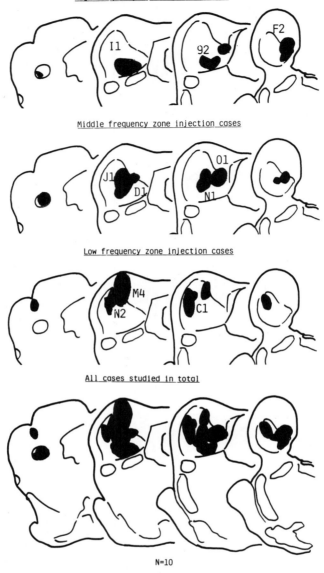

High frequency zone injection cases

Middle frequency zone injection cases

Low frequency zone injection cases

All cases studied in total

N=10

Fig. 5. Superimposed drawings of sites of WGA-HRP injections into central
        nucleus of inferior colliculus in all cases studied (bottom) and
        in three categories according to tonotopical order (upper three).

in the ten cases covered a greater portion of the ICC. For convenience,
they were divided into three categories according to the sites of injections within the ICC in the tonotopical order: 1) 3 low frequency cases
with the dorsolateral sector injections, 2) 4 middle frequency cases
with the middle sector injections and 3) 3 high frequency cases with the
vendromedial sector injections. This categorization of each case has been
reasonably done not only by the sites of the injections in the ICC but
also by estimating appearances of the anterogradely labeled terminals in
the medial geniculate body and the retrogradely labeled neurons in the SOC
and CNC according to their high-to-low frequency organization within each
nucleus.

176

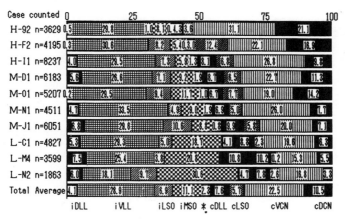

Fig. 6. Percentages of the number of labeled neurons calculated in 10
subcollicular nuclei in all 10 cases studied. Each case bar shows
the ratios of 8 major subnucleus by markings and numerical values
(iDCN and iVCN < 2% in average for each) are taken together and
indicated by an asterisk). Bottom bar shows the ratios of total
average.

A matrix of the number of retrogradely labeled neurons counted in
each subcollicular nucleus in each case has been analyzed statistically.
In the present paper, only a summary report and some interesting aspects
led by the analyzed results will be described.

Comparison of Percentages of the Number of Labeled Neurons among Cases

In agreement with the previous studies (Roth et al., 1978; Adams,
1979; Brunso-Bechtold et al., 1981), the total average bar at the bottom
in Fig. 6 shows that major populations of neurons sending their fibers to
the ICC are in the contralateral CNC (cDCN+cVCN, 33%), the ipsilateral
VLL (iVLL, about 30%) and the SOC (iMSO+iLSO+cLSO, about 25%) and the DLL
(iDLL+cDLL, about 10%) on both sides. When compared the average bar with
other case bars in Fig. 6, it is evident that there are considerable
variation among cases. While some nuclei like iVLL, 28.9% in average and
varying from 18.1% to 33.5%, show small variation, other nuclei like
iMSO, 11.1% in average and varying from 5.4% up to 30.6%, show very large
variation.

Significant difference is detected between the 3 categories in
tonotopical order. In cases with high frequency sector injections (upper
3 bars in Fig. 6), the contralateral components including the cDCN, cVCN
and cLSO occupy a large share, while, in cases with low frequency sector
injections (lower 3 bars in Fig. 6), high proportion of the iMSO is the
most notable. Similar results have been already reported by other investigators (Brunso-Bechtold et al., 1981; Aitkin and Schuck, 1985). On the
other hand, variation of the NLL (iVLL, iDLL and cDLL) does not appear to
be influenced by the tonotopical order.

Ipsi-/Contra-Lateral Ratio in Bilateral Projection Nuclei

In the bilateral projection nuclei such as the DLL, LSO, DCN and
VCN, the proportion of the numbers of labeled neurons on both sides is
not necessarily constant from case to case, but it has been revealed
to vary in order. The ipsi-/contra-lateral ratio calculated and shown

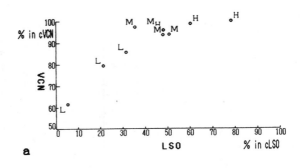

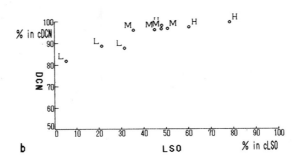

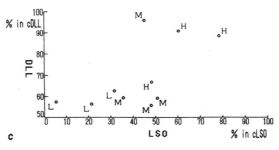

Fig. 7. Plots of ipsi- and contra-lateral ratios in bilateral projection
nuclei and their correlations between LSO and VCN (a), between
LSO and DCN (b) and between LSO and DLL (c) in cases with high-
(H), middle- (M) and low (L) frequency sector injections. See
text for detail.

in Fig. 7 is the number of labeled neurons on the contralateral side
divided by the number of labeled neurons on both sides.

    In the LSO, the ratio varies from 5% to 80% according to the low-to-
high frequency organization in the ICC, when plotted the ratio of individual case with a high- (H), middle- (M) or low- (L) frequency sector
injection (Fig. 7a, b, c). This is in good accordance with Glendenning
and Masterton's "acoustic chiasm" story that crossed projection neurons
are dominant in the medial high frequency sectors of the LSO, while uncrossed projection neurons are dominant in the lateral low frequency
sectors of the LSO (1983, Glendenning et al., 1985). In the present
study, the number of labeled neurons have been counted not separately for
each frequency sector within the LSO but totally whatever the frequency
sectors are. In general, it can be said that crossed components are
dominant in high frequency sector, while uncrossed components are dominant

in low frequency sector in the LSO-ICC projection system.

In the VCN and DCN, about 95% of the labeled neurons appear on the contralateral side in many cases (Fig. 7a, b). However, in the cases with low frequency injections, the ratio decreases from 80% down to 63% in the VCN (Fig. 7a) and from 89% to 82% in the DCN (Fig. 7b). These results suggest that there are tonotopically ordered ipsi- vs. contra-lateral organizations not only in the LSO-ICC system but also in the CNC-ICC system.

By contrast, the ratio in the DLL varies most uniquely (Fig. 7c). Although it can vary between 60% and 80%, neither correlation between the ratios in the DLL and LSO nor any tonotopical order in the variation of the ratio in the DLL is likely to exist.

Inter-Relationships among the Subcollicular Nuclei

It would be reasonably assumed that some of the interrelationships among the subcollicular nuclei in terms of their ICC projections can be led by the present numerical data.

The simplest way is finding out a possible high correlation of the number of labeled neurons between any two nuclei. The first example shown in Fig. 8a is the correlation between the iVLL and the cDCN. This shows very high correlation coefficiency (0.92), which is even higher than the value between the cVCN and the cDCN. The second example shown in Fig. 8b is a possible negative correlation between the iMSO, which is in charge of low frequency, and the cDCN, which is in charge of relatively high frequency. As expected, the tendency that the low frequency injection cases contain more neurons in the iMSO and less neurons in the cDCN than the high frequency injection cases makes the negative correlation as a whole.

Table 1 shows a correlation matrix, in which correlation coefficiencies of all combinations from 10 individual nuclei are calculated. The question is how we can deal with these variability that characterize so many inter-relationships among the nuclei. One realistic way we tried to treat the variation simultaneously and to get objective interpretations from the numerical data is the "multivariate analysis" technique by the use of a computer (Shiba, 1979). Table 2 shows results of the

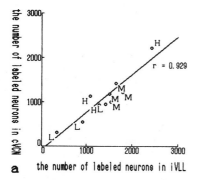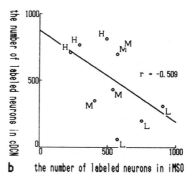

Fig. 8. Correlations of the number of labeled neurons between iVLL and cVCN (a) and between iMSO and cDCN (b) in cases with high-(H), middle-(M) and low-(L) frequency sector injections.

Table 1. Correlation matrix of all combinations.

	iDLL	iVLL	iLSO	iMSO	iVCN	iDCN	cDLL	cLSO	cVCN	cDCN
iDLL	1.000									
iVLL	0.492	1.000								
iLSO	0.477	0.768	1.000							
iMSO	0.439	-0.061	0.020	1.000						
iVCN	0.399	-0.204	-0.286	0.670	1.000					
iDCN	0.279	0.071	0.069	0.782	0.708	1.000				
cDLL	0.857	0.795	0.748	0.401	0.194	0.313	1.000			
cLSO	0.078	0.762	0.789	-0.441	-0.522	-0.158	0.407	1.000		
cVCN	0.415	0.929	0.614	-0.226	-0.273	-0.175	0.695	0.721	1.000	
cDCN	-0.198	0.634	0.414	-0.509	-0.542	-0.214	0.205	0.772	0.707	1.000

"factor analysis" using centroid method. In this statistical procedure,
mathematically assumed factors 1, 2, 3 and 4 (F1, F2, F3 and F4, respectively), which would explain the whole variation, are extracted in a
large eigenvalue sequence. That the contribution rates of F1, F2, F3 and
F4 are 40.1%, 36.3%, 8.4% and 4.6%, respectively, means that the whole
variation can be explained 76% by F1 and F2 combined or explained 90% by
F1, F2, F3 and F4 combined.

Table 2. Factor matrix computed out by using centroid method (Shiba,
1979).

	Factor 1	Factor 2	Factor 3	Factor 4
iDLL	0.704	-0.388	-0.512	0.033
iVLL	0.879	0.437	-0.060	0.107
iLSO	0.756	0.349	-0.082	-0.431
iMSO	0.319	-0.801	0.110	-0.179
iVCN	0.146	-0.815	0.144	0.209
iDCN	0.422	-0.647	0.582	-0.096
cDLL	0.940	-0.124	-0.313	-0.073
cLSO	0.549	0.746	0.114	-0.197
cVCN	0.745	0.538	-0.159	0.345
cDCN	0.350	0.760	0.247	0.146
Eigenvalue	4.010	3.632	0.844	0.468
Contribut.	40.1 %	36.3 %	8.4 %	4.6 %

    Fig. 9, in which the values after the varimax rotation are plotted
in a F1-F2 coordinate (a) and in a F1-F2 coordinate (b), shows the final
results of the factor analysis. In Fig. 9a, if the iDLL, cDLL, iLSO and
cLSO are ignored, three major components of the cDCN, cVCN and iVLL are
seen to form a group with high F1 value and low F2 value, while iMSO is
located away from the three major components with high negative F2 value
and low F1 value. Two minor components of iDCN and iVCN are closely
associated with the iMSO in Fig. 9a. In Fig. 9b, the cDLL and iDLL are
seen to form a group with high negative F3 value, while the cLSO and
iLSO form another group with high negative F4 value. These results suggest
that there could be four groups having something in common in their
variation of the number of labeled neurons, and the commonality could be
reasonably interpreted as follows:

1) iVLL, cDCN and cVCN:
    Injection volume depending variation.
    Diffuse projection type (iVLL type terminal pattern as shown in Fig.
    4 lower).

2) iMSO, iDCN and iVCN:
    Injection site depending variation.
    Nucleotopical projection type (iMSO type terminal pattern as shown
    in Fig. 3 lower).

3) cDLL and iDLL:
    Injection site depending variation.
    Nucleotopical projection type (DLL type terminal pattern as shown
    in Fig. 4 upper and middle).

4) cLSO and iLSO:
    Injection site depending variation.
    Nucleotopical projection type (LSO type terminal pattern as shown
    in Fig. 3 upper and middle).

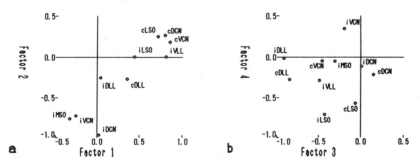

Fig. 9.   Results of factor analysis. Factor values of variables (sub-
          nuclei) after varimax rotation plotted in a F1-F2 coordinate (a)
          and a F3-F4 coordinate (b) after varimax rotation.

It now appears that the three major nuclei of the iVLL, cDCN and
cVCN are distinguished in the diffuse pattern of termination in their ICC
projections, regardless of the difference that the organization of the
projections of the cDCN and cVCN are clearly tonotopical, whereas that
of the iVLL is non-tonotopical. On the other hand, the ICC projections
of the MSO (iMSO), LSO (cLSO, iLSO) and DLL (cDLL, iDLL) should be called
"nucleotopical" in that they have their own distribution patterns of
termination topographically as has shown in the first section of this
paper. It is interesting to note that the first group of the iVLL, cDCN,
and cVCN are functionally monaural, while the second group of the MSO,
LSO and DLL are binaural. Also interesting is that the iDCN and iVCN are
closely associated not with the cDCN and cVCN but with the iMSO.

CONCLUDING REMARKS

Among a variety of auditory, visual and somato-sensory relay centers,
the IC is most unique in its highly convergent manner of the ascending
afferents of different origins as has been demonstrated in this paper.
The IC is unique also in that only one representation of sensory receptor
(i.e. the cochleotopical organization) is laid in a single, large and
structurally rather homogeneous neuronal mass (i.e. the ICC). Regardless of a lot of attempts to parcellate the ICC into several structural
and functional subunits, the ICC appears to be a single inseparable func-

tional unit, so far examined by the terminal patterns of the lateral
lemniscal afferents as well as by mappings of morphological and physiological properties of neurons of the ICC. Although there are apparent
regional variability within the ICC anatomically and physiologically,
it would be better to map this variability objectively without imposing
upon it the subjective criteria for formal subdivision.

ACKNOWLEDGMENT

This research project was supported by the grant No. 61770039 from
the Ministry of Education, Science and Culture of Japan.

REFERENCES

Adams, J. C., 1979, Ascending projections to the inferior colliculus,
J. Comp. Neurol., 183:519-538.

Aitkin, L. M. and Schuck, D., 1985, Low frequency neurons in the lateral
central nucleus of the cat inferior colliculus receive their input
predominantly from the medial superior olive, Hearing Res., 17:87-93.

Brunso-Bechtold, J. K., Thompson, G. C. and Masterton, R. B., 1981, HRP
study of the organization of auditory afferents to the central
nucleus of the inferior colliculus in the cat, J. Comp. Neurol., 197:
705-722.

Elverland, H. H., 1978, Ascending and intrinsic projections of the
superior olivary complex in the cat, Exp. Brain Res., 32:117-134.

Glendenning, K. K. and Masterton, R. B., 1983, Acoustic chiasm: efferent
projections of the lateral superior olive, J. Neurosci., 3:1521-1537.

Glendenning, K. K., Brunso-Bechtold, J. K., Thompson, G. C. and Masterton,
R. B., 1981, Ascending auditory afferents to the nuclei of the
lateral lemniscus, J. Comp. Neurol., 197:673-703.

Glendenning, K. K., Hutson, K. A., Nudo, R. J. and Masterton, R. B., 1985,
Acoustic chiasm II: anatomical basis of bilaterality in lateral
superior olive of cat, J. Comp. Neurol., 232:261-285.

Henkel, C. K. and Spangler, K. M., 1983, Organization of the efferent
projection of the medial superior olivary nucleus in the cat as
revealed by HRP and autoradiographic tracing methods, J. Comp.
Neurol., 221:416-428.

Kudo, M., 1981, Projections of the nuclei of the lateral lemniscus in the
cat: an autoradiographic study, Brain Res., 221:57-69.

Merzenich, M. M. and Reid, M. D. 1974, Representation of cochlea within
the inferior colliculus of the cat, Brain Res., 77:379-415.

Morest, D. K. and Oliver, D. L., 1984, The neuronal architecture of the
inferior colliculus in the cat: defining the functional anatomy of
the auditory midbrain, J. Comp. Neurol., 222:209-236.

Nudo, R. J. and Masterton, R. B., 1984, 2-deoxyglucose studies of
stimulus coding in the brainstem auditory system of the cat, Contrib.
to Sens. Physiol., 8:79-97.

Oliver, D. L., 1984, Dorsal cochlear nucleus projections to the inferior
colliculus in the cat: a light and electron microscopic study,
J. Comp. Neurol., 224:155-172.

Oliver, D. L. and Morest. D. K. 1984, The central nucleus of the inferior
colliculus in the cat, J. Comp. Neurol., 222:237-264.

Roth, G. L., Aitkin, L. M., Andersen, R. A. and Merzenich, M. M., 1978,
Some features of the spatial organization of the central nucleus of
the inferior colliculus of the cat, J. Comp. Neurol., 182:661-680.

Servie're, J., Webster, W. R. and Calford, M. B., 1984, Iso-frequency
labelling revealed by a combined ^{14}C -2-deoxyglucose, electrophysiological, and horseradish peroxidase study of the inferior
colliculus of the cat, J. Comp. Neurol., 228:463-477.

Shiba, S., 1979, "In-shi bunseki-ho", Tokyo Univ. Press, Tokyo (in
Japanese).

Whitley, J. M. and Henkel, C. K., 1984, Topographical organization of the
    organization of the inferior collicular projection and other connec-
    tions of the ventral nucleus of the lateral lemniscus in the cat,
    J. Comp. Neurol., 229:257-270.

DIFFERENT BINAURAL INPUTS SUBDIVIDING ISOFREQUENCY PLANES IN CHICK

INFERIOR COLLICULUS: EVIDENCE FROM 2-DEOXYGLUCOSE

Peter Heil and Henning Scheich

Institute of Zoology, Technical University

Schnittspahnstr. 3, D-6100 Darmstadt, F.R.G.

Recently, Scheich (1983) has mapped the functional organization of field L, the avian analogue of the mammalian auditory cortex, in monaural domestic chicks using the 2-deoxyglucose technique (Sokoloff et al., 1974). He provided evidence that the field contains two columnar systems with different input from the two ears. One of these systems consists of three bands, a rostral, an intermediate, and a caudal one, all oriented orthogonal to the course of isofrequency contours. This system therefore resembles the organization described in cat auditory cortex (Abeles and Goldstein, 1970). The second system concerns the intermediate band only. This band shows multiple alternating columns of ipsi- vs. contralateral dominant input, which are roughly parallel to the course of isofrequency contours (Heil and Scheich, 1985). In order to clarify whether these two columnar systems are generated in the forebrain or whether they can be observed at levels below field L, we have analyzed 2-deoxyglucose patterns in the lower auditory system of monaural chicks, and we found evidence for segregated areas of different binaural interactions in the midbrain nucleus mesencephalicus lateralis, pars dorsalis (MLD), the avian inferior colliculus homologue.

One day to four weeks after unilateral cochlea removal chicks were injected with 18 μCi of (14C) 2-fluoro-2-deoxy-D-glucose (2DG). Eight monaural and five non-operated control chicks were exposed to continuous white noise in a sound-proof chamber. Two additional controls were stimulated with pure tones. For one chick tones alternated between 0.2 and 1.0 kHz, for the other between 1.0 and 2.5 kHz. Alternating tones were presented at a rate of 4/s. After exposure times of 75 min chicks were decapitated, brains quickly removed, and frozen cross-sections cut at 30 um intervals. Processing of sections for autoradiography and details of the computerized densitometric analyses of autoradiograms are described elsewhere (Heil and Scheich, 1986).

Since MLD is the first nucleus in the ascending auditory system of birds which contains appreciable amounts of various binaural and monaural cell types (Coles and Aitkin, 1979; Moiseff and Konishi, 1983), namely EE-, i.e. units excited by stimulation of either ear, EI-, i.e. units excited by stimulation of the contralateral and inhibited by the ipsilateral ear, and EO-cells, i.e. units excited only by the contralateral ear, with no effect of the ipsilateral cochlea, this account concentrates on the chick auditory midbrain.

In 2DG-autoradiographs of cross-sections through MLD of monaural chicks clear differences of 2DG labeling patterns between the two sides are seen in all cases (Fig. 1). On the side contralateral to the intact

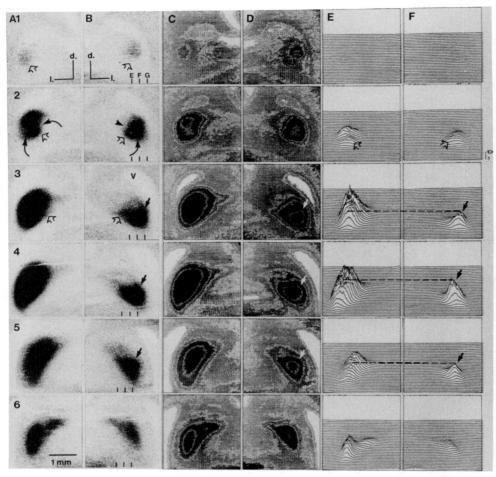

Fig. 1. 2DG patterns in the contralateral (column A) and ipsilateral (column B) nucleus mesencephalicus lateralis, pars dorsalis (MLD) of a monaural chick. Cross-sections from rostral (1) to caudal (6) levels at 240 μm intervals. Note the ventral area of high 2DG uptake in the ipsilateral MLD (solid arrows). Intensity of labeling is approximately identical to the corresponding ventral area of the contralateral MLD (see dashed lines in columns E and F, rows 3-5). Open arrows point to an area- dorsolateral boundary is indicated by bent arrows- also labeled in bilaterally deafened and intact chicks, which is probably not primarily auditory. Lateral scales E, F, and G refer to the parasagittal reconstruction planes of Fig. 2. E, F, and G. Columns C and D show isolabeling contours, and columns E and F densitometric profiles of A and B, respectively.

ear the MLD shows high 2DG uptake throughout its rostrocaudal extent (Fig. 1 A, C, E). These patterns closely resemble those seen in noise-

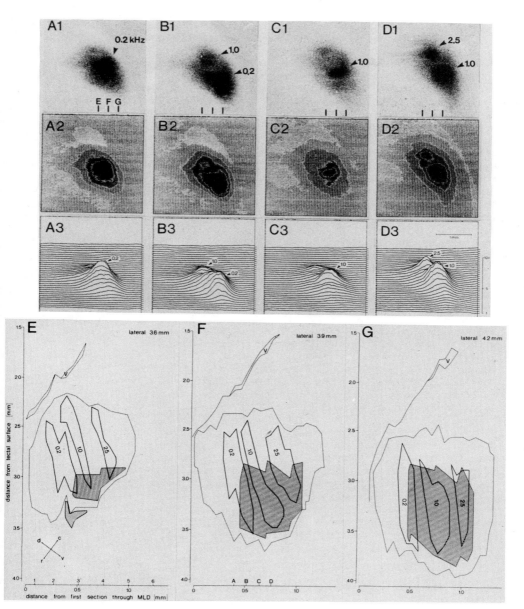

Fig. 2.   Tonotopic and interaural organization of MLD. Row 1 shows auto-
           radiographs of cross-section through the right MLD from rostral
           (A) to caudal (D) at 120 μm intervals. A and B are taken from
           a control chick stimulated with tones alternating between 0.2
           and 1 kHz, C and D from a chick stimulated with tones alternating
           between 1 and 2.5 kHz. 2DG isofrequency contours are identified
           by appropriate numbers (in kHz). Rows 2 ans 3 show isolabeling
           contours and densitometric profiles of corresponding 2DG auto-
           radiographs. E-G: Reconstruction of isofreqency contours in
           three parasagittal planes 3.6 (E), 3.9 (F), and 4.2 mm (G) lateral
           to the midline. Levels are also indicated by appropriate scales
           in row 1 and column B of Fig.1. Scales 1-6 on the abscissa in E
           refer to rostrocaudal levels 1-6 of Fig. 1 and scales A-D in F
           to rostrocaudal levels in upper autoradiographs. Hatched (cont'd)

187

Fig. 2.   (cont'd)   areas indicate the region of strong 2DG uptake
            in the ipsilateral MLD of noise-stimulated monaural chicks
            (Fig. 1 B).

---

stimulated controls. In contrast, the ipsilateral MLD shows strong 2DG
accumulation only at central and ventral locations (solid arrows in Fig.
1 B, D, F, rows 3-5). However, as is evident from the computer-generated
isolabeling contours (Fig. 1 C, D, rows 3-5) the medial to lateral extent
of this area of highest isointensity labeling in the ipsilateral MLD
is smaller than on the corresponding contralateral side, mainly due to
decreased labeling in the lateral area of MLD. The 2DG uptake in this
ventral core is approximately identical in intensity to the corresponding
area of the contralateral MLD, as is indicated by the dashed lines
connecting the peaks in the densitometric profiles (Fig. 1 E, F, rows 3-5).
The dorsal part of the central ipsilateral MLD, separated by a nearly
straight line from the symmetrically labeled ventral area (see isolabeling contours in Fig. 1 D, rows 3-5), the lateral area, and the dorsorostral and caudal poles show markedly less 2DG incorporation than on
the corresponding contralateral side.

     In an attempt to spatially relate these symmetrically labeled ventral
areas of MLD to the tonotopic organization of this nucleus (Coles, 1977),
we used control chicks stimulated with tones that alternated between 0.2
and 1 kHz in one chick, and between 1 and 2.5 kHz in the other during
2DG experiments (see above). Each MLD shows two peaks of 2DG uptake,
with the 1 kHz-peak being identical in both cases (Fig. 2 A-D). Since
labeling of isofrequency contours in field L in these cases was in
agreement with previous findings (Heil and Scheich, 1985), the tonotopic
labeling of MLD was considered reliable. Positions of peaks in MLD shift
in consecutive sections from dorsal in rostral sections to ventral
locations in caudal sections (Fig. 2 A - D).

     For a better survey of the tonotopic organization the isofrequency
contours in the right MLD (corresponding to the ipsilateral MLD of
monaural chicks) of each case were reconstructed in three parasagittal
planes 3.6, 3.9, and 4.2 mm lateral to the midline and with respect to
the dorsocaudal tectal surface. Reconstructed isofrequency contours of
both experiments were superimposed in the appropriate parasagittal
planes (Fig. 2 E-F). The course and locations of these reconstructed
isofrequency contours are in good agreement with electrophysiological
maps of tonotopy in the MLD of this species (Coles, 1977), if the
different planes of sectioning are taken into account (see compass in
Fig. 2 E).

     In the same manner the ventral area of strong and symmetrical 2DG
uptake in the ipsilateral (right) MLD of noise-stimulated monaural
chicks was reconstructed at the same parasagittal levels (scales E-G
in Fig. 1 B) and superimposed onto the tonotopic frame (hatched areas
in Fig. 2 E-G). The reconstructions reveal that the ipsilaterally
labeled (hatched) areas of MLD occupy defined sectors of isofrequency
contours, viz. the most ventral tip of such planes at a medial level (E),
the ventral half at a more lateral level (F), and at level G nearly the
complete dorsoventral extent. Furthermore, the dorsomedial borders of
the hatched areas are oriented roughly orthogonal to the course of isofrequency contours.

     These results clearly demonstrate that there are at least two
bands with different interaural integration mechanisms orthogonal to
the course of isofrequency contours in MLD, similar to one of the systems
in field L (Scheich, 1983). 2DG uptake mainly stems from synaptic
activity in axon terminals and dendrites, thus reflecting input, whether

excitatory or inhibitory, and not from discharging neurones (axonal spikes), i.e. output (e.g. Auker et al., 1983; Heil and Scheich, 1986; Nudo and Masterton, 1986). The consequences of removing one cochlea will differ for the various interaural cell types and thus 2DG uptake. Cells receiving binaural excitatory input (EE-units) should show approximately symmetrical 2DG uptake, since they retain one E-input on either side. EO-units should be labeled only contralaterally, since on the ipsilateral side no active input is available after cochlea removal. For EI-units the situation is more complex. At the first level of E/I-convergence the ipsilateral I-input may produce a 2DG uptake similar to the contralateral E-input (Nudo and Masterton, 1986). One step up, however, the reduced ipsilateral input, resulting from the active I-input one step down should show as reduced 2DG label. Thus, EI-inputs in a comparison of the two sides of the brain will show asymmetric labeling at the second and higher levels, just as EO-inputs. These considerations imply that in MLD of monaural chicks the symmetrically labeled ventral area indicates the location of binaural excitatory input, which will predominantly lead to EE-cell characteristics. This assumption is supported by a comparison of the 2DG patterns in auditory brainstem and lemniscal nuclei (Heil and Scheich, 1986) with the interaural cell types found in these nuclei (Moiseff and Konishi, 1983). As a consequence of these considerations we assume that EE-cells are concentrated in restricted ventrolateral sectors of isofrequency contours. It cannot be excluded that the symmetrically labeled areas may also contain EI-cells, whose binaural properties are generated at the level of the MLD.

The asymmetrically labeled areas of MLD in contrast, should correspond to the loci of EO-cells and EI-cells, whose binaural properties are transferred from lower levels, e.g. from the posterior subdivision of the ventral nucleus of the lateral lemniscus (Moiseff and Konishi, 1983). Whether they are similarly located in restricted sectors of isofrequency contours cannot be derived from our data. This is not unlikely however, since some recent work in mammals provides good evidence for the spatial segregation of cells with different interaural response properties in isofrequency contours of the inferior colliculus (e.g. Wenstrup et al., 1985). A spatially segregated or columnar distribution of the various interaural cell types in the central nucleus of MLD may be essential for the generation of the columnar systems seen in field L (Scheich, 1983) and the establishment of the spatiotopic organization in the external nucleus of MLD (Knudsen and Konishi, 1978).

ACKNOWLEDGEMENT

This study was supported the DFG-SFB 45.

REFERENCES

Abeles, M. and Goldstein, M. H., Jr., 1970, Functional architecture in cat primary auditory cortex, J. Neurophysiol., 33:172-187.

Auker, C. R., Meszeler, E. M. and Carpenter, D. O., 1983, Apparent discrepancy between single unit activity and (14C) deoxyglucose labeling in optic tectum of the rattlesnake, J. Neurophysiol., 49: 1504-1516.

Coles, R. B., 1977, Physiological and anatomical studies of auditory units in midbrain areas of the domestic fowl (Gallus gallus), Thesis, Monash University.

Coles, R. B. and Aitkin, L. M., 1979, The response properties of auditory neurones in midbrain areas of the domestic fowl (Gallus gallus) to monaural and binaural stimuli, J. Comp. Physiol., 134:241-251.

Heil, P. and Scheich, H., 1985, Quantitative analysis and two-dimensional reconstruction of the tonotopic organization of field L in the chick from 2-deoxyglucose data, Exp. Brain Res., 58:532-543.

Heil, P. and Scheich, H., 1986, Effects of unilateral and bilateral cochlea removal on 2-deoxyglucose patterns in the chick auditory system, J. Comp. Neurol., 252:279-301.

Knudsen, E. I. and Konishi, M., 1978, A neural map of auditory space in the owl, Science, 202:778-780.

Moiseff, A. and Konishi, M., 1983, Binaural characteristics of units in the owl's brainstem auditory pathway: Precursors of restricted spatial receptive fields, J. Neurosci., 3:2553-2562.

Nudo, R. J. and Masterton, R. B., 1986, Stimulation-induced (14C)2-deoxyglucose labeling of synaptic activity in the central auditory system, J. Comp. Neurol., 245:553-565.

Scheich, H., 1983, Two columnar systems in the auditory neostriatum of the chick: Evidence from 2-deoxyglucose, Exp. Brain Res., 51:199-205.

Sokoloff, L., Reivich, M., Patlak, C. S., Pettigrew, K. D., DesRosiers, M. and Kennedy, C., 1974, The (14C)deoxyglucose method for the quantitative determination of local cerebral glucose consumption, Trans. Am. Soc. Neurochem., 5:85.

Wenstrup, J. J., Ross, L. S. and Pollak, G. D., 1985, A functional organization of binaural responses in the inferior colliculus, Hearing Res. 17:191-196.

FUNCTIONAL ORGANIZATION OF THE VENTRAL AND MEDIAL DIVISIONS OF

THE MEDIAL GENICULATE BODY (MGB) OF THE CAT

C. Rodrigues-Dagaeff, E. M. Rouiller, Y. de Ribaupierre,
G. Simm, A. Villa and F. de Ribaupierre

Dept. of Physiology
Rue du Bugnon 7
CH-1005 Lausanne, Switzerland

INTRODUCTION

The medial geniculate body (MGB) has been divided by Morest (1964) in 3 main regions:the dorsal, medial and ventral divisons. In each division, nuclei corresponding to specific neuronal populations were defined:the pars lateralis (LV) and ovoidea (OV) in the ventral division, the pars magnocellularis (M) in the medial division and the dorsal (D), deep dorsal (DD) and suprageniculate (SG) nuclei in the dorsal division. Electrophysiological studies have demonstrated that these different nuclei differ by their functional properties (Calford, 1983, Toros-Morel et al., 1981). In LV, most units were sharply tuned and arranged along tonotopic sequences whereas D is not tonotopically organized and units were either unresponsive to tones or poorly tuned. LV and D are the main thalamic equivalents of the 2 segregated systems of connection, the tonotopic and diffuse systems respectively, involving distinct subdivisions of the midbrain, thalamus and cortex (Andersen et al., 1980). The pars magno-cellularis (M) appears functionally more heterogeneous and its pattern of connectivity seems to overlap the tonotopic as well as the diffuse systems of connection. The aim of the present work was to investigate the detailed functional organization of the ventral and medial divisions of the MGB. Functional properties of single units such as responsiveness to clicks, noise and tone bursts, response patterns and latencies, charac-teristic frequency (CF) and sensitivity to repetitive acoustic pulses were systematically analyzed as a function of the position of their recording site in LV, OV or M.

METHODS

The present data are based on the discharge properties of 2892 single units recorded in the MGB of 46 Nembutal and nitrous oxide anesthetized cats (1781 in LV, 371 in OV and 740 in M). The surgical preparation, acoustic stimulation and analysis of single unit activity have been previously described (Rouiller et al., 1979). Electrode tracks were histologically reconstructed on corresponding Nissl frontal sections of the MGB. The identification of one or two electrolytic lesions done on each electrode penetration was used to position each recording site. Borders of MGB subdivisions were determined on the same histological sections using both morphological (cell size and density) and functional (abrupt changes of response properties) criteria, as described by Calford

(1983). Finally, reconstructions on individual sections were adjusted onto a standard MGB, in order to cumulate data from several experiments.

RESULTS

Preliminary data suggested the existence of systematic changes of discharge properties along the rostro-caudal axis of LV (Toros et al., 1979). Such functional gradients were investigated by distributing single units within four 800 microns thick consecutive frontal slices going from A 3.5 to A 6.5 in LV and OV. In M, only 3 such slices were considered along its anterior three quarters (A 4.1 to A 6.5 mm), the number of units in the most posterior quarter of M being insufficient to be considered in the present analysis.

I. PARS LATERALIS (LV)

Responsiveness and Response Patterns

Effects of acoustical stimuli on the activity of MGB single units were tested by their responsiveness to broad band stimuli (clicks and noise bursts) and pure tones. As illustrated in Fig. 1 (left column), the proportion of units sensitive to both tone and broad band stimuli (N+C+T) progressively increased from caudal to rostral; posteriorly, the majority of cells were either not sensitive (NO RESP.) to simple acoustic stimuli or they responded to pure tone exclusively (T). Consistent changes in the distribution of response patterns to the 3 types of stimuli were observed along the rostro-caudal axis. As illustrated for pure tones in the middle column of Fig. 1, the proportion of responses with inhibitory components (dark bins) progressively decreased from posterior to anterior, whereas the relative number of excitatory responses (through, on) increased (chi-square, $p < 0.01$).

Response Latencies

The distribution of response latencies to noise, click and tones systematically varied along the rostro-caudal axis of LV. As illustrated for tones in Fig. 1 (right column), response latencies were on average longer and more variable in the most posterior slice. Going rostrally, the response latencies progressively decreased and were significantly less variable (analysis of variance, $p < 0.01$). These changes of latencies were preserved when they were corrected for the cochlear delay computed with the travelling time model proposed by Gibson et al. (1977), as well as when excitatory or inhibitory responses were considered separately.

Tonotopic Organization and Frequency Selectivity

Within each individual frontal slice, the distribution of single units as a function of their CF was consistent with an increasing CFs from lateral to medial, as previously reported (Aitkin and Webster, 1972; de Ribaupierre and Toros, 1976; Calford and Webster, 1981; Imig and Morel, 1985). The quality of this tonotopic arrangement was estimated by a tonotopic factor (Morel, 1980), computed on 59 track segments in LV, ranging between 0 for a random arrangement and 1 for a perfect tonotopic organization. As shown in Fig. 2A, the tonotopic factor increased progressively from caudal to rostral (coefficient of correlation r=0.48, $p < 0.01$), indicating that a smaller CF local disparity was encountered anteriorly than posteriorly. This variation of the tonotopicity from posterior to anterior might be correlated with changes of frequency selectivity. The distribution of tuning curves indicated that the majority of units located caudally were broadly tuned or responded with multiple frequency ranges, whereas anteriorly most units displayed primary-like or narrow tuning curves.

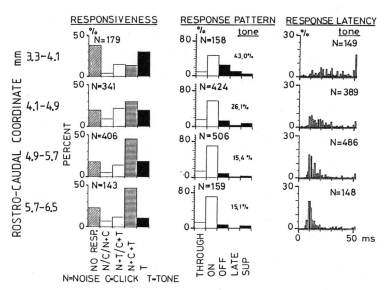

Fig. 1. Distribution of responsiveness to simple acoustic stimuli (left
column), of response patterns to tone bursts (middle column)
and of response latencies to tone bursts (right column) for single
units located in LV as a function of their caudo-rostral
coordinate (4 frontal consecutive slices, from top to bottom).
In each distribution, the sum of all bins is 100%. N indicates
the number of units distributed.

Responses to Repetitive Clicks

The temporal coding properties of MGB single units were analyzed
by their discharge characteristics in response to trains of clicks
presented at increasing repetition rates. Units were classified as
"lockers" when their discharges were precisely synchronized to the

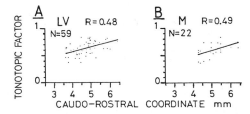

Fig. 2. Degree of tonotopic organization estimated by the tonotopic
factor (see text) computed on electrode track portions crossing
LV (left) or M (right) as a function of the corresponding
caudo-rostral coordinate. N indicates the number of electrode
track portions considered in each plot.

individual clicks, as "groupers" when they showed poor time-locking, as
"special responders" when they presented transient responses at the
onset or offset of the trains and as "non-responders" when insensitive
(Rouiller et al., 1981). The present analysis showed a progressive

increase of the proportion of "lockers" from caudal to rostral where the highest rates of synchronization (200 to 400 Hz) were the most frequently observed. On the other hand, the proportion of "special responders" and "non-responders" was higher posteriorly than anteriorly.

II PARS OVOIDEA (OV)

Previous observations indicated that units located in OV had discharge properties very similar to those seen in LV (Calford, 1983; Toros-Morel et al., 1981). This was also true for the distribution of response properties along the rostro-caudal axis. As in LV, going from posterior to anterior, responsiveness to broad band stimuli and the proportion of excitatory responses increased whereas response latencies decreased and were less variable. The number of units in OV tested for the other response properties was insufficient to allow a comparison with the observations made in LV.

III PARS MAGNOCELLULARIS (M)

As previously reported, units in M are characterized by very heterogeneous discharge properties showing the whole range of possible responses observed in the MGB (Calford, 1983). On the contrary to LV, excitatory responses to tones were more frequent caudally (chi-square, $p < 0.02$) and no systematic change of responsiveness to simple acoustic stimuli and response patterns to noise and click occured going from caudal to rostral (A 4.1 - A 6.5 mm). As in LV, the response latencies to noise, click and tone were, on average, shorter and less variable rostrally (analysis of variance, $p < 0.001$) and the proportion of "lockers" increased going anteriorly as well as their upper frequency limits of locking, whereas the other responses to click trains were more frequent caudally.

Tonotopic Organization

Consistent progressions of unit CFs were observed in the present study along electrode track portions crossing M in the frontal plane. These progressions were consistent with a high to low CF organization going from dorso-medial to ventro-lateral. Although the tonotopicity appeared somewhat less strict than in LV, the tonotopic factor also increased significantly ($p < 0.05$) going from caudal to rostral, as illustrated in Fig. 2B. Frequency selectivity, on the other hand, did not vary significantly along the rostro-caudal axis in M. This tonotopic organization in M was supported by the observed spatial distribution of labeled cells in M resulting from injections of WGA-HRP in CF defined loci in the primary auditory cortex (AI). Following injections in low, middle and high CF regions of AI, clusters of labeled neurons in M were progressively shifted from ventral to dorsal respectively.

DISCUSSION

The two principal nuclei of the ventral division, LV and OV showed a statistically significant progression of response properties along the rostro-caudal axis, perpendicular to the tonotopic gradient of LV. Anteriorly, units generally had short response latencies with discharge characteristics suggesting a rather good preservation of the discharge properties observed in brainstem auditory nuclei (responsiveness to broad band stimuli, sharp frequency tuning, majority of excitatory response patterns and synchronization of discharges to repetitive clicks). Going caudally, a progressively increasing proportion of units in LV and OV were characterized by more variable and longer response latencies, an increasing proportion of inhibitory responses, a broader tuning, a loss of precise time-locking and a poorer responsiveness to broad band stimuli. This functional gradient along the rostro-caudal axis may be correlated with the spatial distribution of the reciprocal connections of LV with the various fields of the auditory cortex. Following WGA-HRP injections in the primary (AI) or anterior (AAF) auditory cortical fields, retrogradally labeled neurons and anterogradally marked corticofugal terminals were found in majority (90%) in the anterior half of LV and OV. The posterior part of LV and OV, showed only rare connections with AI and AAF, and none with the secondary auditory cortex (AII). Injection of WGA-HRP in the posterior auditory field (PAF), however, revealed that it is interconnected with a wider rostro-caudal extent of LV and OV than AAF or AI, including part of the posterior half of these 2 nuclei. These functional and anatomical observations suggest that the neurons located in the anterior half of LV and OV might be principally relay neurons transferring to AAF and AI a well preserved auditory information coming from the periphery. On the other hand, the posterior part of LV and OV, characterized by more complex response properties, might participate to a more elaborated information processing in cooperation with PAF and possibly other cortical fields that need to be determined as well as the precise origins of its inputs from the midbrain.

For the pars magnocellularis (M), the present work revealed its tonotopic organization consisting in a CF gradient oriented from dorsal to ventral, slightly tilted from medial to lateral. Although the CF local disparity appeared more important than in LV, this observation suggests that the anterior part of M might participate to the tonotopic system of connection. The tracing WGA-HRP experiments reported above, however, showed that M is interconnected with the tonotopic cortical fields (AAF principally, AI and PAF) as well as with the non-tonotopically organized area (AII). This suggests that M might functionally act as an interface between the 2 largely segregated tonotopic and diffuse systems of connection.

REFERENCES

Aitkin, L. M. and Webster, W. R., 1972, Medial geniculate body of the cat:organization and responses to tonal stimuli of neurons in the ventral division, J. Neurophysiol., 35:365-380.

Andersen, R. A., Knight, P. L. and Merzenich, M. M., 1980, The thalamocortical and corticothalamic connections of AI, AII, and the anterior auditory field (AAF) in the cat: evidence for two largely segregated systems of connections, J. Comp. Neurol., 194: 663-701.

Calford, M. B. and Webster, W. R., 1981, Auditory representation within principal division of cat medial geniculate body: an electro-physiological study, J. Neurophysiol., 45:1013-1028.

Calford, M. B., 1983, The parcellation of the medial geniculate body of the cat defined by the auditory response properties of single units, J. Neuroscience, 3:2350-2364.

De Ribaupierre, F. and Toros, A., 1976, Single unit properties related to the laminar structure of the MGN, Exp. Brain Res., Suppl. 1, 503-505.

Gibson, M. M., Hind, J. E., Kitzes, L. M. and Rose, J. E., 1977, Estimation of travelling wave parameters from the response properties of cat anteroventral cochlear nucleus neurons, in: "Psychophysics and Physiology of Hearing", E. F. Evans and J. P. Wilson, eds., London, Academic Press, 57-68.

Imig, T. J. and Morel, A., 1985, Tonotopic organization in the ventral nucleus of the medial geniculate body in the cat, J. Neurophysiol., 53: 309-340.

Morel, A., 1980, Codage des sons dans le corps genouille median du chat:

evaluation de l'organisation tonotopique de ses differents noyaux, These, Universite de Lausanne. Juris Druck Verlag, Zurich.

Morest, D. K., 1964, The neuronal architecture of the medial geniculate body of the cat, J. Anat., 98:611-630.

Rouiller, E., De Ribaupierre, Y. and De Ribaupierre, F., 1979, Phase-locked responses to low-frequency tones in the medial geniculate body, Hearing Res., 1:213-226.

Rouiller, E., De Ribaupierre, Y., Toros-Morel, A., and De Ribaupierre, F., 1981, Neural coding of repetitive clicks in the medial geniculate body of cat, Hearing Res. 5, 81-100.

Toros, A., Rouiller, E., De Ribaupierre, Y., Ivarsson, C., Holden, M. and De Ribaupierre, F., 1979, Changes of functional properties of medial geniculate body neurons along the rostrocaudal axis, Neurosc. Letters Suppl. 3, S5.

Toros-Morel, A., De Ribaupierre, F. and Rouiller, E., 1981, Coding properties of the different nuclei of the cat's medial geniculate body, in: "Neuronal Mechanisms of Hearing". J. Syka and L. Aitkin, eds., Plenum Press, New-York/London, 239-243.

TOPOGRAPHY OF THE THALAMOTELENCEPHALIC PROJECTIONS IN THE AUDITORY

SYSTEM OF A SONGBIRD (STURNUS VULGARIS)

Udo Häusler

Institut für Zoologie
Technische Universität München
Garching, Lichtenbergstrasse 4

INTRODUCTION

The main auditory center in the telencephalon of birds is field L
of Rose (1914). Field L is located in the medio-rostral part of the
neostriatum and receives its main input from nucleus ovoidalis (ov) of
the thalamus (Karten, 1968; Kelly and Nottebohm, 1979). Field L, the
surrounding auditory neostriatum and the hyperstriatum ventrale in its
vicinity are tonotopically organized (Zaretsky and Konishi, 1976; Bonke
et al., 1979; Leppelsack, 1981; Müller and Leppelsack, 1985; Rubsamen
and Dorrscheid, 1986). The main tonotopic gradient runs in the dorsoventral direction while isofrequency layers have been reported to lie
perpendicular to field L. Field L can be subdivided into three layers
L1-L3 where L2 has been shown to be the terminating area of the ov
fibers (Bonke et al., 1979). Whereas L2 has high unit activity in multi-
unit recordings and simple tuning properties of the neurons, L1 and L3
exhibit more complex tuning and lower unit activity (Scheich et al.,
1979). Comparable results have been reported for the starling by Müller
and Leppelsack (1985). Ov in the pigeon receives afferent fibers from the
nucleus mesencephalicus lateralis p. dorsalis (MLD), which is the avian
homologue of the inferior colliculus, of both sides of the brain (Karten,
1969). The nucleus ovoidalis has been shown to be tonotopically organized
and composed of neurons with relatively simple tuning properties comparable
to those of field L (L2) (Häusler, 1984; Bigalke et al., 1987).

The aim of this study was to establish the link between tonotopic
organization of the forebrain and the organization of the auditory
thalamus, using a combination of electrophysiological recordings and
retrograde tracer experiments.

METHODS

The experiments were carried out on 14 Starlings (4 male, 10 female).
For measurement of response properties, multiunit recordings were made
with metal electrodes. Pure tones and bandlimited noise with different
center frequencies and 250 ms duration were presented to the birds.
Best frequencies were obtained during the recording session by audio-
visual inspection and verified off line with the aid of a computer.
After electrophysiological characterization of the auditory

area, an injection, as described below, was made into the recorded regions. The areas were selected to show the same simple tuning properties and strong phasic-tonic responses both to pure tones and narrowband noise. Such regions were considered to represent the termination area of the ov fibers. This was verified later, after histological processing of the brain. After the electrophysiological recordings, the platinum iridium electrodes were withdrawn and replaced by a glass micropipette filled with a 3% aqueous solution of weat germ agglutinin conjugated horse-radish peroxidase (WGA-HRP) (Sigma). Small quantities between 0.3 and 5 nl of WGA-HRP were injected at the recording site. In the first two animals a 30% solution of standard HRP (Sigma type VI) instead of WGA-HRP was used. After perfusion of the animal the brain was removed from the skull and sectioned in the frontal plane with the aid of a freezing microtome. Sections were processed with the tetramethylbenzidine method according to Mesulam (1978) and counterstained with neutral red. The label was visualized using polarisation microscopy, and camera lucida drawings of the labelled areas in the injection zone, field L and the ov were made.

RESULTS

After injections of standard HRP into field L, label was found in the ov relatively independently of the injection site. No differences could be seen between injections into high frequency and low frequency areas. The latero-dorsal area of ov, which contains the larges neurons with the most dense Nissl substance, showed the strongest label. Neurons in the ventral part were less intensely labelled. Additional clusters of labelled neurons were found in the tractus of the nucleus ovoidalis (tov) in two areas, one about 0.9 mm below ov and the other about 2.0 mm below. Counting of labelled and unlabelled neurons in ov showed that the proportion of labelled neurons was of the order of 99%. There were about 3 to 8 unlabelled neurons in the entire ov, which in addition were always found at or near the ventral border of ov. In this location it is difficult to determine whether a neuron belongs to ov or not. This clearly indicates the lack of a significant population of local interneurons in ov and is in good agreement with other experiments which failed to demonstrate such neurons in the ov (Häusler, 1983).

In contrast to injections of standard HRP, WGA-HRP application resulted in distinct distribution patterns of labelled neurons in the ov. These patterns show a dependence on the location and volume of the injection. WGA-HRP injections are therefore classified according to injection volume into large injections (more than 5nl) and small injec-tions (about 0.5 nl). In experiments with large injection volume, a large part of ov is labelled but some areas are completely free of labelled cells. After large central injections into field L at the site correspond-ing to 2 kHz, the very medial and the latero-ventral areas are excluded. After injections into ventral areas, where best frequencys of about 5kHz are represented, the lateral and latero-ventral areas in ov were not labelled.

After these large WGA injections, additional label was found in locus coeruleus (Loc). A projection from Loc to the caudal neostriatum has previously been bescribed for the pigeon (Kitt and Brauth, 1986), but the projection appears to be more intense in the starling than in the pigeon.

Small injections of WGA-HRP (between 0.4 and 0.7 nl) show labelling patterns in the ovoidalis that are highly dependent on the injection site. The pattern consists of a number of labelled neurons restricted to specific areas of ov. With these injections, a detailed topography of the thalamo-telencephalic projections was determined. Injections into

the high frequency area in the medio-ventral part of field L, results in
a cluster of labelled neurons in the very medial part of ov (Fig. 1 site
a). Some label extends slightly laterally at the dorsal and ventral
borders of ov. At the other extreme, injections into the laterodorsal
area of field L, where low frequencies are represented, labelled neurons
could be found in the ventral parts of ov extending down into the proximal
part of the tov (Fig. 1 site c). Injections with a location intermediate
to la. and lc. result in labelled neurons whose location is also intermediate. (Fig. 1 site b). The more the injections extend into latero-
dorsal areas in field L, the more label lies latero-ventrally in ov. The
injections into the center of field L result additionally in the largest
number of labelled cells in ov. After injections into field L at a level
more rostral than site b in Fig. 1., but at the same center frequency,
resulted in a latero-dorsal labelling pattern in ov (Fig. 1 site d). These
findings suggest a tonotopic organization of ov, with a frequency gradient
from medial to lateral and from dorsal to ventral as well. High frequencies
lie medial and low frequencies ventro-lateral, with the most intense
representation in the 2 kHz region.

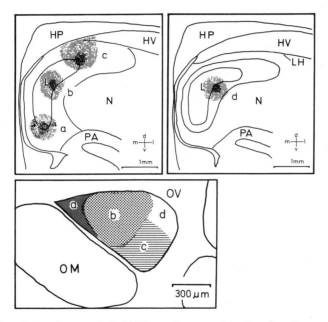

Fig. 1.   Injection sites of 4 different experiments (a, b, c, d) and
          corresponding labelled areas in the ov. Upper part: Frontal
          sections trough field L, right one is 300 microns rostral as the
          left. Lower part:Frontal section through ov. Best frequencies
          at the injection sites are 5 kHz at a, 2 kHz at b and d and
          about 200 Hz at c. Abbreviations: L: field L, HP: hippocampus,
          HV: hyperstriatum ventrale, LH: lamina hyperstriatica, N:neo-
          striatum, PA: paleostriatum, OM: tractus occipitomesencephalicus,
          OV: nucleus ovoidalis. m: medial, l: lateral, d: dorsal,
          v: ventral.

     After small injections, labelled cells could in some cases be found
in the tov. The areas in the tov do not show an organization which is as
clear as the pattern found in ov. After injections into low frequency
areas, label was found in the neurons of the tov directly below ov. After

injections into midfrequency areas at lkHz, label was accumulated mainly
in the clusters more distant from ov. However, as it is difficult to
define the clear borders of these clusters with only a few labelled cells
it is not possible to derive a clear-cut organization of these areas.
Cells in the tov often showed labelled dendrites which are oriented in
the direction of the acoustic fibers running up to the nucleus ovoidalis,
and did not extend to surrounding brain areas.

DISCUSSION

These experiments have, for the first time, demonstrated a lack of
local interneurons in the thalamic auditory nucleus of birds. This finding
is in contrast to similar experiments in mammals. In the cat about 10%
of the neurons of the ventral division of the medial geniculate body
(MGB) were found unlabelled after massive injection of HRP into the
auditory cortex (Winer, 1984). This is interpreted as an indication of
the existence of local thalamic interneurons in the MGB. Although this
interpretation has to be made with care, (in tracer experiments negative
evidence does not prove complete absence of a connection), it may be
taken as an indication of structural differences between the mammalian
and avian auditory pathways. This would mean that the avian auditory
thalamus has a more homogeneous structure and may therefore have less
complex processing capacity than that of mammals. The homogeneity of the
cell population in ov is also supported by other anatomical evidence
which demonstrated a lack of Gaba-ergic neurons in the ov (Müller, 1987).
This suggests that the ov is composed of a population of excitatory
projection neurons simply transmitting information received from the
midbrain to field L in the forebrain. This is in good agreement with
electrophysiological evidence which shows the simple tuning properties
of these neurons (Bidermann-Thorson, 1970; Häusler, 1983; Bigalke et al.,
1987) which are found in the auditory midbrain nucleus MLD.

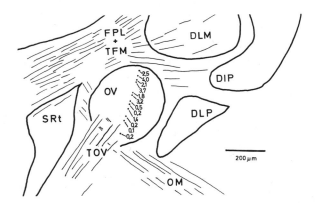

Fig. 2.   Sagittal section through the thalamus of the starling.
          Locations of single unit recordings of one microelectrode
          penetration and corresponding best frequencies are represented
          by the dots inside ov. Abbreviations: DIP: nucleus dorso-
          intermedius posterior thalami, DLM: n. dorsolateralis
          anterior thalami pars medialis, DLP: n. dorsolateralis
          posterior thalami, FPL+TFM: fasciculus prosencephali
          lateralis, SRt: n. subrotundus, TOV: tractus ovoidalis.

The second finding of this study revealed a tonotopic organization
of the ov. A frequency gradient from medial to lateral and from dorsal
to ventral was demonstrated. Electrophysiological recordings from ov

clearly support this view. Fig. 2 shows a sagittal section through nucleus ovoidalis with the locations of single unit recordings and their corresponding best frequencies (Häusler, 1983). Despite some scatter, a decrease of the best frequency from dorsal to ventral is obvious. The frequency gradient from medial to lateral has also been demonstrated by electrophysiology (Bigalke et al., 1987).

In addition to the simple tonotopicity, it was possible to demonstrate that locations in field L differing in a rostrocaudal level but not in frequency representation, receive their input from different locations in ov. This shows that the terminals of the ov neurons do not diverge onto complete isofrequency planes, but on the contrary project point to point from nucleus ovoidalis to field L even within isofrequency planes.

ACKNOWLEDGMENTS

Supported by the Deutsche Forschungs Gemeinschaft SFB 204

REFERENCES

Biederman-Thorson, M., 1970, Auditory evoked responses in the cerebrum and ovoid nucleus of the ring dove, Brain Res., 24:235-245.

Bigalke-Kunz, B., Rübsamen, R. and Dörrscheid, G. J., 1987, Tonotopic organisation of the auditory thalamus in a songbird, the European starling, J. Comp. Physiol. A, 161:255-265.

Bonke, D., Scheich, H. and Langner, G., 1979, Responsiveness of units in the auditory neostriatum of the guinea fowl (numida meleagris) to species specific calls and synthetic stimuli. I. Tonotopy and functional zones, J. Comp. Physiol., 132:243-255.

Bonke, B. A., Bonke, D. and Scheich, H., 1979, Connectivity of the auditiory forebrain nuclei in the guinea fowl (Numida meleagris), Cell Tiss. Res., 200:101-121.

Häusler, U., 1983, Histologische und elektrophysiologische Unter-suchungen an einzelnen Neuronen des Nucleus ovoidalis in Zwischen-hirn des Staren. (Sturnus vulgaris L.) Diplomarbeit, Lehrstuhl für allgemeine Zoologie, Ruhr-Universitat Bochum.

Karten, H. J., 1967, The organization of the ascending auditory pathway in the pigeon (Columba livia). I. Diencephalic Projections of the inferior colliculus (nucleus mesencephali lateralis, pars dorsalis), Brain Res., 6:409-427.

Karten, H. J., 1968, The ascending auditory pathway in the pigeon (Columba livia). II. Telencephalic projections of the nucleus ovoidalis thalami, Brain Res., 11:134-153.

Kelly, D. B. and Nottebohm, F., 1979, Projections of a telencephalic auditory nucleus field L - in the canary, J. Comp. Neurol., 183: 455-470.

Kitt, C. A. and Brauth, S. E., 1986, Telencephalic projections from midbrain and isthmal cell groups in the pigeon. I locus coeruleus and subcoeruleus, J. Comp. Neurol., 247:69-91.

Leppelsack, H. J., 1981, Antworteigenschaften auditorischer Vorderhirnneurone eines Singvogels unter besonderer Berücksichtigung des Gesangslernens. Habilitationschrift, Abteilung für Biologie, Ruhr-Universitat Bochum.

Mesulam, M. M., 1978, Tetramethyl benzidine for horseradish peroxidase neurohistochemistry: A noncarcinogenic blue reaction-product with superior sensitivity for visualizing neural afferents and efferents, J. Histochem. Cytochem., 26:106-117.

Müller, C. M., 1987, Strukturelle und funktionelle Aspekte inhibi-torischer interaktionen vermittels Gaba auf den oberen Hörbahn-stationen des Huhnen (Gallus gallus). Dissertation Fachbereich Biologie, Technische Hochschule Darmstadt.

Müller, C. M. and Leppelsack, H. J., 1985, Feature extraction and tonotopic organisation in the avian auditory forebrain, Exp. Brain Res., 59:587-599.

Rose, M., 1914, Über die Cytoarchitektonische Gliederung des Vorderhirns der Vogel, J. f. Psychol. u. Neurol., 2:278-352.

Rübsamen, R. and Dörrscheid, G. J., 1986, Tonotopic organisation of the auditory forebrain in a songbird, the european starling, J. Comp. Physiol. A., 158:639-646.

Winer, J. A., 1984, Identification and structure of neurons in the medial geniculate body projecting to primary auditory cortex (AI) in the cat, Neurosc., 13:395-413.

Zaretsky, M. D. and Konishi, M., 1979, Tonotopic organization in the avian telencephalon, Brain Res., 111:167-171.

STUDY WITH HORSERADISH PEROXIDASE (HRP) OF THE CONNECTIONS BETWEEN

THE COCHLEAR NUCLEI AND THE INFERIOR COLLICULUS OF THE RAT

F. Collia, D. E. Lopez, M. S Malmierca, and M. Merchan

Chair of Histology, Faculty of Medicine
University of Salamanca
C/Fonseca, 37007, Salamanca, Spain

INTRODUCTION

The connections between the inferior colliculus and the cochlear nuclei evaluated by injection of horseradish peroxidase (HRP) into the colliculus were analyzed in the cat (Adams, 1979), (Cant, 1982, HRP and electron microscopy) and in the rat (Beyerl, 1978). These authors reported the existence of labelled neurons in the ipsilateral ventral and dorsal cochlear nuclei. General consensus exists concerning the absence of label in the octopus neurons in the PVCN and in the bushy neurons of the AVCN.

In the present work we performed massive injections of HRP into the inferior colliculus of the rat in order to quantify the maximum number of neurons in the cochlear nuclei that project their axon to the colliculus.

MATERIAL AND METHODS

In the present work 30 Wistar rats weighing 300 g were studied. The animals were injected using 1 μl Hamilton syringes with 0.4 μl of Sigma type VI peroxidase at 40% dissolved in phosphate buffer, pH 7.2-7.4.

Injection was performed in the inferior colliculus taking the stereo-taxis atlas coordinates of Paxinos and Watson (1982) (interaural 0.22 mm, lateral 1.5 mm, depth, 4.5 mm). Injection time was 20 min. With this dose massive labelling of the whole colliculus was achieved.

After a survival period of 36-48 h the animals were perfused with glutaraldehydeformaldehyde according to the technique of Mesulam (1982). The brains were submerged in a 0.2 M solution of phosphate buffer with 10% sucrose for 2-24 h. The zone between the superior colliculus to behind the dorsal cochlear nucleus was serially sectioned at 50 μm in a freezing microtome. The cuts were processed according to the method of Mesulam (1982) and were contrasted with neutral red. The serial sections of the cochlear nuclei were drawn with a camera lucida and the unequivocally labelled neurons were represented on those drawings.

RESULTS

The results obtained are shown schematically in Fig. 1, representing

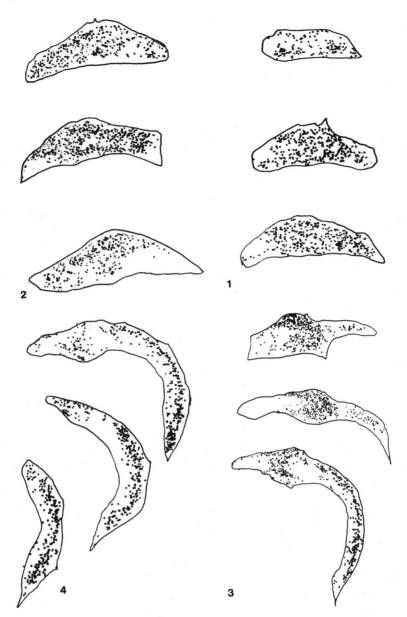

Fig. 1. Camera lucida schemes of a series of transverse sections
after HRP injection into the inferior colliculus showing the
topographical distribution of the labelled neurons in the
contralateral cochlear nuclei.

12 sections that are characteristic of the contralateral cochlear nuclei
of one animal from the experimental series. The total number of labelled
neurons of the cochlear nuclei was 10882 of which 8575 were found in the
DCN. Regarding the neuronal types labelled, pyramidal cells were clearly
identified in the DCN and stellate cells in the VCN. The classic area
of the nucleus of the octopus cells was seen to be free of HRP-positive
label.

In the ipsilateral cochlear nuclei the maximum number of labelled

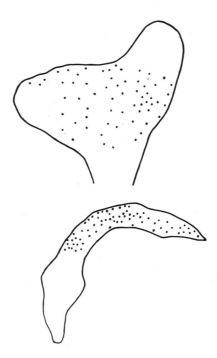

Fig. 2.   Scheme of transverse and parasagittal sections after HRP
          injection into the inferior colliculus showing the topographical
          distribution of the labelled neurons in the ipsilateral cochlear
          nuclei.

neurons was 131; 68 in the VCN and 63 in the DCN. Labelling intensity in
the ipsilateral cochlear nuclei was less pronounced than in the contralateral nuclei and neuronal identification was only possible in the case
of the pyramidal neurons of the dorsal cochlear nucleus.

REFERENCES

Adams, J. C., 1979, Ascending projections to the inferior colliculus, J.
     Comp. Neurology., 183:159-538.
Beyerl, B. D., 1978, Afferent projections to the central nucleus of the
     inferior colliculus in the rat, Brain Research, 145:209-223.
Cant, N. B., 1982, Identification of cell types in the anterolateral
     cochlear nucleus that project to the inferior colliculus, Neuroscience
     Letters, 32:241-246.

PHYSIOLOGICAL PROPERTIES OF UNITS IN THE COCHLEAR NUCLEUS

ARE ADEQUATE FOR A MODEL OF PERIODICITY ANALYSIS IN THE AUDITORY MIDBRAIN

Gerald Langner

Zoological Institute, TH - Darmstadt

Schnittspahnstr. 3, 61 Darmstadt, FRG

INTRODUCTION

The temporal patterns of neuronal discharges in the auditory nerve code the information about the periodicities of acoustic signals. It is now widely accepted that the auditory system has to make use of this information and that it needs for that purpose central mechanisms for periodicity analysis. Particularly, periodic envelopes which are typical for many communication sounds evoke in human subjects the perception of periodicity pitch (Schouten et al., 1962). This phenomenon is directly related to similar pitch effects, for example, the "missing fundamental": Harmonic sounds are composed of a fundamental and additional harmonic components. Their envelope has the period of the fundamental even after passing a filter as, for example, the cochlea. This holds as long as the filter output contains at least two subsequent harmonic components. This seems to be the reason for a pitch corresponding to the fundamental even when it is missing. Very often AM-signals with a sinusoidal carrier and modulation are used for investigating the processing and perception of periodic signals. In the harmonic case the three components of such signals, the carrier frequency and its two sidebands, correspond to components of a harmonic sound and the modulation frequency to the (missing) fundamental.

TEMPORAL CODING OF PERIODICITY IN THE AUDITORY MIDBRAIN

Pattern models of periodicity pitch require a neuronal spectral representation of such signals ("central spectrum") with sufficiently high resolution of the frequency components (Terhardt et al., 1982; Goldstein, 1978; Wightman, 1973). In contrast to such expectations, it was demonstrated in the Guinea fowl (Numida meleagris) that information relevant for periodicity pitch is represented and processed in the time domain at the level of the midbrain (Langner, 1981, 1983). For a temporal representation of envelope period at least some of the components of a harmonic sound have to remain spectrally unresolved in order to stimulate auditory neurons as a complex up to the level of processing. The neuronal responses to AM-signals suggest that the auditory system performs a kind of cross-correlation analysis between the modulation and the carrier frequency with the neurons in the midbrain functioning as coincidence detectors. Consequently these units are tuned not only to the carrier (i.e. to their center frequency, CF) but also to a "best modulation frequency" (BMF), or a "missing fundamental". The (rate) tuning of the neurons to temporal information at their input does not imply that their output is highly synchronized, although in the bird some neurons were found which synchronize to modulation frequencies up to 1300 Hz. The suggested correlation theory may explain certain details of pitch perception in humans including the "second effect of pitch shift" (Schouten et al., 1962). The conclusion is that periodicity pitch in humans may be based on similar temporal mechanisms (Langner, 1985).

Neurons tuned to envelope periodicities relevant for pitch perception were also observed in the midbrain of the cat (Schreiner and Langner, 1984) and of various other animals (Rose and Capranica, 1983; Rees and Moller, 1983). Many details of the neuronal response properties in the cat were found to be similar to those observed in the Guinea fowl and could be described by the same model. In addition, in the cat the maxima of the modulation transfer functions (BMF) were found to be mapped concentrically on isofrequency laminae in the central nucleus of the inferior colliculus (ICC) with the highest BMF as a focus in the center.

A NEURONAL MODEL FOR PERIODICITY ANALYSIS

When AM signals are used as stimuli, one of the prominent effects observed in the midbrain of birds and cats is that the BMF of a tuned unit varies as a function of the carrier as described by the "periodicity equation" (Langner, 1983). Another effect is that during the first 30 to 50 ms after stimulus onset synchronization to modulation frequencies below the BMF increases and, in contrast, decreases for modulation frequencies above the BMF. These and other details of the responses could be simulated by a computer model.

The model is composed of a trigger unit, an oscillator circuit, a reducer circuit, and a coincidence unit (Fig. 1). The trigger unit detects the zero crossings of the modulating waveform and triggers the oscillator and the reducer circuit. The oscillator generates several action potentials with regular intervals, thereby coding the periodicity of the envelope with delayed responses. Such delays are necessary for a temporal correlation analysis. Corresponding oscillatory responses with intervals independent of the stimulus periods and preference for multiples of 0.4 ms were observed in birds and cats. In the oscillator circuit of the model this preference is produced by one or more synapses with an adequate delay. Some influence of the stimulus amplitude on the oscillator intervals is modelled by auditory nerve input.

The observed effects of the carrier frequency on the responses of coincidence units may be explained by one input that codes the carrier in a "reduced" way, i.e. by intervals multiple to its period. The reducer circuit receives - besides the trigger input - additional input from the auditory nerve. The membrane potential of the output unit of the reducer circuit is kept at a certain value below threshold by an inhibitory input of a unit activated tonically by the auditory nerve. The trigger unit initiates an inhibition of this inhibitory unit after which the output unit can start integrating over its input from the auditory nerve. In the range of phase coupling, below 5 kHz, the threshold is reached after a certain number of steps. As a result, the output unit synchronizes to the modulating waveform with a phase delay corresponding to an integer multiple of the carrier period. The circuit is reset after each response of the output unit. A constant phase delay results when spike rate in the auditory nerve responses saturate at a low intensity, usually 20 to 50 dB above threshold (Kiang et al., 1965).

The transient response property of the auditory nerve during the

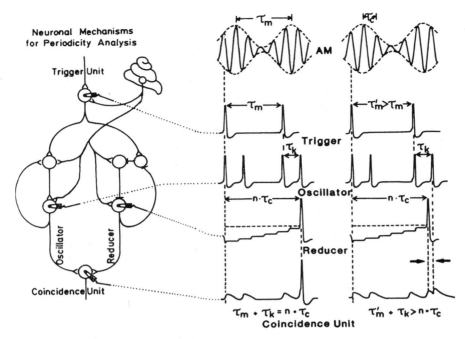

Fig. 1.　Left: A neuronal model for periodicity analysis. The auditory
nerve projects to trigger, reducer, and oscillator units.
Oscillator and reducer are triggered by the trigger unit and project to a coincidence unit. Right: A response scheme of the
components of the periodicity model. The trigger unit synchronizes
to four envelope cycles of AM-signals, on the right side with a
greater modulation period τ_m. The oscillator unit generates in
this example two spikes with an intrinsic interval τ_k. The
reducer unit integrates after the first trigger spike over input
activity synchronized to the carrier. It reaches threshold after
a certain number (n) of carrier cycles. The coincidence unit
responds to coinciding oscillator and reducer spikes. This
condition, as expressed by the periodicity equation (bottom), is
fulfilled in the first example (left side) but not in the second
(right side), due to a small change in modulation frequency.
(Additional synaptic and conductive delays were considered as
being irrelevant.)

first 30 to 50 ms impairs the phase delay and as a result also the temporal
response pattern of the coincidence unit.

Coincidence units respond when they receive coinciding inputs from
the oscillator and the reducer circuit. This is described in terms of
the stimulus periods by the periodicity equation. Action potentials
coincide after each modulation cycle when the phase delay of the reducer
equals one modulation period plus a delay introduced by an oscillation.
The modulation frequency with that period will elicit the best response
(BMF) in that unit. Consequently, the BMF will be a function of the
phase delay and thereby of the carrier period. Different time courses of
synchronization for different modulation frequencies (see above) are due
to the transient responses of the reducer which result from the transient
time courses of the auditory nerve.

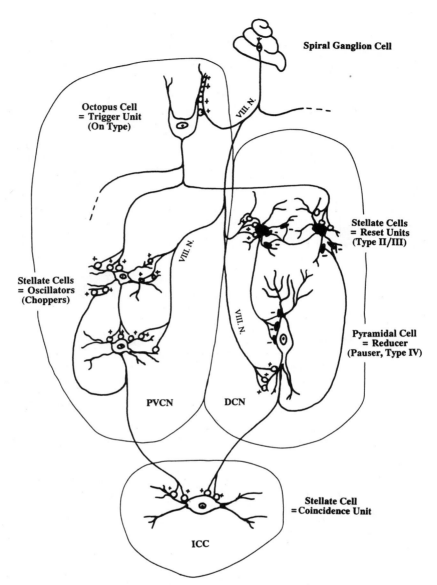

Fig. 2. <u>A neuronal periodicity model including neurons of the cochlear</u>
<u>nucleus</u>. It is suggested that an octopus cell of the PVCN syn-
chronizes to envelope modulations and triggers the oscillatory
responses of stellate cells in the PVCN and pauser type responses
of a pyramidal cell in the DCN. The pyramidal cell starts to
integrate over its eight-nerve input only after its initial
inhibition by a stellate cell ('black neuron') is inhibited. In
order to function as part of a correlation network the stellate
cells of the PVCN and the pyramidal cells of the DCN are
supposed to project onto the same neurons in the central nucleus
of the inferior colliculus. A computer simulation of this network results in temporal response pattern of the coincidence
neurons as observed in actual neurons of the midbrain.

210

ANATOMICAL AND PHYSIOLOGICAL CORRELATES
FOR THE COMPONENTS OF THE PERIODICITY MODEL

The coincidence unit of the computer model simulates in great detail response patterns observed in units of the midbrain in birds and cats while the response properties of the remaining neurons in the model resemble those of real neurons investigated in detail in the posterior ventral (PVCN) and the dorsal (DCN) cochlear nucleus of mammals. It is therefore suggested that these neurons may play a role in periodicity analysis by preprocessing temporal information in a way adequate for the correlation analysis completed by the coincidence units in the midbrain (Fig. 2).

Octopus cells in the PVCN have on-type responses as required in the model and are able to follow high repetition rates of acoustic signals which suggests that they are important for the processing of periodicity pitch (Møller, 1970). However, the role of the octopus cells as trigger units requires projections to the stellate cells in PVCN and DCN which remain to be demonstrated.

It is known that the ICC receives direct input from the PVCN and DCN (Adams, 1979). Projection neurons are stellate neurons in PVCN and pyramidal cells in DCN. Stellate cells of the PVCN display chopper responses. They respond with regular spike intervals independent of the stimulus frequencies similar to the oscillator neurons of the model (Rhode et al., 1983) and may indeed be involved in periodicity processing (Frisina et al., 1985). Input from the auditory nerve is considered to have some influence on the chopper intervals.

Fusiform or pyramidal cells in the DCN were identified physiologically as type-IV or pauser units (Rhode et al., 1983), which exhibit complicated patterns of response to tone-burst stimulation. In spite of their inhibitory response even at CF, as soon as the amplitude exceeds values of 20 to 30 dB above threshold, they respond to broadband signals even at higher intensities (Young and Brownell, 1976). They synchronize to the envelope of periodic signals in a way compatible with the reducer unit of the model (Schreiner and Snyder, pers. com.) and also have inhibitory connections like proposed in the model (Mugnaini, 1985). A further prediction of the model is that stellate cells of the PVCN and pyramidal cells in the DCN converge onto the same units in the ICC and produce the observed coincidence effects.

SUMMARY

The suggested model for periodicity analysis includes a functional role for most of the important neuron types of the cochlear nucleus which do not project into the complex of the olivary nuclei and may therefore be involved in other than localization tasks. Besides several well-known physiological properties and anatomical connections of certain neuron types in mammals, the model includes a number of predictions and connections which are in principle testable. Not every aspect of the model could be discussed in this paper and it is still open how these predictions transfer to the bird brain. It should be noticed that the model can probably be extended from periodicity analysis to general frequency or even intensity coding.

Supported by the Deutsche Forschungsgemeinschaft.

REFERENCES

Adams, J. C., 1979, Ascending projections to the inferior colliculus, J. Comp. Neurol., 183:519-538.

Frisina, R. D., Smith, R. L. and Chamberlain, S. C., 1985, Differential encoding of rapid changes in sound amplitude by second order auditory neurons, Exp. Brain Res., 60:417-422.

Goldstein, J. L., 1978, Mechanisms of signal analysis and pattern perception in periodicity pitch, Audiology, 17:421-445.

Kiang, N. Y. S., Watanabe, T., Thomas, E. C. and Clark, L. F., 1965, Discharge patterns of single fibers in the cat's auditory nerve. MIT, Cambridge

Langner, G., 1981, Neuronal mechanisms of pitch analysis in the time domain, Exp. Brain Res., 44:450-454.

Langner, G., 1983, Evidence for neuronal periodicity detection in the auditory system of the guinea fowl:implications for pitch analysis in the time domain, Exp. Brain Res., 52:333-355.

Langner, G., 1985, Time coding and periodicity pitch, in: "Time Resolution in Auditory Systems", A. Michelsen, ed., Springer, Berlin, 108-121.

Møller, A. R., 1970, Two different types of frequency selective neurons in the cochlear nucleus of the rat, in: "Frequency analysis and periodicity detection in hearing", R. Plomp and G. F. Smoorenburg, eds., Sijthoff, Leiden, 168-174.

Mugnaini, E., 1985, GABA neurons in the superficial layers of the rat dorsal cochlear nucleus: light and electron microscopic immunocytochemistry, J. Comp. Neurol., 235:61-81.

Rees, A. and Møller, A. R., 1983, Responses of neurons in the inferior colliculus of the rat to AM and FM tones, Hearing Res., 10:301-330.

Rhode, W. S., Oertel, D. and Smith, P. H., 1983, Physiological response properties of cells labeled intracellularly with horseradish peroxidase in cat ventral cochlear nucleus, J. Comp. Neurol., 213: 448-463.

Rose, G. J. and Capranica, R. R., 1983, Temporal selectivity in the central auditory system of the leopard frog, Science, 219:1087-1089.

Schouten, J. F., Ritsma, R. J. and Cardozo, B. L., 1962, Pitch of the residue, J. Acoust. Soc. Am., 34:1418-1424.

Schreiner, C. and Langner, G., 1984, Representation of periodicity information in the inferior colliculus of the cat, Soc. Neurosci. Abstr., 10:395.

Terhardt, E., Stoll, G. and Seewann, M., 1982, Pitch of complex signals according to virtual pitch theory-tests, examples, and predictions, J. Acoust. Soc. Am., 71:671-678.

Wightman, F. L., 1973, The pattern-transformation model of pitch, J. Acoust. Soc. Am., 54:407-416.

Young, E. D. and Brownell, W. E., 1976, Responses to tones and noise of single cells in dorsal cochlear nucleus of unanesthetized cats, J. Neurophysiol., 39:282-300.

MONAURAL PHASE SENSITIVITY IN NEURONS FROM THE CAT'S AUDITORY SYSTEM

E. A. Radionova

I. P. Pavlov Institute of Physiology

Leningrad, USSR

The present paper deals with some neurophysiological processes underlying brain sensitivity to changes in the phase structure of sound signals (for references see: Buunen, 1976; Kubovy and Jordan, 1979; Radionova, 1982; Raiford and Schubert, 1971; Rose et al., 1974).

Frequency following response (FFR) and single neuron activity of the cochlear nucleus (CN) and inferior colliculus (IC) were studied using mainly two-tone harmonic signals of rectangular envelope with different phase relations of the harmonics. Electrostatic earphones were used. The harmonic frequencies varied from 0.1 to 20 kHz, their intensities usually did not exceed 80 dB SPL. The signal duration amounted to 20 ms or more. While changing the phase of the signal harmonics the wave-form of the CN FFR usually showed a good correspondence with the wave-form of the sound signal. However, under certain conditions and within a restricted range of phase values nonlinear effects were pronounced, when the first harmonic could be partly or even fully suppressed (Fig. 1, $+30°$). Both the reproduction of the signal wave-form and the suppression effect could be observed in the cochlear microphonics as well (Altman and Radionova, 1974). The IC FFR also changed its wave-form following phase variations in the two-tone or more complicated sound signals. However, there was no correspondense between the FFR wave-form and the wave-form of the signal.

Unlike the summed activity, in single neurons it was necessary to adjust appropriate frequencies and intensities of the two-harmonics in order to obtain a phase effect. Altogether, of the 54 tonic neurons from the CN only 35 (or 65%) showed phase sensitivity (Kotelenko and Radionova, 1975; Radionova, 1986). To simplify the experiment, single tone signals with varying initial and end phases were used, and IC neurons with phasic ON-OFF responses were studied. Similar results were obtained for both the ON- and OFF- responses. It was found (Fig. 2) that in case signal frequency F < CF, the maxima of the response lay at or near 90 and 270°. The same relations were true for the summed ON- and OFF-responses (the evoked potentials) at the IC, CN and the auditory nerve levels. It was suggested that the observed phenomena were connected with elastic and inertial forces at the cochlear partition level which differently control oscillations below and above the resonant frequency (Radionova, 1979). Thereafter it proved possible to predict the OFF-response value for two-tone signals ending at different moments, i.e. at different phase relations of its components (Radionova, 1979).

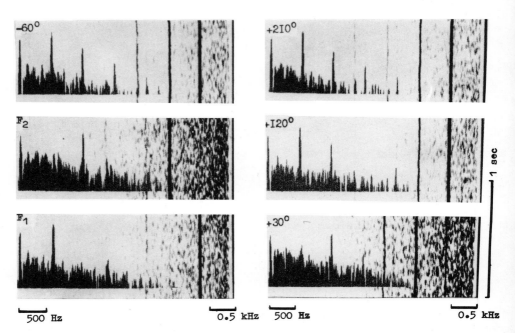

Fig. 1. Spectrograms (on the left) and dynamic spectrograms (on the right) of the FFRs from the CN following stimulation with the 1st harmonic (F_1=0.5 kHz), the 2nd harmonic (F_2=1.0 kHz) and both in all other cases, at different phase values of the second harmonic (figures). The first peak in each spectrogram is an artifact.

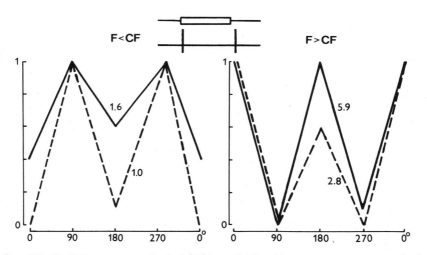

Fig. 2. IC ON-OFF neuron. Probability of the OFF-response vs. end phase of tone signals of different frequencies (figures, kHz). CF means the characteristic frequency (2.3 kHz).

The present data pointed to the significance of the neuron characteristic frequency for the phenomena of phase sensitivity. It was found that in tonic low frequency neurons (CF < 5 kHz) the φ-function showed one maximum within the 360° range of the Ψ-values of the second harmonic (Fig. 3, A). In high frequency neurons (CF > 11 kHz) this function showed two maxima (Fig. 3, C), and in neurons of the intermediate group the φ-

function was of an intermediate shape (Fig. 3, B). It was suggested that the different character of the Ψ-function is determined by the different character of oscillations of the cochlear partition at the basal and apical ends of the cochlea, the apical end oscillations would be asymmetrical (Dallos, 1986; Kiang, 1965; Rose et al., 1969 and others), whereas the basal end oscillations would be rather symmetrical, with both polarities effective for excitation of the auditory fibers (Kiang, 1965; Palmer and Russell, 1986).

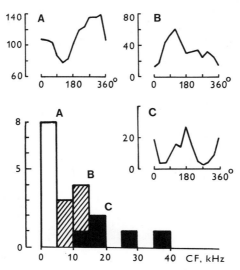

Fig. 3.   Graphs: numbers of impulses in discharge per 10 trials vs. phase value of the 2nd harmonic in the two-tone signal for three tonic CN-neurons with low, mid- and high CFs (A, B, C). A, B, C, CFs: 3.9, 7 and 11 kHz; signals: 2+4 kHz, 3.8+7.6 kHz, 5.5+11 kHz. Below: number of neurons with different CFs.

To examine the above suggestion the FFRs from 32 locations in the CN were studied on 9 cats with click series of opposite polarities. As might be expected, in the high frequency regions of the CN, clicks of both polarities were equally effective, whereas in the low frequency regions they evoked different responses.

Thus, it may be concluded that phase sensitivity of the central auditory neurons is determined to a great extent by processes at the level of the cochlear partition.

REFERENCES

Altman, J. A. and Radionova, E. A., 1974, Frequency following response of the cochlear nuclei to sound stimuli consisting of two harmonic components, Physiol. J. USSR, 60:1796-1805 (in Russian).
Buunen, T. J. F., 1976, On the perception of phase differences in acoustic signals, Delft, W. D. Meinema B. V.
Dallos, P., 1986, Neurobiology of cochlear inner and outer hair cells: intracellular recordings, Hearing Res., 22:185-198.
Kiang, N. Y.-S., 1965, Discharge patterns of single fibers in the cat's auditory nerve, Cambridge, Massachusetts, MIT Press.
Kotelenko, L. M. and Radionova, E. A., 1975, On the phase sensitivity of neurons of the cat's auditory system, J. Acoust. Soc. Amer., 57:979-982.

Kubovy, M. and Jordan, R., 1979, Tone-segregation by phase:On the phase sensitivity at the single ear, J. Acoust. Soc. Amer., 66:100-106.

Palmer, A. R. and Russell, I. J., 1986, Phase-locking in the cochlear nerve of the guinea pig and its relation to the receptor potential of inner hair cells, Hearing Res., 24:1-15.

Radionova, E. A., 1979, Phase sensitivity of neurons of the inferior colliculus, Biophysics,23:533-538 (English translation).

Radionova, E. A., 1982, Neurophysiological phenomena of the monaural phase sensitivity in the auditory system, in: "Sensory systems. Audition". Leningrad: Nauka, 72-86 (in Russian).

Radionova, E. A., 1986, Monaural phase sensitivity of auditory system in the cat., J. Evolutionary Biochem. Physiol., 310-317 (English translation).

Raiford, C. . and Schubert, E. D., 1971, Recognition of phase changes in octave complexes, J. Acoust. Soc. Amer., 50:559-567.

Rose, J. E., Brugge, J. F., Anderson, D. J. and Hind, J. E., 1969, Some possible neural correlates of combination tones, J. Neurophysiol., 32:402-423.

Rose, J. E., Kitzes, L. M., Gibson, M. M. and Hind, J. E., 1974, Observations on phase-sensitive neurons of anteroventral cochlear nucleus of the cat: nonlinearity of cochlear output, J. Neurophysiol., 37: 218-253.

Sanotskaja, N. N., 1977, Cochlear microphonics in cat evoked by two-tone harmonic sound signals with different phase spectra, Physiol. J. USSR, 63:976-983, (in Russian, English summary).

COCHLEAR POTENTIALS AND RESPONSES FROM STRUCTURES OF

AUDITORY PATHWAY INFLUENCED BY HIGH-INTENSIVE NOISE

M. Biedermann, E. Emmerich, H. Kaschowitz, and F. Richter

Institute of Physiology

Friedrich-Schiller-University, Jena, G. D. R.

Former acute investigations on guinea pigs showed the cochlear damaging effects of impulsive noise. Biedermann et al. (1977) reported on alterations of input-output functions of cochlear microphonics in guinea pigs exposed to 10 noise impulses of varied intensities between 139 and 164 dB SPL. These electrophysiological findings were confirmed by morphological investigations of Meyer's group (1980, 1985), who found large parts of destruction or total hair cell loss in the organ of Corti of the impulse noise exposed guinea pigs. The acute experimental technique on guinea pigs that were narcotized only allowed us to track a time range in each single animal. To compensate for this we turned to a method giving the possibility of continuous long time observation of alterations not only in the cochlea but also in other nuclei of the auditory system. Besides, the chronic method also opened the possibility for repeating noise exposures with increasing intensities in the same animal.

Our present study gives an overview of the chronic effects of impulse noise on the auditory system of awake rabbits. To that end we recorded cochlear microphonics (CM), auditory evoked brainstem potentials (ABR) and evoked potentials (EP) from inferior colliculus and medial geniculate nucleus. The aim of our study was to compare peripheral and central electrophysiological alterations after impulse noise trauma using noise intensities that occur in human environment and service.

The experiments were made on awake white New Zealand male rabbits with chronically implanted electrodes using the technique described by Biedermann and Emmerich (1982). For recording CM and ABR, Ag-AgCl electrodes were implanted near the cochlea on the petrosal part of the temporal bone. For recording EP, chrome-nickel wire electrodes were positioned by means of a stereotaxic headholder in medial geniculate nucleus and inferior colliculus.

CM were stimulated by sine waves of 1.25; 2.50; 5.00 and 10.00 kHz. ABR were evoked by half a sine wave (clicks) of 3.1 kHz and alternated polarization to eliminate the CMs. Test stimuli for recording of EP were burst waves of 2.1; 4.0 and 6.3 kHz and 1 ms duration. Noise impulses were produced by means of a spark noise generator with an intensity of 164 dB SPL. By using cloth-covering for attenuation this intensity could be reduced to 153 and 144 dB SPL. The experimental design is given in Fig. 1A.

The awake rabbits were trained to the experimental design, then they were exposed to 10 impulses each at intervals of 15 s. Alterations in amplitudes of the potentials and changes of peak latencies of ABR and EP were tracked. The results were compared graphically with values prior to impulse noise trauma.

Impulse Noise Effects on CM

The CM input-output functions before noise exposure had a uniform parabolic course (Richter and Biedermann, 1987). After 10 impulses of 144 dB SPL only a small decrease of amplitudes was observed. The amplitudes reached the control level 2 h after noise exposure and a complete recovery of the CM functions occurred.

Impulses of 153 dB SPL produced amplitude reductions that were not predictable concerning time and extent of restitution process. Only in a few of the rabbits examined was a complete recovery seen; one rabbit experienced a total loss of CM 90 min after noise trauma.

In the animals exposed to impulses of 164 dB SPL a uniform decline of all CM functions occurred. The CM loss was seen to be complete 2 h after noise trauma. No restitution of CM was seen in any of these cases. The characteristic effects of the three impulse intensities are depicted in Fig. 1B.

These results are in agreement with acute experiments on chinchillas (Henderson et al., 1974) and on guinea pigs (Biedermann et al., 1977). The rabbit might be less susceptible to impulse intensities of 144 and 153 dB SPL than is the guinea pig. The threshold in rabbits for irreversible CM alterations could be higher.

Impulse Noise Effects on ABR

In the rabbits replicable, stable ABR could be recorded prior to impulse noise exposure. Mean values were averaged and used to compare it with results after noise exposure. The ABR mostly consisted of 4 waves, wave I had a peak latency of 2.5 ms, the other waves followed at intervals of about 1.5 ms.

After exposure to 144 dB SPL impulses small changes of ABR amplitudes were seen, temporarily reaching 40%-70% of controls. Peak latencies increased significantly, greatest changes were seen 2 h after noise exposure. 7 days after this the ABR amplitudes were at the control level, whereas the increased peak latencies did not recover and remained above the control level.

10 impulses of 153 dB SPL caused decrease of ABR amplitudes. Peak latencies further increased, but with lesser tendency to restitution. In two of the cases the ABR became unrecordable 2 h after noise exposure.

After exposure to impulses of 164 dB SPL the ABR became unrecordable in any of the rabbits. No recovery was seen at all.

Immediately after impulse noise trauma ABR peak latencies increased, whereas the intervals between the waves remained nearly constant. These increases almost always affected wave I which corresponds to the gross-action potential of the cochlear nerve. This leads to our opinion that cochlear alterations caused the increase in ABR peak latencies and the delay of the gross-AP. Using higher impulse intensities the time intervals between waves II - IV were altered, now showing a participation of brain stem divisions generating these waves.

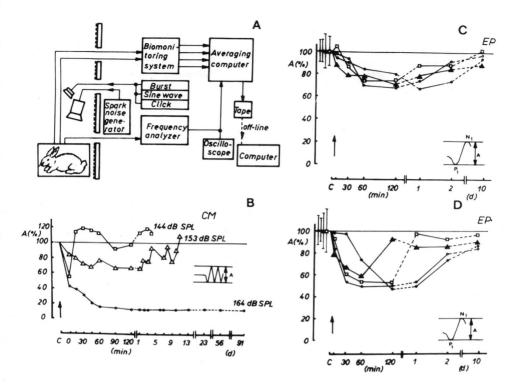

Fig. 1. A. Block diagram of our experimental design. B. Percentage CM
amplitude loss after impulse noise exposure to 10 impulses each
of 144, 153 and 164 dB SPL. Abscissae show time after noise
exposure (arrow). Ordinates give CM amplitudes in % equated to
control before exposure. Test tone frequency was 2.50 kHz, its
intensity was 70 dB SPL. C. Alterations of EP amplitudes from
medial geniculate nucleus after exposure to 10 impulses of 153 dB
SPL. Test stimuli were burst waves of 4.0 kHz and 1 ms duration.
Symbols represent 4 different rabbits. Abscissae give percentage
alterations of amplitudes equated to control before noise
exposure. Symbols are given as mean values with standard deviation (control) for each rabbit. Ordinates show time lapse of
alterations, arrow marks noise exposure. D. Alterations of EP
amplitudes from inferior colliculus after 10 impulses of 153 dB
SPL. For explanation of the diagram see Fig. 1C.

Impulse Noise Effects on EP

Noise impulses of 144 dB SPL caused temporarily observable reductions
of EP amplitudes. The time course of restitution resembled recovery of
cochlear potentials. Extents of decrease of EP amplitudes varied, larger
reductions were observed in the inferior colliculus than in medial geniculate nucleus. Peak latencies of P_1 and N_1 were nonsignificantly prolonged.

Effects of 153 dB SPL impulses on EP are shown in Fig. 1C/D.
Amplitudes were reduced to 50% - 70% of controls. Peak latencies were
prolonged significantly. These changes were observed to be irreversibly
only in a few of the rabbits. After impulses of 164 dB SPL amplitude loss
was seen to be complete within 2 h after noise trauma. Within 4 months no
recovery of EP could be observed.

These electrophysiological findings indicated that cochlear alterations caused by impulse noise acted on structures of the auditory pathway, too. We conclude that the described changes of auditory evoked potentials from medial geniculate nucleus and inferior colliculus were caused by altered afferent inputs coming from damage in the organ of Corti. Influences of continuous noise on central structures have been given by Syka and Popelář (1982) and Popelář and Syka (1978, 1982). Our results are in agreement with their findings concerning the larger extent of noise effects on inferior colliculus than on medial geniculate nucleus.

REFERENCES

Biedermann, M., Büttner, G., Kaschowitz, H. and Pingel, T., 1977, Die Wirkung von Einzelimpulsen auf das Mikrophonpotential der Meer-schwinchencochlea, Acta Biol. Med. Ger., 36:1097-1105.

Biedermann, M. and Emmerich, E., 1982, Die Implantation von Langzeitel-elektroden zur Erfassung von Cochleapotentialen beim Kaninchen, in: "Kongressbericht Cochlea-Forschung 1980", Wissenschaftliche Zeit-schrift der Martin-Luther-Universität Halle/Wittenberg, 166-171.

Henderson, D., Hamernik, R. P. and Sitler, R. W., 1974, Audiometric and histological correlates of exposure to 1 msec noise impulses in the chinchilla, J. Acoust. Soc. Am., 56:1210-1221.

Meyer, Ch. and Biedermann, M., 1980, Immediate alterations in the impulse noise exposed organ of Corti of the guinea pig. Acta Otolaryngol. (Stockh), 90:250-256.

Meyer, Ch. and Biedermann, M., 1985, Früh- und Spätreaktionen der Meer-schweinchencochlea auf Impulsschall, Anat. Anz., 158:5-12.

Popelář, J. and Syka, J., 1978, Effect of high-intensity sound on cochlear microphonics and activity of inferior colliculus neurons in the guinea pig, Arch. Otorhinolaryngol., 221:115-122.

Popelář, J. and Syka, J., 1982, Noise impairment in the guinea pig. II Changes of single unit responses in the inferior colliculus, Hearing Res., 8:273-283.

Richter, F. and Biedermann, M., 1987, Input-output functions of cochlear microphonics in chronic experiments on awake rabbits, Arch. Otorhinolaryngol., 244:91-92.

Syka, J. and Popelář, J., 1982, Noise impairment in the guinea pig. I. Changes in electrical activity along the auditory pathway, Hearing Res. 8:263-272.

AUDITORY CORTEX

AUDITORY CORTEX: MULTIPLE FIELDS, THEIR ARCHITECTONICS AND CONNECTIONS

IN THE MONGOLIAN GERBIL

H. Steffen, C. Simonis, H. Thomas, J. Tillein and H. Scheich

Institute for Zoology
Technical University Darmstadt
Schnittspahnstr. 3, D-6100 Darmstadt, FRG

Tonotopic organization of auditory cortical fields has been described, so far, in a number of different species, like cat (Merzenich et. al., 1975; Knight, 1977; Reale and Imig, 1980), monkey (Imig et al., 1977; Brugge, 1982; Aitkin et al., 1986), ferret (Kelly et al., 1986) and in rabbit (McMullen and Glaser, 1982). At least two tonotopically organized fields are known in rodents, like in the grey squirrel (Merzenich et al., 1976), mouse (Stiebler, 1987) and guinea pig (Hellweg et al., 1977; Redies et al., 1987). Due to its unusual specialization in the low frequency range (Ryan, 1976), the mongolian gerbil may be a good model for auditory research in that range, including aspects of human speech processing. We have engaged in the functional analysis of the auditory cortex of this animal using electrophysiological, 2-deoxyglucose and pathway tracing techniques in parallel.

ELECTROPHYSIOLOGY

Fig. 1 is a reconstruction of the results of a fine grain electrophysiological analysis in 15 adult gerbils. A lateral view of the surface of the temporal cortex is depicted, rostral left, dorsal up. The penetration was roughly tangential to the pial surface, stepping from dorsal to ventral regions (step size = 50 μm). Sinus tone bursts (250 ms burst and 500 ms pause) from a loudspeaker close to the contralateral ear were used as stimuli to the anaesthetized animals (Urethane). To date, four different tonotopically organized fields could be distinguished: Two fields with a mirror-image tonotopy had adjacent high frequency representations, the low frequency representations were rostral in a smaller anterior field (AAF) and caudal in a larger field (AI). More caudally, additional fields were found, a dorsal field (DP) and a ventral posterior field (VP). In field VP, low frequency responsiveness was situated rostrally, adjacent to the low frequency representationm of AI, high frequency representations could be found in the caudal part of the field. Field DP is peculiar in the sense that its tonotopical organization was concentrical, high frequencies in the center being surrounded by a belt of lower frequencies. In AI, for which the best maps were obtained, isofrequency contours ran roughly dorso-ventrally with frequencies below 4 kHz covering at least two thirds of the field.

1. Cytoarchitecture: A horizontal section through the brain of a gerbil at the level of the auditory cortical fields shows - in the Nissl stain - cortical regions of different cell-packing density as well as soma diameters (Fig. 2B). Rostrally, a cell dense band marks layer IV of the somatosensory cortex (SS). More caudally, this band is interrupted (?) and appears again more caudally in layer IV. In comparison with the results of electrophysiology (Fig. 2D) and 2-Deoxyglucose (2-DG) mapping of auditory cortical activity (Fig. 2C), the zone with the cell dense band in layer IV of the temporal cortex corresponds to fields AAF and AI.

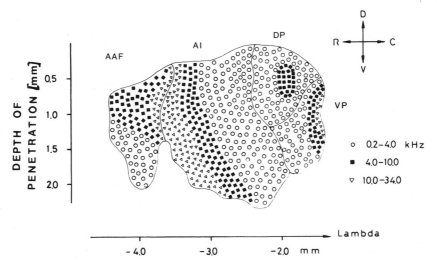

Fig. 1. Reconstruction of the results of the electrophysiological exploration of the auditory cortex in 15 adult gerbils. The direction of the penetration was roughly tangential to the pial surface, stepping from dorsal to ventral loci. Step size: 50 μm; distance between penetrations: 200 - 300 μm. Stimulus: tone bursts via loudspeaker (60 dB) close to contralateral ear. Four auditory fields with a mirror-image tonotopy with respect to their common borders are clearly distiguishable. Note the relative large representation of the low frequency range (< 4 kHz) in AI.

2. Fiberarchitecture: In a fiber stain (Liesegang; Fig. 2A), this zone is characterized by a very dense fiber meshwork throughout all supra-granular layers including layer IV; in addition, there is a fiber-dense zone visible in the extreme caudal part of the cortex, right at the border of the rhinal fissure. Tracing experiments with HRP show that cells in this zone project to cortical auditory fields (Fig. 3).

3. 2-Deoxyglucose: Fig. 2C shows the autoradiography of a horizontal section through the right hemisphere of a gerbil that has been exposed to alternating tones (1.4 and 5.5 kHz, 3/s alternation) during 90 minutes after injection of 14-C-Fluoro-2-deoxyglucose. Two days prior to the injection, a best frequency determination in AI was made electro-physiologically at 5.5 kHz, and the electrode's position was marked. Isofrequency contours of 1.4 and 5.5 kHz (dark stripes) are clearly visible. The interpretation of the multiple labeled contours is as follows: contour 1: activity of 1.4 and 5.5 kHz neurons fuses to one

contour in field AAF. This field has less spatial frequency resolution
than field AI (see Fig. 1). Contours 2+3: frequency contours of 5.5 kHz
(2) and 1.4 kHz (3), respectively, in field AI. Note the lesion caused
by the electrode at 5.5 kHz in contour 2. Contours 4+5: Two isofrequency
contours of 1.4 kHz in field DP, the isofrequency contours of 5.5 kHz
being hardly visible in between. The 1.4 kHz contour marks twice because
of the concentric nature of the frequency map in this field (see Fig. 1).
The interpretation of these contours is based on a complete reconstruction of this case from serial sections and on a large number of other
cases using different stimulus frequencies (Scheich et al., 1986).

ROSTRAL

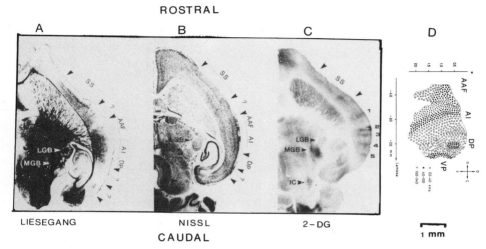

CAUDAL

Fig. 2.   Comparison of photomicrographs of the right hemispheres of three
          different gerbils. Horizontal sections at corresponding dorso-
          ventral levels show the cyto- and fiberarchitectonical organiza-
          tion that underlies the functional regions in the auditory
          cortex as revealed by 2-DG- (C) and electrophysiological mapping
          experiments (D), respectively. A: In the fiber stain, a cortical
          region that corresponds to the electrophysiologically defined
          fields AAF, AI and DP stands out by its dense fiber plexus
          throughout layers IV, V and VI. At the extreme caudal end of the
          temporal cortex, near the rhinal fissure, there is another fibre-
          dense zone, still unexplored electrophysiologically. Cells in
          this zone project to the auditory fields AI and DP as revealed
          by HRP injections into these fields (see also Fig. 3A+B). B:
          A cell-dense band in layer IV of the temporal cortex indicates
          the rostro-caudal extent of the auditory cortical fields AAF and
          AI. C: 2-DG-autoradiography as a result of a two-tone experiment:
          alternating tones, 1.4 and 5.5 kHz, respectively, were presented
          during 90 minutes to an awake gerbil. In the cortex, black
          stripes indicate contours of enhanced cellular activity. For
          interpretation of these contours see text. D: Insert of Fig. 1
          at the corresponding magnification and location for comparison
          with the fields in the adjacent histological sections.

4.   HRP - tracing: Single injections into the frontal field AAF reveal a
complex pattern of columns of labeled cells in AI and DP/VP, respectively,
that project to a restricted part of field AAF (Fig. 3A). Individual
columns of each field have converging projections onto AAF. Injections
into field AI, in turn, reveal only one strong projection from and to AAF

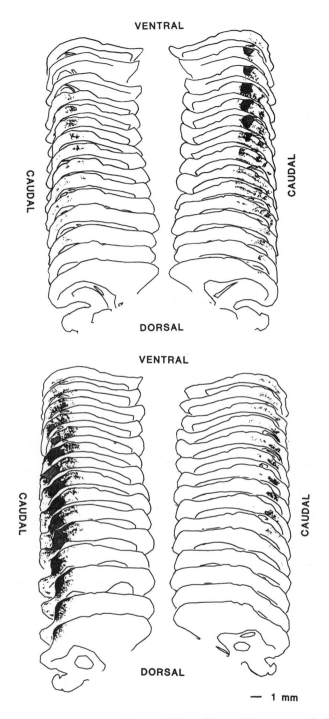

Fig. 3.   Reconstructions of single injections of WGA-HRP (Sigma, type
VI, 1%) into cortical auditory fields. Series of horizontal
sections are shown (computer-aided digitalization of sections;
distance between sections: 200 μm). Each black dot indicates
the position of one labeled cell. The injection site is marked
with grey-fill. Left and right hemispheres are split and (cont'd)

Fig. 3. (cont'd) depicted separately left and right.
All intracortical projections are reciprocal; for reasons of
clarity, regions of anterograde transport are not marked.
A: Injection of HRP into the frontal auditory field (AAF). Ipsilateral projections to this field come from three distinct
cortical areas, situated caudally to the injection site:
fields AI, DP and one extreme caudal field of an hitherto
unknown function. Note the heterotopic dorso-caudal projection
(left side) from contralateral field AI besides the homotopic
one.
B: Injection into field AI and part of DP. Ipsilateral cortical
projections come from field AAF and a still more caudal field
of unknown function. Like AAF, AI also receives a heterotopic
rostro-ventral projection from contralateral field AAF besides
the homotopic one.

and a projection from one location in a more caudal field, presumably
field DP, onto AI (Fig. 3B). The projections from the contralateral
auditory cortex to fields AI and AAF are generally much wider than what
is known from other sensory cortices. Also, there is - besides the
homotopic projection - a pronounced heterotopic projection from contralateral field AAF to ipsilateral field AI and vice versa. Other ipsilateral connections of the fields are reciprocal with striatum and medial
geniculate body (at least three nuclei), terminate in ipsilateral
inferior colliculus and originate in lateral amygdala, loc. coeruleus and
raphe nuclei.

SUMMARY

2-DG-experiments as well as electrophysiological mapping show that the
auditory cortex of the gerbil is composed of at least four tonotopically
organized fields: AAF, AI, DP and VP. DP and VP have a common border of
low frequency representation with each other and with AI. In field DP,
an island of high frequency representation (4 - 10 kHz) is surrounded
by lower frequency representations. AAF and AI show a mirrorimage
tonotopy with low frequency representation caudal in AI and rostral in
AAF and high frequency representation at their common border, the isofrequency contours being oriented roughly dorso-ventrally in AI. The
cyto- and fiberarchitectonics of these cortical regions reflects this
parcellation: differences in cell-packing densities (Nissl) and fiber
densities (Liesegang) match well with the rostro-caudal extent and
locations of the physiologically defined fields. Furthermore, tracer
injections (WGA-HRP) into each one of these fields reveal reciprocal
connections with distinct parts of the other fields.

ACKNOWLEDGEMENTS

Supported by Deutsche Forschungsgemeinschaft SFB 45.

REFERENCES

Aitkin, L. M., Merzenich, M. M., Irvine, D. R. F., Clarey, J. C.,
and Nelson, J. E., 1986, Frequency representation in auditory cortex
of the common marmoset (callithrix jacchus jacchus)
J. Comp. Neurol., 252:175-185.
Brugge, J. F., 1982, Auditory cortical areas in primates, in: "Cortical
sensory organization. Vol. 3: Multiple auditory areas", Woolsey,
C. N., ed., Humana Press, Clifton, N. Jersey, 59-69.
Hellweg, F. C., Koch, R. and Vollrath, M., 1977, Representation of the
cochlea in the neocortex of guinea pigs, Exp. Brain Research 29:
467-474.

Imig, T. J., Ruggero, M. A., Kitzes, L. M., Javel, E. and Brugge, J. F., 1977, Organization of auditory cortex in the owl monkey (Aotus trivirgatus), J. Comp. Neurol., 171:111-128.

Kelly, J. B., Judge, P. W. and Phillips, D. P., 1986, Representation of the cochlea in primary auditory cortex of the ferret (Mustela putorius) Hear. Res., 24:111-115.

Knight, P. L., 1977, Representation of the cochlea within the anterior auditory field (AAF) of the cat, Brain Res., 130:447-467.

McMullen, N. T. and Glaser, E. M., 1982, Tonotopic organization of rabitt auditory cortex, Exp. Neurol., 75:208-220.

Merzenich, M. M. and Kaas, J. H., 1980, Principles of organization of sensory-perceptual system in mammals, in: "Progress in psychobiology and physiological psychology", J. M. Sprague and A. N. Epstein, eds., Academic Press N. Y. Vol 9: 1-42.

Merzenich, M. M., Kaas, J. H. and Roth, G. L., 1976, Auditory cortex in the grey squirrel: tonotopic organization and architectonic fields, J. Comp. Neurol., 166:387-402.

Merzenich, M. M., Knight, P. L. and Roth, G. L., 1975, Representation of the cochlea within primary auditory cortex in the cat, J. Neurophys., 38:231-249.

Reale, R. A. and Imig, T. J., 1980, Tonotopic organization in auditory cortex of the cat, J. Comp. Neurol., 192:265-291.

Redies, H., Sieben, U. and Creutzfeldt, O. D., 1986, Tonotopic organization of the auditory cortex of the guinea pig, Neurosc. Lett. Suppl., 26:S407.

Ryan, A., 1976, Hearing sensitivity of the mongolian gerbil (Meriones unguiculatus), J. Acoust. Soc. Am., 59:1222-1226.

Scheich, H., Heil, P. and Langner, G., 1986, Tonotopic organization of gerbil auditory cortices mapped with 2-deoxyglucose, Soc. Neurosci. Abstr. Vol. 12, Part 2:1274.

Stiebler, I., 1987, A distinct ultrasound-processing area in the auditory cortex of the mouse, Naturwissenschaften, 74:96-97.

Woolsey, C. N., 1971, Tonotopic organization of the auditory cortex, in: "Physiology of the auditory system". M. B. Sachs, ed., Natl. Educational Consultants Inc. Baltimore, 271-281.

GOLGI STUDIES ON THE HUMAN AUDITORY CORTEX

H. L. Seldon

Univ. HNO Klinik Köln,

Josef Stelzmannstr. 9, 5000 Köln 41, F.R.G.

For over 100 years it has been known that speech perception is dependent on the superior temporal cortex in the left hemisphere in most people. Since then we have not made much progress in finding out why this is so - this is the question I wish to address but will not answer. An obvious subsequent question is whether there are any anatomic structures peculiar to this area, which could at least provide a basis for its language capabilities.

Economo and Koskinas (1925) and Economo and Horn (1930) showed that the surface (planum temporale and transverse gyri) in the dominant left hemisphere is larger than on the right and has a different gyrus composition. Their findings were confirmed by Geschwind and Levitsky in 1968 and since then by several others. Cytoarchitectonic studies (Economo and his colleagues, also Galaburda et al., 1978-a-b, and Galaburda and Sanides, 1980) have shown that especially area TA (on the lateral edge of the gyrus and the upper bank of the superior temporal sulcus), the secondary auditory cortex, is significantly larger in the left hemisphere than in the right one. Economo and Koskinas (1925) also emphasized that the somata especially in areas TA and TB show one of the most impressive columnar organizations of the neocortex, aptly described as "organ pipes".

However, all these findings show that the left hemisphere auditory areas are larger, but do not explain why they can do something that the right hemisphere cannot. Let us go a step further and examine the wiring in the auditory areas. For this we use Golgi impregnation to examine individual neuronal structure. Ramon y Cajal (1902) did this around the turn of the century and provided us with a qualitative catalog of cell types. I did this some years ago and stored the 3-D coordinates of the neuron trees in a computer (Seldon, 1985). When we examine the structures, we notice immediately a strong tendency toward a columnar organization of the dendrites of pyramidal as well as nonpyramidal cells, just like the cell somata columns. Recently DeCarlos et al. (1987), using Golgi impregnation of samples from newborns who had succumbed of hypoxia, found that some of the beaded, sparsely-spined nonpyramidal cells are actually chandelier cells.

It is difficult to study axons in Golgi-impregnated adult human samples. By accident I had a specimen which showed largely axon impregnation with few cells. The most remarkable characteristic of this was the

229

macroscopically visible "columns" about 350-500 μm wide - i.e., roughly the width of ocular dominance columns in the visual cortex or binaural versus suppression columns in cat auditory cortex. Up to now there has been insufficient material to make hemispheric comparisons.

We still have not answered the question of what makes the Wernicke region so particular. Visually the neurons look the same in both hemispheres, but if we do some statistical comparisons, we find some interesting differences. First we find that the center-center distance between somata columns is distinctly larger in the left hemisphere. We also find that the tangential extent of dendrites on the left is greater than on the right (Seldon, 1985). (On the other hand, the vertical component on the right is often greater.) However, and this is the crux of the matter, the ratio of tangential extent to inter-column intervals is clearly less on the left - this implies that dendrites from one column in the left hemisphere actually overlap with fewer neighboring columns than those in the right hemisphere. This in turn implies that left hemisphere columns share relatively fewer afferent inputs with their neighbors and thus, at least theoretically, are more capable of independent responses. Note that since columns are considered as functional units, what interests us is not the input to a neuron but the input to a column. If we make a simple estimate of the many-to-many mapping in the auditory areas, we note that left hemisphere columns generally get more input due to their larger tangential size, but that an afferent on the left contacts fewer columns due to their greater spacing.

Thus, as we see, there are certain anatomic features which imply that the dominant auditory cortex may be capable of actions which the other side cannot perform. From neurolinguistics and neurophysiology we can get a cue as to what these actions might be. Steinschneider et al. (1982) have shown that information on individual acoustic features of signals is already coded in the cortical afferents of trained monkeys. Cortical neurons show different responses to a given vowel, dependent on the preceding consonants, i.e., they differentiate combinations. A similar phenomenon was demonstrated by Zeki (1983) in responses of visual cortex cells to colors. Hebben (1986) found that left hemisphere brain lesions did not change basic phonetic feature differentiation and concluded that the phonetic code is either represented in cortical and subcortical structures or redundantly in the cortex. Dichotic hearing tests, e.g. Studdert-Kennedy (1976), Blumstein et al. (1977) imply a categorical, encoded perception for certain consonants (20-60 msec stop consonants such as ba-pa) and for truncated vowels (40 msec) in the dominant hemisphere. As long ago as 1914 von Monakow mentioned that sensory aphasics had more difficulties with con-sonants than with vowels (p. 628). Recently Tanaka et al. (1987) described a case of pure word deafness and amusia after bilateral lesions of the temporo-parietal junction - linguistic tests showed the word deafness to be due to a deficiency in discriminating consonants. This implies that the particular feature of the Wernicke region is its capability for reacting to and distinguishing combinations including very short linguistic signals. This correlates nicely with our inference of more inputs/column and of relatively greater column separation here as compared to the right hemisphere.

To test the model we have to locate functional columns. Up to now, lesion analysis in aphasia has generally only confirmed the location of sensory linguistic perception to the dominant superior temporal gyrus. A few cases have implied spatially separate mappings for different semantic categories (e.g., Warrington and McCarthy, 1983; McCarthy and Warrington, 1985; Hart et al., 1985; Goodglass et al., 1986). However, the performance of the few patients tested has been quite variable, as have been the CT-localizations of lesions. Lesions in the Wernicke area do not preclude recovery of single-word comprehension but are related to sentence comprehension (Selnes et al., 1984).

So to test the model we have to go to acoustic evoked potentials, magnetoencephalography (MEG) and "brain mapping". Here we have two possibilities, dipole localization (e.g., Romani et al., 1982; Scherg and von Cramon, 1985) or current source density (CSD) analysis (e.g., MacKay, 1984). Although CSD is an elegant, direct application of Maxwell's equations (V.D = p and V.J = -dp/dt where D = electric displacement, J = current density, p = charge density, t = time - Jackson, 1962, p. 177), current difficulties in humans are that with recording electrodes in the temporal region only a 2-D map can be made and dipole sources parallel to the scalp can only be detected by polarity reversals (MacKay, 1984). With the dipole localization method, subjectively the most appropriate model, i.e., that which best fits the anatomy, would be an adaptation of the annular sector dipole layer (Darcey, 1979) including a sector for the planum temporale and transverse gyri, one for the upper bank of the superior temporal sulcus, and a strip of radial dipoles for the lateral gyral surface. This, however, leads to a set of equations with a large number of unknown variables, which immediately makes even an approximated solution unlikely. A single equivalent dipole could be located, but it subjectively does not satisfy our model.

For MEG, Cuffin and Cohen (1977) showed analytically that the magnetic field perpendicular to the surface of a sphere (e.g., head) or an infinite plane is due exclusively to an underlying tangential dipole (i.e., not to secondary volume currents). Romani et al. (1982) used this to demonstrate tonotopic dipole organization in primary auditory cortex. Although elegant, the MEG also locates single equivalent dipoles; it also does not detect strictly radial dipoles such as on the lateral gyrus.

REFERENCES

Blumstein, S. E., Cooper, W. E., Zurif, E. B. and Carmazza, A., 1977, The perception and production of voice-onset time in aphasia, Neuropsychology, 15:371-383.
Cajal, S. R., 1902, Studien uber die Hirnrinde des Menschen, Nr. 3, J. A. Barth, Leipzig
Darcey, T. M., 1979, Methods for the localization of electrical sources in the human brain and applications to the visual system, Ph. D. Thesis, Cal. Inst. of Tech., Pasadena
DeCarlos, J. A., Lopez-Mascaraque, L., Cajal-Agueras, S. R. and Valverde, F., 1987, Chandelier cells in the auditory cortex of monkey and man - a Golgi study, Exp. Brain Res., 66:295-302.
Economo, C. v. and Koskinas, G. N., 1925, Die Cytoarchitektonik der Hirnrinde des erwachsenen Menschen, Springer, Berlin.
Economo, C. v. and Horn, L., 1930, Über Windungsrelief, Masse und Rindenarchitektonik der Supratemporalflache, ihre individuellen und ihre Seitenunterschiede, Z. Gesamte Neurol. Psychiatr., 130: 678-757.
Galaburda, A. M., LeMay, M., Kemper, T. L. and Geschwind, N., 1978a, Right-left asymmetries in the brain, Science, 199:852-856.
Galaburda, A. M., Sanides, F. and Geschwind, N., 1978b, Human brain: cytoarchitectonic left-right asymmetries in the temporal speech region, Arch. Neurol., 35:812-817.
Galaburda, A. M. and Sanides, F., 1980, Cytoarchitectonic organization of the human auditory cortex. J. Comp. Neurol., 190:597-610.
Geschwind, N. and Levitsky, W., 1968, Human brain: left-right assymmetries in temporal speech region, Science, 161:186-187.
Goodglass, H., Wingfield, A., Hyde, M. R. and Theurkauf, J. C., 1986,

Category specific dissociations in naming and recognition by aphasic patients, Cortex, 22:87-102.

Hart, J., Berndt, R. S. and Caramazza, A., 1985, Category-specific naming deficit following cerebral infarction. Nature, 316:439-440.

Hebben, N., 1986, The role of the frontal and temporal lobes in the phonetic organization of speech stimuli: a multi-dimensional scaling analysis, Brain and Lang., 29:342-357.

Jackson, J. D., 1962, Classical Electrodynamics, J. Wiley and Sons, New York.

MacKay, D. M., 1984, Source density analysis of scalp potentials during evaluated action. I. Coronal distribution, Exp. Brain Res., 54:73-85.

McCarthy, R. and Warrington, E. K., 1985, Category specificity in an agrammatic patient - the relative impairment of verb retrieval and comprehension, Neuropsychologia, 23:709-727.

Monakow, C. v., 1914, Die Lokalisation im Grosshirn, Bergmann, Wiesbaden.

Romani, G. L., Williamson, S. J., Kaufman, L. and Brenner, D., 1982, Characterization of the human auditory cortex by the neuromagnetic method, Exp. Brain Res., 47:381-393.

Scherg, M. and Cramon, D. v., 1986, Evoked dipole source potentials of the human auditory cortex, Electroenceph. clin. Neurophysiol., 65: 344-360.

Seldon, H. L., 1985, The Anatomy of Speech Perception: Human Auditory Cortex, in: "Cerebral Cortex", Vol. 4, A. Peters, E. G. Jones, eds., Plenum Press, New York, 273-327.

Selnes, O. A., Niccum, N., Knopman, D. S. and Rubens, G. B., 1984, Recovery of single word comprehension. CT-scan correlates, Brain and Lang., 21:72-84.

Steinschneider, M., Arezzo, J. and Vaughan, H. G., 1982, Speech evoked activity in the auditory radiations and cortex of the awake monkey, Brain Res., 252:353-366.

Studdert-Kennedy, M., 1976, Speech perception, in: "Contemporary Issues in Experimental Phonetics", N. J. Lass, ed., Academic Press, New York, 243-293.

Tanaka, Y., Yamadori, A. and Mori, E., 1987, Pure word deafness following bilateral lesions: a psychophysical analysis, Brain, 110:381-404.

Warrington, E. K. and McCarthy, R., 1983, Category specific access dysphasia, Brain, 106:859-878.

Zeki, S., 1983, Colour coding in the cerebral cortex: the reaction of cells in monkey visual cortex to wavelengths and colours, Neuroscience, 9:741-766.

STRUCTURE OF THE GABA-ERGIC INHIBITORY SYSTEM IN THE CHICKEN

AUDITORY PATHWAY REVEALED BY IMMUNOCYTOCHEMISTRY

Christian M. Müller

Institut für Zoologie, Technische Hochschule
Schnittspahnstrasse 3, 6100 Darmstadt, F.R.G.
Max-Planck-Institut für Hirnforschung
Deutschordenstrasse 46, 6000 Frankfurt/M., F.R.G.

The auditory system of birds has been shown to be an advantageous model to study audition in a higher vertebrate. From a neuroethologists view this notion is supported by the fact that birds, especially song-birds, have evolved a strikingly complex repertoire of communicative sounds which they use for intraspecific communication. The accuracy of sound analysis is also very similar to what is known from studies in mammals, including man (Dooling et al., 1975). The auditory pathway of birds has been shown to share multiple similarities with the mammalian auditory system. This includes not only the presence of comparable auditory nuclei (Boord, 1969), but also similar interconnectivities (Boord, 1969; Bonke et al., 1979). Due to the recent progress in immunocytochemistry it has become feasible to further characterize the structure of nuclei in the central nervous system by labeling neurons with a known transmitter content. Such studies can guide physiological experiments aimed at the function of neuronal circuits. The inhibitory neurotransmitter gamma-aminobutyric acid (GABA) is present in a number of structures involved in sensory processing and has recently been shown to play an important role in telencephalic auditory processing in the chicken (Müller and Scheich 1987). I used the method of immuno-cytochemistry to describe the distribution of GABAergic perikarya and nerve terminals in the entire auditory pathway of the chicken. The results point out the importance of GABA as a neurotransmitter at all levels of auditory processing in the avian auditory pathway.

METHODS

The present report is based on experiments with 10 chickens aged 4 to 8 weeks. Under deep anesthesia the animals were perfused intra-cardially with saline, followed by fixative. Sections of the brains at the level of auditory nuclei were cut on a cryostate or with a vibratome. To stain presumably GABAergic neurons and terminals I used antisera directed against the GABA synthetizing enzyme glutamate decarboxylase (GAD; Oertel et al., 1981) and an antiserum directed against GABA (Storm-Mathisen et al., 1983). The location of the antigen was visualized by the indirect peroxidase-antiperoxidase- or by the avidin-biotin-method using diaminobenzidine as the chromagen. Control sections with the first antiserum replaced by non-immune-serum never revealed staining. Details about the methods of immunostaining can be found in Müller (1987, 1988).

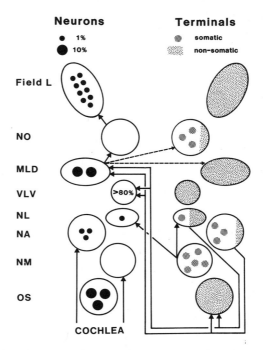

Fig. 1.   GABAergic neurons and terminals in the auditory pathway
of the chicken. Schematic drawing of the interconnectivity
within the auditory pathway. The nuclei are indicated at the
left of the scheme. The percentage of immunolabeled neurons
is given by the symbols in the left part of the figure, while
the presence of labeled terminals is indicated in the right
part of the panel. If numerous immunopositive terminals were
frequently found in direct association with somata in a given
nucleus, this is indicated by the symbol of "somatic" labeling.
"Non-somatic" labeling corresponds to stained terminals which
are scattered in the neuropil without clear association with
somata.

RESULTS

    The findings of the GABA- and GAD-immunostaining are summarized in
Fig. 1. The scheme shows the auditory nuclei (labeled at the left margin
of the figure) and their major interconnectivities (Heil and Scheich,
1986). In addition, the percentage of labeled neurons within each
nucleus is given in the left part of the scheme, while the presence
of terminal labeling is indicated in the right part. Terminal staining
is further divided into somatic (i.e. when numerous immunopositive
terminals were clearly associated with a single cell body) and non-
somatic labeling. In the following I will describe the immunolabeling
observed within the auditory pathway.

GABA in Auditory Brainstem

    The avian cochlear nucleus complex consists of the nucleus magno-
cellularis (NM) and the nucleus angularis (NA). Both nuclei receive
direct afferents from the cochlear ganglion. NM is free of GABA-positive
neurons, but contains punctate labeling indicative of GABAergic axon
terminals. These terminals heavily sourrond the unstained cells of NM.
In NA about three percent of the cells reveal GABA immuno-reactivity. The

stained cells are medium sized and mainly located in the dorsolateral aspects of NA. Stained terminals are less numerous and smaller than those in NM. Although the majority of presumed GABAergic axon endings in NA is directly associated with unstained somata, there are also numerous terminals scattered in the neuropil.

Nucleus laminaris (NL) of birds consists of a monolayer of perikarya through most of its extent and receives afferents both from the ipsi- as well as from the contralateral NM. Cell bodies in the perikaryal layer do not stain with the antibodies but are surrounded by several GABA-positive terminals. In addition, terminal labeling is present throughout the dendritic layers of NL. Few neurons are present in the dendritic layers which are always GABA-positive. This cell population comprises up to about 1% of neurons in the entire NL.

A high density of stained neurons is found both in the nucleus olivaris superior (OS) and in the ventral nuclei of the lemniscus lateralis (VLV). Both nuclei receive afferents from the cochlear nucleus complex and the NL. While up to 30% of the neurons in OS show GABA-immunoreactivity, immunopositive cells amount to more than merely 80% of the neurons in the VLV. In the anterior portion of VLV every cell is labeled with the antiserum to GABA. Besides stained neurons, terminals are also labeled in both nuclei, which are predominantly scattered in the neuropil.

GABA in Auditory Midbrain

Nucleus mesencephalicus lateralis pars dorsalis (MLD) in the midbrain also contains a high proportion of presumably GABAergic neurons. Labeled cells are round to oval and amount to up to 20% of the cell population in the MLD. The cells are homogeneously distributed within the nucleus and their soma diameters do not differ significantly from those of the overall population of cells in MLD. Labeled axon terminals are scattered in the neuropil, whereas single somata are contacted by only a few stained boutons.

GABA in Auditory Thalamus

The thalamic n. ovoidalis (NO) is devoid of immunostained cells. This finding has also been confirmed after large injections of colchicine into the forebrain, in order to block the axonal transport. Despite the lack of stained neurons, the NO is densely covered by stained terminals which surround the unstained somata. In addition, the neuropil also reveals scattered terminal staining.

GABA in Caudal Auditory Telencephalon

In the caudal auditory telencephalon, which consists of the trilaminated field L and the auditory parts of the hyperstriatum ventrale (HV), about 9% of the cell population reveals immunoreactivity for GABA and GAD. The soma diameters of stained neurons are significantly smaller then those of the overall population of neurons. Presumably GABAergic cells are present in all laminae of field L and the HV. The distribution of intercellular distances between neighbouring stained cells reveals two maxima. This pattern of cell distribution is compatible with the notion that the GABAergic system in the auditory forebrain consists of two cell populations, one with a higher cell density than the other (Müller, 1988). The two cell populations may correspond to the two types of neurons with a low spine-density known from Golgi-studies (Saini and Leppelsack, 1981), namely neurons with local axon arborizations and those with laterally projecting axons.

Besides labeled cells immunopositive terminals are also present throughout all laminae of field L and the HV. Terminal labeling is densest in the thalamorecipient lamina L2 of field L. Immunopositive axon terminals are scattered in the neuropil and are only rarely directly associated with somata.

CONCLUSIONS

In summary, using antisera directed against GABA and the GABA synthetizing enzyme GAD, I have identified presumably GABAergic cells and terminals in the auditory pathway of the chicken. The presence of immunopositive terminals in all nuclei of the auditory system suggests that GABAergic inhibitory interactions play an important role in auditory processing at most, if not all, levels of the auditory pathway. As the overall picture of GABA immunoreactivity shares several features with that obtained in mammalian species (e.g. Thompson et al., 1985), the notion is supported that the avian auditory system is a valuable model to study audition in a higher vertebrate.

ABREVIATIONS

MLD-nucleus mesencephalicus lateralis, pars dorsalis; NA-nucleus angularis; NL-nucleus laminaris; NM-nucleus magnocellularis; NO-nucleus ovoidalis; OS-nucleus olivaris superior; VLV-nucleus ventralis lemnisci lateralis.

ACKNOWLEDGEMENT

The author is indebted to Dr. W. Oertel for providing the GAD antiserum and Dr. I. Wulle for the gift of GABA antiserum. Partially supported by the Deutsche Forschungsgemeinschaft (SFB 45).

REFERENCES

Bonke, B. A., Bonke, D. and Scheich, H., 1979, Connectivity of the auditory forebrain nuclei in the Guinea Fowl (Numida meleagris), Cell Tiss. Res., 200:101-121.
Boord, R. L., 1969, The anatomy of the avian auditory system, Ann. N. Y. Acad. Sci., 167:186-198.
Dooling, R. J. and Saunders, J. C., 1975, Hearing in the parakeet (Melopsittacus undulatus): absolute thresholds, critical ratios, frequency difference limens, and vocalizations, J. Comp. Physiol. Psychol., 88:1-20.
Heil, P. and Scheich, H., 1986, Effects of unilateral and bilateral cochlea removal on 2-deoxyglucose patterns in the chick auditory system, J. Comp. Neurol., 252:279-301.
Müller, C. M., 1987, Gamma-aminobutyric acid immunoreactivity in brainstem auditory nuclei of the chicken, Neurosc. Lett., 77:272-276.
Müller, C. M., 1988, Distribution of GABAergic perikarya and terminals in the centers of the higher auditory pathway of the chicken, Cell Tiss. Res., (in press).
Müller, C. M. and Scheich, H., 1987, GABAergic inhibition increases the neuronal selectivity to natural sounds in the avian auditory forebrain, Brain Res., 414:376-380.
Oertel, W. H., Schmechel, D. E., Mugnaini, E., Tappay, M. L. and Kopin, I. J., 1981, Immunocytochemical localization of glutamate decarboxylase in rat cerebellum with a new antiserum, Neurosc., 6:2715-2735.
Saini, K. D. and Leppelsack, H.-J., 1981, Cell types of the auditory caudomedial neostriatum of the starling (Sturnus vulgaris), J. Comp. Neurol., 198:209-229.

Storm-Mathisen, J., Leknes, A. K., Bore, A. T., Vaaland, J. L.,
    Edminson, P., Haug, F. M. S. and Ottersen, O. P., 1983, First
    visualization of glutamate and GABA in neurons by immunocyto-
    chemistry, Nature (Lond.), 301:517-520.
Thompson, G. C., Cortez, A. M. and Man-Kit Lam, D., 1985, Localization
    of GABA immunoreactivity in the auditory brainstem of guinea pigs,
    Brain Res., 339:119-122.

FUNCTION OF THE GABA-ERGIC INHIBITORY SYSTEM IN THE CHICKEN

AUDITORY FOREBRAIN

Christian M. Müller

Institut für Zoologie, Technische Hochschule,
Schnittspahnstrasse 3, 6100 Darmstadt, F.R.G.
Max-Planck-Institut für Hirnforschung
Deutschordenstrasse 46, 6000 Frankfurt-M., F.R.G.

The auditory structures in the caudal auditory telencephalon (CAT) of birds consist of the trilaminated field L and the auditory part of the hyperstriatum ventrale (HV). These substructures share a common tonotopic organization, whereas iso-frequency planes cut perpendicular to all laminae of field L and the HV (Scheich et al., 1979b; Müller and Leppelsack, 1985). The input to the CAT arises from the diencephalic n. ovoidalis and terminates in the central lamina (L2) of field L (Bonke et al., 1979a). From there information is further relayed to the adjacent laminae L1 and L3, as well as to the HV. This hierarchical pattern of interconnectivity is paralleled by an increase in neuronal response selectivity. While cells in the thalamorecipient lamina usually respond to pure tones, about 15 to 20% of the neurons in the postsynaptic structures respond only to complex auditory stimuli, e.g. noise bands or natural sounds (Scheich et al., 1979a; Müller and Leppelsack, 1985). In addition, tone responsive units in the higher order laminae often reject broad band sounds although they contain frequencies that are responded to when presented as pure tones. The increase of response selectivity from the input layer L2 to the postsynaptic structures has been attributed especially to inhibitory interactions in the CAT (Langner et al., 1981; Müller and Leppelsack, 1985).

In a recent study I showed that a subpopulation of neurons in the CAT contains both gamma-aminobutyric acid (GABA) and the GABA synthetizing enzyme glutamate decarboxylase (Müller, 1988). Furthermore, iontophoretic application of a GABA antagonist substantially reduces the response selectivity of units in the CAT to natural vocalizations (Müller and Scheich, 1987). To further characterize the influence of inhibitory interactions mediated by GABA on telencephalic auditory information processing, we tested the influence of iontophoretically applied GABA-antagonist bicuculline onto the response characteristics of single units in the CAT to synthetic stimuli (Müller and Scheich subm. a, b). The present report summarizes the findings of the iontophoresis experiments and describes findings from in vitro studies supporting the existence of two GABAergic systems in the CAT of the chicken.

METHODS

Iontophoresis Experimets

Single unit recordings were obtained from chronically prepared, fully awake chickens by means of multibarrel glass electrodes with a tungsten wire inserted in the central barrel (Müller and Scheich, 1987; subm.). The outer barrels were filled with solutions of GABA (1M, pH 7.5), and NaCl (2M, pH 3.0 and 7.0). During the recording sessions the animals were placed in a soundproof booth. Standard techniques for single unit recording and on-line data aquisition were used (Bonke et al., 1979b). Drug ejection was controlled by a Neurophore iontophoresis assembly. Sound stimuli consisted of pure tones, two-tone combinations, and white noise bursts, and were presented either via a loudspeaker or via earphones sealed into the external ear. After termination of the experiments electrode tracks were reconstructed in Nissl stained sections using small electrolytic lesions as references.

In vitro Experiments

Intracellular recordings were obtained from an in vitro slice preparation that includes all laminae of the CAT. Chickens aged 3 to 8 weeks were decapitated during Nembutal anesthesia and frontal sections (400 μm) through the CAT were cut on a vibratome. The sections were transferred to a recording chamber and continuously superfused with oxygenated Ringer solution thermostatically held at a temperature of 31°C. Standard recording and stimulation techniques were employed (for details see Müller, 1987).

RESULTS AND CONCLUSIONS

Iontophoresis Experiments

In order to evaluate the mechanisms by which GABAergic inhibition increases the response selectivity of single units in the CAT we investigated the influence of iontophoretically applied BIC on the response characteristics to tones, two-tone combinations, and dichotic stimuli. Four major effects of BIC iontophoresis were observed:
1) Iso-intensity-response areas obtained with pure tone stimulation widened during the blockade of GABA receptors, i.e. the tuning of single units was reduced. This effect was only seen in units lying postsynaptic to the thalamorecipient lamina L2. In addition, most cells that responded to natural sounds, but did not respond to pure tones before BIC application, displayed clear excitatory response areas to tones thereafter.
2) Two-tone stimuli, consisting of the units best frequency combined with a test frequency increased in 100 Hz steps, revealed areas of two-tone supression that extended from the border of the excitatory response area up to three octaves from the best frequency. During BIC iontophoresis these areas of two-tone supression were reduced to a narrow frequency band just outside the original excitatory response area. Units from the input lamina L2 always had only a very narrow frequency. range capable of inducing two-tone supression, which was BIC insensitive.
3) When tested with tone stimuli with increasing intensity, several units displayed non-monotonically rising intensity-response functions. These functions had either a saturation characteristic (i.e. after an initial rise of the response strength with increasing intensities a plateau was reached), or an optimum characteristic (i.e. the response strength rose with increasing intensities followed by a reduction of the response strength with further increases of the sound pressure). During BIC iontophoresis most of these intensity-response functions changed to a monotonically rising function. This effect was observed in units of all laminae of field L and the HV.

240

4) If both ears were stimulated independently with earphones sealed into the external ears, three types of binaural interactions were observed. The majority of units revealed excitatory responses to stimulation of either ear (EE-units). A second group of cells responded only to stimulation of the contralateral ear and this response was uninfluenced by additional ipsilateral stimulation (EO-units). The third group of cells showed an excitatory response to contralateral stimulation which was reduced or supressed by simultaneous ipsilateral stimulation (EI-units). During BIC-iontophoresis all tested units of the EO-type responded also to ipsilateral stimulation alone, i.e. changed to EE-units. The same effect was also seen in half of the EI-units. The other half of the EI-units had a typical EO-characteristic during BIC, i.e. the interaural inhibition was blocked, but ipsilateral stimulation alone did not excite the cells. These effects were also seen in units from all laminae of field L and the HV.

It is concluded from these findings that GABAergic inhibition is responsible for at least three response characteristics of telencephalic auditory units in the chicken:i. it focuses neuronal responses to narrow frequency ranges by lateral inhibition; ii. it limits the response strength of certain units at high intensity levels; and iii. it mediates inhibitory interaural interactions. While the first effect is usually seen only when the stimulus contains multiple frequency components, the latter two actions are elicited with single frequencies. Because the inhibitory effects differ also in other respects (e.g. the response limitation must be mediated by an inhibitory network with a high threshold), it is worthwhile assuming that these mechanisms reflect the actions of different GABAergic cell populations. Indeed, immunocytochemical studies suggest the presence of at least two types of GABAergic neurons in the auditory forebrain of the chicken, namely local interneurons and laterally projecting interneurons (Müller, 1988). I suggest that the former are the anatomical substrate for response limitation and interaural inhibition, while the latter are responsible for the lateral inhibition.

In vitro Studies

In order to test for the presence of multiple inhibitory actions I used an in vitro slice preparation of the auditory forebrain of the chicken. Intracellular recordings were obtained from neurons located in the lamina L1 of field L and the HV. First the effect of locally applied GABA on the membrane potential was investigated by positioning a GABA-containing pipette (10 mM in Ringer solution) at different locations near to an impaled cell. GABA was then applied by pressure ejection. Two effects were observed: i. If the GABA pipette was positioned remote from the impaled neuron, the effect of GABA application was a depolarization (Fig. 1A, application sites 30 and 150 μm below the surface) paralleled by a decrease of the membrane resistance. ii. If the pipette was positioned closer to the impaled cell, the depolarizing response was overlayed by an additional hyperpolarizing component (Fig. 1a, application sites 60, 90, and 120 μm below the surface). These two components could also be elicited by electrically stimulating the input to a given cell. Hyperpolarizing IPSP's were only seen in units lying in lamina L1 or the HV when a corresponding tonotopic location in lamina L2 was electrically stimulated (Fig. 1B). The reversal potential of this hyperpolarization was negative with respect to the resting potential. When the stimulating electrode was positioned laterally to the impaled cell in the same lamina (L1 or HV), IPSP's were always depolarizing (Fig. 1C) with reversal potentials more positive than the resting membrane potential. It is concluded from these findings that inhibitory neurons, activated by stimulation of lamina L2, induce hyperpolarizing inhibition in neurons lying in corresponding tonotopic locations in the postsynaptic structures. As hyperpolarization is seen only with GABA application close

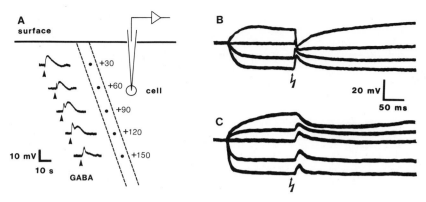

Fig. 1. A:Effect of locally applied GABA on the membrane potential
of a neuron in lamina L1. Pressure application of GABA from a
micropipette elicits a depolarization at all tested positions
(given by the spots along the oblique track; the numbers
correspond to the penetration depth from the surface of the slice
in microns). When GABA is applied close to the soma of the
impaled cell, an additional hyperpolarizing component is visible
with a short delay. B: Inhibitory potential in a neuron from
lamina L1 elicited by electrical stimulation of a corresponding
tonotopic location in L2. The potential is hyperpolarizing at
the cells resting membrane potential and reverses polarity at
slightly hyperpolarized levels. C: Inhibitory potential in a
neuron from L1 after electrical stimulation at a different
tonotopic location (1 mm lateral) in the same lamina. The potential is depolarizing at the cells resting membrane potential
and reverses polarity at depolarized levels. Resting membrane
potentials are -70 mV (A), -68 mV (B), and -72 mV (C).

to the soma of impaled neurons it is concluded that GABAergic cells
mediating hyperpolarizing inhibition preferentially synapse at somata or
proximal dendritic segments. In contrast, electrical stimulation of
remote tonotopic locations within a given structure elicit depolarizing
responses. This is compatible with the hypothesis that lateral inhibition
is mediated by GABAergic neurons synapsing preferentially at distal
portions of dendrites. The inhibitory effect of this action of GABA
probably results from the concomitant reduction of the dendritic membrane
resistance which prevents the propagation of EPSP's along a dendrite
(shunting inhibition).

The results of the iontophoretic and in vitro studies lead to the
model of the GABAergic inhibitory system in the chicken auditory
telencephalon shown in Fig. 2. A given neuron in the higher order laminae
of field L or the HV receives an excitatory input from a corresponding
tonotopic spot in lamina L2. In addition, inhibitory input is provided
by two GABAergic cell populations. One population consists of
local interneurons which receive their excitatory drive from a similar
tonotopic location than their target neuron. These cells synapse
preferentially near to the soma of the postsynaptic cell and mediate
hyperpolarizing inhibition. This inhibitory system is thought to be
involved in response strength limitation and interaural inhibition. A
second GABAergic input originates from laterally projecting interneurons

located in a different tonotopic location within the same lamina. These neurons terminate preferentially on distal dendritic segments, mediate depolarizing inhibition, and are thought to be responsible for lateral inhibition. The two-tone experiments revealed that lateral inhibition can act over three octaves. From quantitative investigations of the tonotopic organization of the chicken auditory forebrain it is known that

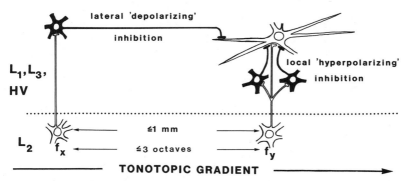

Fig. 2. Model of the GABAergic system in the auditory telencephalon of the chicken. Relay cells are indicated by open symbols, while GABAergic interneurons are represented by the filled symbols. See text for further explanation.

frequencies separated by three octaves are located about 1 mm apart from each other (Heil and Scheich, 1985). Thus it can be concluded that laterally projecting GABAergic interneurons can act over a distance of 1 mm. In summary, I propose that the auditory telencephalon of the chicken contains two GABAergic inhibitory systems with different mechanisms of action and which are involved in different aspects of auditory information processing.

REFERENCES

Bonke, B. A., Bonke, D. and Scheich, H., 1979a, Connectivity of the auditory forebrain nuclei in the Guinea fowl (Numida meleagris), Cell Tiss. Res., 200:101-121.
Bonke, D., Scheich, H. and Langner, G., 1979b, Responsiveness of units in auditory neostriatum of the Guinea fowl (Numida meleagris) to species specific calls and synthetic stimuli. I. Tonotopy and functional zones, J. Comp. Physiol., 132:243-255.
Heil, P. and Scheich, H., 1985, Quantitative analysis and two-dimensional reconstruction of the tonotopic organization of the auditory field L in the chick from 2-deoxyglucose data, Exp. Brain Res., 58:532-544.
Langner, G., Bonke, D. and Scheich, H., 1981, Neuronal discrimination of natural and synthetic vowels in field L of trained mynah birds, Exp. Brain Res., 43:11-24.
Müller, C. M., 1987, Differential effects of acetylcholine in the chicken auditory neostriatum and hyperstriatum ventrale-studies in vivo and in vitro, J. Comp. Physiol. A, (in press).
Müller, C. M. and Leppelsack, H. J., 1985, Feature extraction and tonotopic organization in the avian auditory forebrain, Exp. Brain Res., 59:587-599.
Müller, C. M. and Scheich, H., 1987, GABAergic inhibition increases the neuronal selectivity to natural sounds in the avian auditory forebrain, Brain Res., 414:376-380.
Müller, C. M. and Scheich, H., Interaural inhibitory response

        characteristics are reestablished in the avian auditory forebrain
        by a GABAergic mechanism, (subm.).
Müller, C. M. and Scheich, H., Contribution of GABAergic inhibition to
        the response characteristics of auditory units in the avian
        forebrain, (subm.).
Scheich, H., Langner, G. and Bonke, D., 1979a, Responsiveness of units in
        the auditory neostriatum of the Guinea fowl (Numida meleagris) to
        species specific calls and synthetic stimuli. II. Discrimination of
        iambus-like calls, J. Comp. Physiol., 132:257-276.
Scheich, H., Bonke, B. A., Bonke, D. and Langner, G., 1979b, Functional
        organization of some auditory nuclei in the guinea fowl
        demonstrated by the 2-deoxyglucose technique, Cell Tiss. Res.,
        204:17-27.

RESOLUTION OF COMPONENTS OF HARMONIC COMPLEX TONES BY

SINGLE NEURONS IN THE ALERT AUDITORY CORTEX

R. W. Ward Tomlinson and Dietrich W. F. Schwarz

Otoneurology Research Lab.
Div. Otolaryngology and Dept. Physiology
Univ. British Columbia
Vancouver, Canada

Humans perceive the pitch of a harmonic complex tone without its fundamental as identical with that of the fundamental presented alone (the phenomenon of the missing fundamental) (cf. de Boer, 1976). In psychophysical experiments, rhesus monkeys were able to match harmonic complex tones with fundamentals to the same without fundamentals (Tomlinson and Schwarz, 1987). It was concluded that rhesus monkeys could perceive the pitch of the missing fundamental. This percept is generally thought to arise within the central nervous system (Houtsma and Goldstein, 1973; Terhardt, 1972). We studied the auditory cortex of alert monkeys to determine how stimulus features relevant to pitch perception might be encoded. Single neurons in auditory cortex of unanesthetized monkeys were isolated, the units' sensitivity to pure tones and noise stimuli were determined, and sideband inhibition was assessed, typically with two tones. We then applied various complex tones and compared the responses to those obtained using the previous simpler stimuli.

The pure and complex tones were generated by a 16 bit digital synthesizer with a 70 kHz sample rate. The sounds were presented to the animals from an overhead speaker in a sound attenuated room. A PDP 11/23 PLUS computer was used to control stimuli and collect response patterns.

A stainless steel chamber was surgically implanted vertically above the auditory cortex and the animal was allowed to recover for at least one week before recording sessions began. Epoxylite coated tungsten microelectrodes were lowered through the dura mater in a guard tube and then advanced downwards to the superior temporal plane. We searched for units with noise and tone burst stimuli. The responses were digitized and plotted in dot raster format. Each dot in such a plot represents a spike occurring at the time shown on the abscissa in response to a tone of a frequency identified on the ordinate along a logarithmic scale.

Fig. 1 shows a series of responses of one neuron to pure and complex tones taken at several different intensities, between 10 db SPL and 60 dB SPL. This unit has an extremely narrowly tuned response area for pure tones (right column) with a characteristic frequency of 4.2 kHz. The bandwidth of the response broadens only slightly at higher intensities. In the complex tone rasters (left column) one sees that the cell responds to several discrete ranges of fundamental frequency. Close analysis of the frequency histograms to the right of each raster shows six distinct

PURE TONES

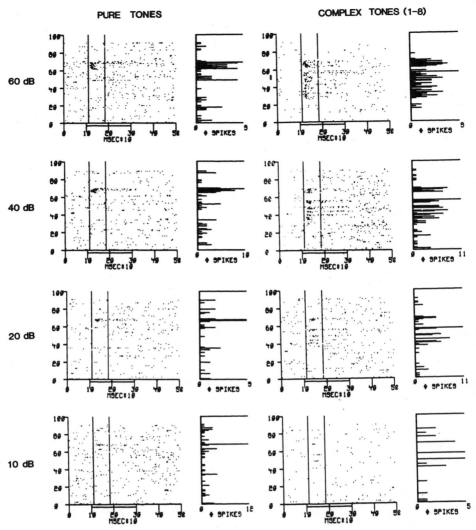

Fig. 1. Dot rasters showing the response of a single neuron to pure tones
and complex tones. Left column: responses to pure tones. The
ordinate represents frequency from 80 (frequency no. 1) to
18000 Hz (frequency no. 91) on a logarithmic scale. The abscissa
represents time (bin width, 2 milliseconds). The bar on the
ordinate shows the presence of a 200 ms tone. Each dot represents
one neural event. Right column: response of the neuron to an
8 component harmonic series. The ordinate represents the
fundamental frequency of the complex on a logarithmic scale from
80 to 18000 Hz. The abscissa is as on the left. The intensity
of each plot is shown on the left. Histograms: Frequency histograms are generated by counting the events between the two lines
on the dot rasters.

peaks, best seen at 40 and 20 dB. At 60 dB several bands merge and at
10 dB the response starts to disappear in the background firing rate. The
upper bands are the more distinct and than the lower bands.

    The complex tone stimuli used were an eight component harmonic series starting with the fundamental (f0). As the fundamental is increased, the first component of a complex to enter the unit's excitatory receptive field will be the eighth harmonic. Since the responses are plotted on a fundamental frequency ordinate, the first response (horizontal row on the dot rasters) should be seen at a position on the y-axis corresponding to one eighth of the unit's best frequency (BF). The next response should be seen at one seventh BF and so on up to the BF itself. Thus frequency histogram peaks are expected at ordinate positions forming a complete subharmonic series of BF. The center frequencies of the bands have been tabulated with the corresponding subharmonics of the BF for the 40 dB response in Table I. The close correspondence of histogram peak values with subharmonics indicates that the first six components of the eight component complexes cause an excitation whenever their frequencies are close to the unit's BF.

Table I. Observed center frequencies of response bands at 40 dB and corresponding subharmonics of the pure tone best frequency.

Observed peak frequencies (Hz)	Predicted frequencies (Hz)
4200	4200
2115	2100
1408	1400
1130	1050
841	840
712	700

Which characteristics of cortical neurons can give rise to such responses? Fig. 2 illustrates the effect of the bandwidth of the a hypothetical neuron's pure tone response area on response patterns to complex tones. We used linear Gaussian filters to simulate the neuronal response at three bandwidths, with center frequencies of 4.2 kHz. The bandwidth employed in the top row is 3 times greater than that of the bottom row. For the bandwidth of the top row (one octave), comparable to the more narrowly tuned cortical units we have seen, only three distinct peaks are visible. Six peaks are present in the middle row complex tone response, whose pure tone response bandwidth (o.5 oct.) is comparable to the 40 dB response of the neuron in Fig. 1, and eight peaks can be distinguished in the bottom row (0.3 oct.). The lowest components are resolved best, as one sees in the neural response. The number of the highest resolvable harmonic is inversely proportional to the filter bandwidth. This is illustrated in the response at 60 dB in Fig. 1, where the broadening of the pure tone response bandwidth causes a decrease in the number of discriminable histogram peaks when compared to the 40 dB response.

The major determining factor in the ability of a neuron to resolve components in a complex is the existence of sufficiently narrow tuning in the pure tone response. The large majority of cortical neurons we have seen (ca. 95%) do not show the discrete responses to the components of a complex tone seen in Fig. 1, since in general they do not possess the requisite sharpness of tuning. Humans can discriminate the presence of up to 8 components of a harmonic complex (Plomp, 1964) with the lowest components being the easiest to resolve. This resolution extends up to levels of 80 dB (eg. Pick, 1977; Scharf and Meiselman, 1977), above which it deteriorates. The best neurons we have seen can resolve 6 components on

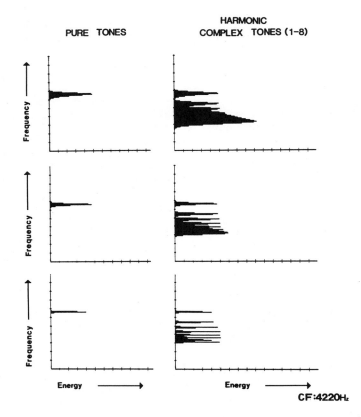

PURE TONES   HARMONIC
                    COMPLEX TONES (1-8)

Energy ⟶   Energy ⟶

CF:4220Hz

Fig. 2.   The response of three gaussian filters to pure tones (left) and
          eight component complex tones (right). Each row shows one filter
          bandwidth. The center frequency of the filters is 4.2 kHz. The
          ordinates are as in Fig. 1. The abscissas represent the amount
          of energy passed by the filters, on a linear scale (arbitrary
          units).

the basis of firing rate, over a dynamic range of approximately 40 dB.

     In order to account for the human capacity on the basis to the
neuron type shown in Fig. 1, an array of units tuned to all relevant
frequencies with narrower bandwidths that those observed in the monkey
cortex would have to be postulated. It is unknown if these monkeys can
resolve more than 6 components.

     In summary, this population of units in the alert auditory cortex
can resolve components of a complex tone almost as well as humans in
psychophysical experiments.

ACKNOWLEDGMENTS

     This research is being supported by the Pacific Otolaryngology
Foundation and the Medical Research Council of Canada.

REFERENCES

de Boer, E., 1976, On the residue and auditory pitch perception, in:
     "Handbook of Sensory Physiology", Vol. V/3, 479-583.
Goldstein, J. L., 1973, An optimum processor theory for the central

formation of the pitch of complex tones, J. Acoust. Soc. Am., 54: 1496-1516.

Houtsma, A. J. M. and Goldstein, J. L., 1972, The central origin of the pitch of complex tones: Evidence from musical interval recognition, J. Acoust. Soc. Am., 51:520-528.

Pick, G. F., 1977, Comment on paper by Scharf and Meiselman, in: "Psychophysics and Physiology of Hearing", P. Wilson and E. F. Evans, eds., Academic Press, London, 233-234.

Plomp, R., 1964, The ear as a frequency analyser, J. Acoust. Soc. Am., 36:1628-1636.

Scharf, B. and Meiselman, C. H., 1977, Critical bandwidth at high intensities, in: "Psychophysics and Physiology of Hearing", P. Wilson and E. F. Evans, eds., Academic Press, London, 221-232.

Terhardt, E., 1972, Zur Tonhoehenwahrnehmung von Klangen, Acustica 26: 173-199.

EFFERENT AUDITORY SYSTEM

PHYSIOLOGY OF THE OLIVOCOCHLEAR EFFERENTS

John J. Guinan Jr.

Electrical Engineering and Computer Science Dept.
and Research Laboratory of Electronics
Massachusetts Institute of Technology; and
Eaton-Peabody Lab, Massachusetts Eye & Ear Infirmary
243 Charles St. Boston MA, 02114, U.S.A.

Our knowledge of auditory efferents has changed greatly over the past few years. A new conception of the anatomical organization of the efferents has been developed and it has been demonstrated that efferents can produce mechanical changes in the cochlea. These, in turn, have motivated experiments intended (1) to determine the cochlear effects produced by each of the new efferent groups and (2) to determine the nature of the mechanical changes produced by efferents. In this paper, I review some recent experiments and the substantial changes which they have produced in our understanding of the physiology of olivocochlear efferents.

There are two major systems of olivocochlear efferents: the medial olivocochlear (MOC) system which originates in the medial part of the superior olivary complex and projects to the outer hair cells (OHCs), and the lateral olivocochlear (LOC) system which originates in or near the lateral superior olivary nucleus and projects to the dendrites of auditory-nerve fibers (Warr and Guinan, 1979; Guinan et al., 1983). Both of these systems have crossed and uncrossed projections. In particular the crossed-MOC system has approximately 2.5 times as many fibers as the uncrossed-MOC system (Warr et al., 1982).

Many of the experiments which have explored efferent effects in the cochlea have excited the efferents with an electrode where the fibers cross at the midline of the floor of the fourth ventricle (Fig. 1). Such stimulation has usually been called crossed-OCB, or COCB, stimulation, but as we shall see, this is probably a misnomer. I will call this "midline-OCB stimulation" or simply "OCB stimulation". At the midline, there are crossed-MOC and crossed-LOC fibers, and uncrossed-MOC fibers pass close by. Another important consideration is that most, or all, of the MOC fibers are myelinated and most, or all, of the LOC fibers are unmyelinated (Guinan et al., 1983). Since unmyelinated fibers have very high thresholds for excitation by extracellular shocks and do not respond well at high stimulus rates (e.g. over 100/sec), the unmyelinated crossed LOC fibers may not be stimulated with the normal midline-OCB stimulation. One goal of the experiments I will describe was to determine whether this conjecture is true. Another goal was to compare the effects produced by uncrossed versus crossed MOC fibers. It has been reported that crossed-OCB fibers produce an increase in the cochlear microphonic but uncrossed-

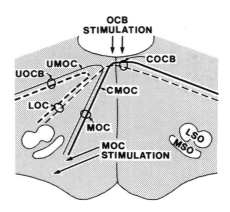

Fig. 1. Schematic brainstem cross-section showing the efferents which
originate on the left side in the cat, and the locations of
stimulating electrodes. Abbreviations for all figures: CMOC:
crossed medial olivocochlear efferents, COCB: crossed olivocochlear bundle, LOC: lateral olivocochlear efferents, LSO:
lateral superior olivary nucleus, MOC: medial olivocochlear
efferents, MSO: medial superior olivary nucleus, OCB: midline
olivocochlear bundle, UMOC: uncrosed medial olivocochlear
efferents, UOCB: uncrossed olivocochlear bundle (Gifford and
Guinan, 1987).

OCB fibers do not (Sohmer, 1966; Fex, 1967).

　　To answer these questions, we stimulated MOC fibers with an electrode
at their brainstem origin (Gifford and Guinan, 1987) where MOC neurons
and processes are physically separate from LOC neurons and processes
(see Fig. 1). Our data are consistent with the interpretation that such
"MOC stimulation" is selective in activating MOC efferents and not
LOC efferents. I will refer to the stimulation as "uncrossed-MOC" (UMOC)
stimulation when the recording electrode was on the same side as the

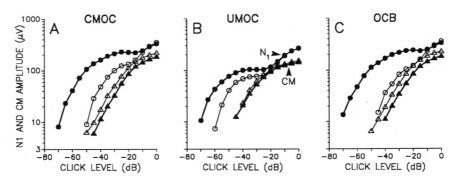

Fig. 2. Typical effects of efferent stimulation on cochlear responses
as functions of click level. Circles: N_1 amplitudes; triangles:
CM amplitude. Filled symbols and thick lines:without efferent
stimulation; open symbols and thin lines: with efferent
stimulation (from Gifford and Guinan, 1987).

stimulating electrode and as "crossed-MOC" (CMOC) stimulation when the recording electrode was on the opposite side. We also tried electrical stimulation at the origin of LOC fibers. In these experiments, we found only small shock-induced effects which could most easily be explained as due to stimulation of distant MOC neurons or fibers (Guinan and Gifford, unpublished).

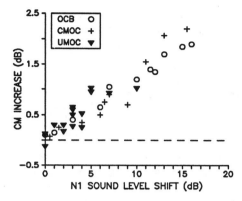

Fig. 3. CM increase vs. N_1 sound-level shift for efferent stimulation at different locations. CM increase was measured at mid-range sound levels and N_1 shift was measured at 10 dB above the N_1 threshold. Efferent shock levels were adjusted to produce points with different N_1 sound-level shifts (Gifford and Guinan, 1987).

Fig. 2 shows the effects of MOC and midline-OCB stimulation on N_1 and CM. Efferent stimulation at each location reduced amplitude of N_1 and increased the amplitude of CM. To look in more detail at the UMOC-induced CM increase, we compared effects on CM with effects on N_1 for electrical stimulation at each location (Fig. 3). This figure shows that points from UMOC stimulation (triangles) fall along the same line as the points for stimulation at the other locations. We do note that the UMOC-induced CM increases are small, averaging 1/2 dB, and could easily have been missed in previous experiments in which the responses were not averaged. In any event, these data show that CMOC and UMOC stimulation, and midline-OCB stimulation, all produce qualitatively similar effects.

Stimulation at the different sites, however, produced effects with different amplitudes (Fig. 4). The average N_1 level shift with UMOC stimulation was 5 1/2 dB, and the average shift with CMOC stimulation was 14 1/2 dB. Thus, CMOC stimulation produced an average effect which was about 2 1/2 times stronger than the average effect of UMOC stimulation. This is approximately equal to the ratio of the number of MOC neurons which project to the opposite cochlea compared to the number of MOC neurons which project to the ipsilateral cochlea (Warr et al., 1982). Thus, it appears that, on a per neuron basis, stimulation of CMOC or UMOC neurons produces equivalent effects, both qualitatively and quantitatively.

Fig. 4 also shows that the average N_1 shift with midline-OCB stimulation was 20 dB. This is almost exactly equal to the sum of the effects of both CMOC and UMOC stimulation. Considering the amount of variation from cat to cat, the exactness of the sum must be partly by chance. However, considering that many UMOC fibers course close to the midline-OCB electrode, it seems almost certain that many UMOC fibers are also stimulated by the midline-OCB electrode. This is one reason for calling this stimulation "midline-OCB stimulation" and not "COCB stimulation".

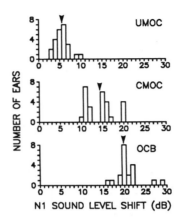

Fig. 4.   The distribution of the N_1 sound-level shifts for UMOC, CMOC and midline-OCB stimulation. The bar heights represent the number of ears with the corresponding shifts. Data from 52 ears in 31 cats. Arrows indicate the average shifts (Gifford and Guinan, 1987).

Another reason, for avoiding the term "COCB stimulation" is that crossed-LOC fibers do not appear to be stimulated, or if they are, there is no effect which seems attributable to them. Both the inhibition of N_1 and the increase in CM produced by midline-OCB stimulation are attributable to the effects of MOC fibers. In addition, intravenous injections of strychnine blocked the effects of stimulation at all three locations to the same degree and with the same time course (Gifford and Guinan, 1987) which suggests that similar mechanisms are involved in all of these. Finally, stimulation at all three locations also produced a decrease in endocochlear potential which, like the other effects, was largest for midline-OCB stimulation and smallest (but definitely still present) for UMOC stimulation (Gifford and Guinan, 1987). This decrease in endocochlear potential is attributable to MOC synapses on OHCs (see Fex, 1967).

All of the effects we have presented so far have been on "gross" potentials from the cochlea. We have done similar experiments in which we determined the effects of MOC stimulation on the firing patterns of single auditory-nerve fibers. We obtained sound-level functions with and without efferent stimulation for tone-bursts at the characteristic frequency (CF). One pair of level functions showing a large depression of plateau rate is shown in Fig. 5. From level-function pairs, we measured the efferent-induced shift of the rising phase along the sound-level axis, ΔL, and the efferent-induced change in the plateau (saturation) rate, ΔR_p (Fig. 5).

Fig. 6 shows the efferent-induced sound-level shift, ΔL, as a function of fiber CF, for efferent stimulation in the three different locations. There are two things to note from this figure. First, the effect is largest at mid-range CFs, approximately 2-10 kHz, and is much less at 20 kHz. We will return to this point later. Second, UMOC stimulation produced smaller effects, on the average, than CMOC stimulation which,

in turn, produced smaller effects than midline-OCB stimulation. It would
be nice to be able to make a quantitative comparison of these effects,
but there are cat to cat amplitude variations (see Fig. 4), and any
comparison must take into account the strong dependence of the effect

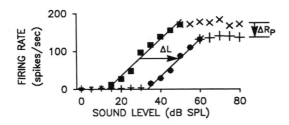

Fig. 5.   Rate versus sound-level functions from a single auditory-nerve
          fiber with (+) and without (x) CMOC stimulation. The filled
          circles and squares show the points chosen by a computer program
          to be in the rising phases; regression lines were fitted to these
          points. The sound-level shift, ΔL, was measured between these
          lines at the middle of the rate range. Points in the plateau
          are shown with larger + or x symbols. The change in the plateau
          rate, ΔR_p, was measured by averaging the spikes only from the
          sound levels at which both plateaus overlapped (Guinan and
          Gifford, 1988a).

on CF which is difficult because the CF range varied from cat to cat. We
can say, however, that these single-fiber data appear to be consistent
with the hypotheses that individual CMOC and UMOC fibers produce similar
effects and that the effects of midline-OCB stimulation are due to both
CMOC and UMOC fibers.

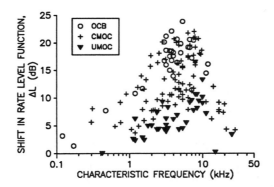

Fig. 6.   Efferent-induced sound-level shift, ΔL, versus CF. Data from
          161 fibers from 7 cats (Guinan and Gifford, 1988a).

     We turn now to efferent-induced changes in plateau rate, ΔR_p. These
show a dependence on fiber CF which, as a first approximation, is similar
to the CF dependence of ΔL shown in Fig. 6. However, in contrast to ΔL
which shows only a small dependence on fiber spontaneous rate (SR), ΔR_p
depends strongly on spontaneous rate. Fig. 7 shows that the depression
of the plateau rate is largest, on the average, for low- and medium-SR
fibers. To compare ΔR_p's from stimulation in different places, we must

keep in mind that ΔR_p's are small and in many cases the expected measurement error due to the randomness of the spikes is comparable in magnitude to the measured changes. If Fig. 7 is restricted to the most accurate points, the scatter of points from each stimulus location is much less (most points shown in Fig. 7 as efferent-induced increases in ΔR_p disappear). All considered, the most we can say is that, on the

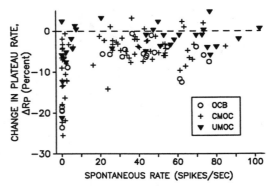

Fig. 7. Efferent-induced change in plateau rate, ΔR_p, versus spontaneous rate. 130 fibers from 7 cats (Guinan and Gifford, 1988a).

average, UMOC stimulation evoked the smallest changes, CMOC stimulation was next, and midline-OCB stimulation evoked the largest changes. This is consistent with the other effects produced by stimulation in these locations.

It is important to note that the finding of plateau shifts is not evidence for the involvement of LOC fibers. First, it is possible to suggest plausible mechanisms by which synapses on OHCs might produce plateau depressions (e.g. efferent-induced electrical changes in the cochlea such as those discussed later might be very important in this). Second, the plateau depressions are larger for CMOC stimulation than for UMOC stimulation, and they are still larger for midline-OCB stimulation. This is what we should expect if the plateau depression was produced by MOC neurons. In contrast, LOC neurons are far more numerous on the ipsilateral side than on the contralateral side. Finally, the plateau shifts are largest for auditory-nerve fibers with mid-range CFs. This pattern is most similar to the pattern of innervation of MOC fibers on OHCs. In contrast, uncrossed-LOC fibers project relatively evenly over the length of the cochlea, and crossed-LOC fibers project predominantly to the apex (Guinan et al., 1984). Thus, the data are consistent with the efferent-induced plateau shifts being produced by the MOC synapses on OHCs and are not evidence for LOC neuron involvement. However, we can not rule out that the plateau shifts are due to efferent synapses on the dendrites of auditory-nerve fibers.

All of the efferent effects presented appear to be due to MOC neurons. Considering our data, and all of the data in the literature, there is no known effect which can be definitely attributed to crossed or to uncrossed LOC neurons, even though LOC neurons are more numerous than MOC neurons. Perhaps this is because LOC neurons have unmyelinated fibers which are not stimulated or perhaps LOC neurons only produce very slow effects in the cochlea.

MECHANISMS

We have concluded that all known efferent effects are likely to be due to MOC synapses. In the cochlea, MOC fibers correspond to the "thick efferents" and, most, if not all, MOC synapses end on OHCs (see Guinan et al., 1983; Liberman and Brown, 1986; Ginzberg and Morest, 1983; Spoendlin, 1966; Dunn, 1975; Liberman, 1980). For the purposes of this paper, I will consider only MOC synapses on OHCs (Fig. 8). This will emphasize that all of the efferent effects described could be produced by MOC synapses on OHCs.

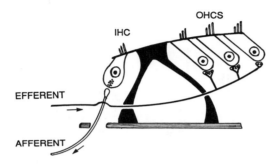

Fig. 8.   Schematic showing the relationship of MOC efferents and radial afferents.

The evidence is now quite compelling that N_1 and the normal recordings from single auditory-nerve fibers are from the myelinated radial afferent fibers which innervate inner hair cells (IHCs) (Fig. 8), and not from the unmyelinated spiral afferent fibers which innervate OHCs (see Liberman, 1982). The question to be considered in the rest of this paper is how do efferents on OHCs inhibit afferents which innervate IHCs?

Some insight into this question was obtained from intracellular recordings in guinea-pig IHCs by Brown and Nuttal (1984). They measured IHC receptor potentials in response to CF tones as a function of sound pressure level both with and without midline-OCB stimulation. Brown and Nuttal found that efferent stimulation shifted the rising phases of the level functions to higher sound levels, and that it usually produced a small depression in the receptor potential in the plateau region. These effects were found without there being any efferent-induced change in the IHC conductance which means that the changes were not due to efferent synapses directly on the IHCs. Furthermore, the IHC receptor potential comes too early in the chain of events for synapses on dendrites of the auditory-nerve fibers to cause the observed changes. Thus, these data provide additional evidence for the hypothesis that the effects of midline-OCB stimulation are due to MOC synapses on OHCs. These data show that efferents act at a site which is earlier than the IHC receptor potential, but they do not indicate the nature of the coupling mechanism from OHCs to IHCs.

From measurements of sound pressures in the earcanal, Mountain (1980) and Siegel and Kim (1982) have shown that midline-OCB stimulation produces mechanical changes in the normal cochlea. This has suggested that the coupling between OHCs and IHCs might be mechanical. Brownell et al. (1985) provided the first direct evidence that OHCs can be mechanically active. Fig. 9 shows some measurements I have made of efferent-induced cochlear mechanical changes monitored in the ear canal by the efferent-evoked sound-pressure change, Poc, with a single tone sound stimulus (Guinan, 1986). Presumably, part of the sound which travels into the cochlea is "reflected" within the cochlea, travels backward through the middle ear, and appears in the earcanal. Such a

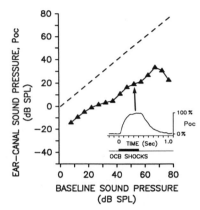

Fig. 9.   Ear-canal sound pressure, Poc, evoked by midline-OCB stimulation as a function of the baseline sound level. Poc is the sound which when added to the baseline sound produces the measured sound. The dashed line shows the baseline sound level. The inset shows the time course of the Poc magnitude at one baseline sound level (data from Guinan, 1986).

sound in the earcanal has been called a "synchronous evoked emission" (Kemp and Chum, 1980). If efferent activity changed synchronous evoked emissions, then changes in earcanal sound pressures exactly like the Poc's we have measured would be produced. Whatever the exact mechanism, the fact that efferent activity can change earcanal sound pressures means that efferents must produce mechanical changes within the cochlea. We note that relative to the sound put into the earcanal (the dashed line in Fig. 9) the biggest efferent-evoked Poc's were at low sound levels (that is, the two curves are closest at low sound levels). This is consistent with the mechanical change in the cochlea being produced by MOC synapses on OHCs, and current theories which suggest that OHCs produce their greatest change on cochlear mechanics at low sound levels.

Data such as the above have led to the hypothesis that efferent synapses on OHCs change the mechanics of the cochlea such that a given sound produces less mechanical drive to IHCs. This would produce the sound-level shifts seen both in IHC receptor potentials and in the firing of auditory-nerve fibers. The mechanical change could come about in either of two ways: (1) With no change in the motion of the basilar membrane, the efferents could decrease the coupling of this motion to the IHC stereocilia, or (2) The efferents might decrease the motion of the basilar membrane. Both of these mechanical coupling hypotheses are compatible with the data presented so far.

We will now examine in more detail some effects of efferent stimulation on the firing of single auditory-nerve fibers to see whether the mechanical-coupling hypotheses can account for all of the phenomena observed. Fig. 10 is a summary of the data I will consider. Fig. 10A shows

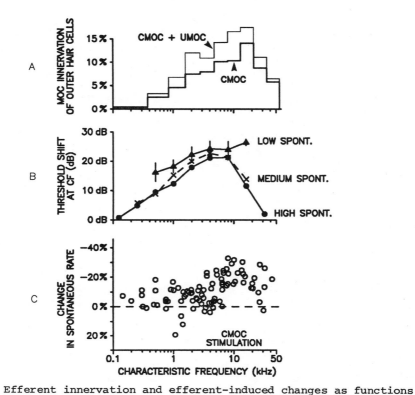

Fig. 10. Efferent innervation and efferent-induced changes as functions
of CF.

A: the density of OHC innervation from MOC fibers derived from
anterograde transport (Fig. 4b of Guinan et al., 1984). Thick
line: CMOC data, thin line: CMOC+UMOC data. Data replotted as
a function of CF using the cochlear place to CF map (based
on afferent CFs) of Liberman (1982).

B: efferent-induced threshold shift at CF, as a function of
fiber CF, for fibers in the three spontaneous rate (SR) classes
of Liberman (1978). Each point shows the average over an octave
band. Error bars: +/- one s.e.m. (high-SR error bars are mostly
smaller than the point size, medium-SR error bars (omitted for
clarity) are intermediate). Data from 400 auditory-nerve fibers
(286 with midline-OCB stimulation, 114 with CMOC stimulation).
To weigh all cats evenly before averaging, data from each cat
were normalized to an N_1 shift of 20 dB, i.e. for a cat with
an N_1 shift of 15 dB, all of the threshold shifts were multi-
plied by 20/15 (from Guinan and Gifford, 1988c).

C: The change in spontaneous rate in the first second after
the onset of CMOC stimulation as a function of fiber CF (data
from Guinan and Gifford, 1988b).

that the peak density of the MOC innervation of OHCs is for CFs of about
10-20 kHz and that the density is high from about 2 to 30 kHz. This plot
should represent the density of efferent endings on OHCs which are
activated by CMOC or midline-OCB stimulation, since there is no reason
to think that such stimulation should be biased toward any frequency
region.

Fig. 10B shows the average efferent-induced threshold shift at CF as a function of fiber CF, for fibers in the three spontaneous rate classes of Liberman (1978). We are using the threshold shift rather than the level shift, ΔL, (Fig. 6) because we have more data on threshold shifts. Both measures are nearly the same because efferent activity produces a nearly parallel shift in the rising phase of level functions in the cat (Wiederhold, 1970; Gifford and Guinan, 1983). Fig. 10B shows that in every CF region, low-SR fibers had larger threshold shifts than high-SR fibers.

The threshold shifts for high-SR fibers appear to be displaced to lower CFs relative to the MOC innervation of OHCs (Fig. 10A, B). The peak is at a lower CF. Particularly striking is the difference at 16 kHz where the innervation is at its maximum but the high-SR threshold shifts are only half of their maximum. The Comparison is quite different for low-SR fibers. The threshold shifts of these fibers appear to peak at 16 kHz which is near the peak of the MOC innervation. Unfortunately, we do not have sufficient low-SR data in the 32 kHz octave.

Finally, Fig. 10C shows that CMOC stimulation inhibited spontaneous rate by at most 30% in data from four cats[1]. The maximum efferent-induced change in spontaneous rate was in the 10-20 kHz range. This is close to the CF range with the maximum density of CMOC innervation (Fig. 10A), but is distinctly higher than the CF region which shows the maximum threshold shift of high-SR fibers (Fig. 10B).

We now consider the mechanisms which might account for the data in Fig. 10. First, an efferent-induced reduction of the sound drive to IHCs cannot explain the efferent-induced reduction in spontaneous rate (Fig. 10C) because there is no sound in this case. We propose that this efferent inhibition of spontaneous rate is due to electrical coupling between OHCs and IHCs as is shown in Fig. 11. Normally, the positive endocochlear potential (EP) drives a resting current through the OHCs. Activation of MOC synapses on OHCs increases the OHC conductance which, in turn, increases the OHC current and decreases EP (e.g. Fig. 11C). Such an efferent-induced decrease in EP has been observed many times (Fex, 1967; Konishi and Slepian, 1971; Teas et al. 1972; Brown and Nuttall, 1984; Gifford and Guinan, 1987). Now consider the effect of this decrease in EP on the IHCs. Normally, the positive EP also drives a resting current through the IHCs which depolarizes the IHCs. An efferent-induced reduction of EP will reduce the resting current and hyperpolarize the IHCs (Fig. 11B). If the "true spontaneous activity" of auditory-nerve fibers is due to spontaneous release of transmitter by IHCs (see Siegel and Dallos, 1986) then a hyperpolarization of IHCs should decrease spontaneous activity. Such a mechanism is suggested by the data of Sewell (1984) who found that furosemide-induced reductions of EP reduced spontaneous rate by approximately 10% for every 1 mV reduction in EP. This fits well quantitatively with the typical 2-3 mV changes found in EP from CMOC stimulation and the maximum 30% CMOC-induced reduction of spontaneous rate (Fig. 10C). Thus, we propose that MOC stimula-

---

[1] Our experiments included one other very-sensitive cat which had much larger inhibitions, up to 70%, with the largest inhibitions in the 3-7 kHz CF range. We interpret the data from this cat as showing large inhibitions of spontaneous activity because some of the "apparent spontaneous" firings were actually responses to background sounds and that these were inhibited strongly by CMOC stimulation (such a hypothesis was first proposed by Wiederhold and Kiang, 1970). To restrict our data as much as possible to the effects of CMOC stimulation on "true spontaneous activity", the data from this one aberrant cat have been excluded from Fig. 10C.

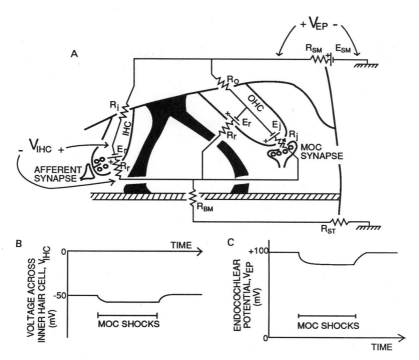

Fig. 11. Suggested mechanism for the suppression of spontaneous activity
by MOC efferents. A: A model of the cochlea showing the electrical connection between OHCs and IHCs (symbols as in Geisler,
1974, from which this was adapted). B: Hypothetical IHC voltage
change produced by MOC shocks (exaggerated in amplitude). C:
Endocochlear potential change produced by MOC shocks (based
on measurements, e.g. Gifford and Guinan, 1987, but exaggeraged
for clarity).

tion reduces spontaneous activity through a hyperpolarization of IHCs
produced by the efferent-induced reduction of EP. Variations of this idea
have been proposed before (e.g. Geisler, 1974; Sewell, 1984).

Let us now consider the distribution along the length of the cochlea
of the efferent inhibition which might be produced by such electrical
coupling. First, we consider the effects which would be produced if
efferents innervating only a narrow region of OHCs could be activated
(Fig. 12A). Electrically coupled effects would spread with the electrical
space constant as shown in Fig. 12B. Although the spread may not be as
symmetric as shown, there should be appreciable spread toward both the
apex and the base. Efferent-induced changes in the mechanics of the organ
of Corti might spread very differently. A mechanical change which reduced
the coupling of basilar membrane motion to IHC stereocilia might be
expected to spread little from the region of the OHCs producing the
mechanical change (not illustrated in Fig. 12). On the other hand, if
the efferent-induced mechanical change caused a reduction in the motion
of the basilar membrane, this reduction in motion would be carried apically by the traveling wave so that the efferent effect would spread apically
from the active OHC synapses (Fig. 12C). Cochlear models suggest that
OHC amplification of basilar membrane motion would be carried apically
in the cochlea (see Neely and Kim, 1983; Kemp, 1986; Kim, 1986).

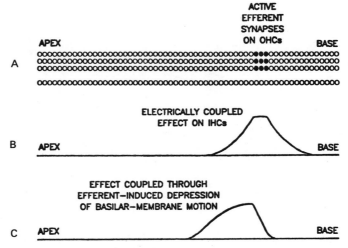

Fig. 12. Hypothetical distribution along the length of the cochlea of effects produced by activation of a narrow band of efferent synapses on OHCs. See text.

Let us apply these ideas to interpreting the data shown in Fig. 10. With our hypotheses, the CMOC-induced change in spontaneous activity is maximum in the 10-20 kHz region (Fig. 10C) because this is the region which receives the greatest CMOC innervation (Fig. 10A) and the change is coupled to IHCs roughly symmetrically about the activated region (Fig. 12A,B). We do not know why the inhibition of spontaneous activity appears to be relatively low at 2-5 kHz whereas the innervation decreases evenly with decreasing CF.

The apical displacement of the efferent-induced threshold shifts of high-SR fibers (Fig. 10B) does not fit with the hypothesis that an efferent-induced mechanical change decreases the coupling of basilar membrane motion to IHCs because such an effect would be expected to produce a CF distribution similar to the MOC innervation of OHCs (Fig. 10A). We interpret the efferent-induced threshold shifts of high-SR fibers as being due to an efferent-induced reduction in the motion of the basilar membrane. This accounts for the apical displacement of the high-SR threshold shifts compared to the MOC innervation of OHCs (Fig. 10 A,B) because changes in basilar membrane motion are carried apically relative to their place of origin (Fig. 12A,C). High-SR fibers have the lowest thresholds (in dB SPL) and might therefore be expected to be most affected by efferent-induced changes in basilar membrane motion since these changes are probably greatest (on a dB scale) at low sound levels (see Fig. 9).

The efferent-induced inhibition of low-SR fibers (which have the highest thresholds in dB SPL) may not be displaced at all relative to the efferent innervation pattern (Fig. 10 A,B). Considering that (1) the hypothesized electrical coupling (Fig. 11) should operate at all sound levels, (2) low-SR fibers might be more sensitive than other fibers to small voltage changes in IHCs (Sewell, 1984), and (3) efferent-induced mechanical changes may be relatively small at higher sound levels (see Fig. 9), it seems reasonable to suggest that much of the efferent-induced inhibition of low-SR fibers might be due to electrical effects such as those shown in Fig. 11. Of course, other factors may also play a role such as efferent-induced local mechanical effects and efferent-induced

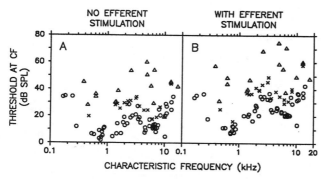

Fig. 13. The overall effect of midline-OCB stimulation on thresholds at CF in one cat. Δ :low-SR fibers. X: medium-SR fibers. O: high-SR fibers (Guinan and Gifford, 1988c).

changes in the currents around the dendrites of auditory-nerve fibers.

The effects of efferent stimulation on the thresholds of auditory-nerve fibers in one cat are shown in Fig. 13. In contrast to the common supposition that efferent effects are only present at low sound levels, Fig. 13 shows that efferent stimulation can have substantial effects over a wide range of sound levels. These data support the idea that one role of the MOC efferents is to help extend the useful dynamic range of the auditory system. It seems likely that the efferent inhibition at low sound levels is due primarily to an efferent-induced decrease in basilar membrane motion, whereas the efferent inhibition at high sound levels is due primarily to other mechanisms which probably include electrical effects produced by the efferent synapses on OHCs and coupled through EP.

SUMMARY CONCLUSIONS

1. Individual CMOC and UMOC efferent fibers produce similar effects.
2. The effects of midline-OCB stimulation probably include contributions from both CMOC and UMOC efferents.
3. The effects of midline-OCB stimulation are due only to MOC efferents.
4. There are no known effects which can be definitively attributed to LOC efferents. Every effect which has been reported can be explained as being produced by MOC efferents.

SUMMARY HYPOTHESES

1. The MOC-induced hyperpolarization of OHCs is electrically coupled to IHCs and spreads both apically and basally.
2. MOC synapses on OHCs reduce the motion of the basilar membrane at the location of the synapses and more apically.
3. The combined effects of the two OHC to IHC coupling mechanisms produces efferent inhibition of auditory-nerve responses over a wide range of sound levels; mechanically at low sound levels, and electrically at high sound levels.

ACKNOWLEDGEMENTS

Much of the work described here was done with M. L. Gifford. I thank the members of the Eaton-Peabody Laboratory for help throughout this work, particularly Drs. J. J. Rosowski, and M. C. Liberman for their comments on the manuscript. This work was supported by NINCDS grants RO1 NS-20269 and PO1 NS13126.

REFERENCES

Brown, M. C., and Nuttall, A. L., 1984, Efferent control of cochlear inner hair cell responses in the guinea-pig, J. Physiol., 354:625-646.

Brownell, W. E., Bader, C. R., Bertrand, D. and de Ribaupierre, Y., 1985, Evoked mechanical responses of isolated cochlear outer hair cells, Science, 227:194-196.

Dunn, R. A., 1975, A comparison of Golgi-impregnated innervation patterns and fine structural synaptic morphology in the cochlea of the cat. Ph. D. Dissertation, Harvard University.

Fex, J., 1967, Efferent inhibition in the cochlea related to hair cell dc activity: study of postsynaptic activity of the crossed olivocochlear fibers in the cat, J. Acoust. Soc. Am., 41:666-675.

Geisler, C. D., 1974, Model of crossed olivocochlear bundle effects, J. Acoust. Soc. Am., 56:1910-1912.

Gifford, M. L., and Guinan, J. J., Jr., 1983, Effects of crossed-olivo-cochlear-bundle stimulation on cat auditory nerve fiber responses to tones, J. Acoust. Soc. Am., 74:115-123.

Gifford, M. L., and Guinan, J. J., Jr., 1987, Effects of electrical stimulation of medial olivocochlear neurons on ipsilateral and contralateral cochlear responses, Hearing Res., 29:179-194.

Ginzberg, R. D. and Morest, D. K., 1983, A study of cochlear innervation in the young cat with the Golgi method, Hearing Res., 10:227-246.

Guinan, J. J., Jr., 1986, Effects of efferent neural activity on cochlear mechanics, Scand. Audiol. Suppl., 25:53-62.

Guinan, J. J., Jr. and Gifford, M. L., 1988a, Effects of Electrical Stimulation of Efferent Olivocochlear Neurons on Cat Auditory-Nerve Fibers. I. Rate versus sound level functions. (Submitted to Hearing Res.)

Guinan, J. J., Jr. and Gifford, M. L., 1988b, Effects of Electrical Stimulation of Efferent Olivocochlear Neurons on Cat Auditory-Nerve Fibers. II. Spontaneous Rate. (Submitted to Hearing Res.)

Guinan, J. J., Jr., and Gifford, M. L., 1988c, Effects of electrical stimulation of efferent olivocochlear neurons on cat auditory-nerve fibers. III. Tuning curves and threshold at CF. (in prep.)

Guinan, J. J., Warr, W. B. and Norris, B. E., 1983, Differential olivo-cochlear projections from lateral versus medial zones of superior olivary complex, J. Comp. Neurol., 221:358-370.

Guinan, J. J., Warr, W. B. and Norris, B. E., 1984, Topographic Organization of the Olivocochlear Projections from the Lateral and Medial Zones of the Superior Olivary Complex, J. Comp. Neurol., 226:21-27.

Kemp, D. T., and Chum, R. A., 1980, Observations on the generator mechanism of stimulus frequency acoustic emissions - two tone suppression, in: "Psychophysical, Physiological and Behavioural Studies in Hearing", van den Brink, G. and Bilsen, F. A., eds., Delft Univ. Press, 34-42.

Kemp, D. T., 1986, Otoacoustic emissions, travelling waves and cochlear mechanisms, Hearing. Res., 22:95-104.

Kim, D. O., 1986, Active and nonlinear cochlear biomechanics and the role of outer-hair-cell subsystem in the mammalian auditory system, Hearing. Res., 22:105-114.

Konishi, T., and Slepian, J. Z., 1971, Effects of the electrical stimula-tion of the crossed olivocochlear bundle on cochlear potentials recorded with intracochlear electrodes in guinea pigs, J. Acoust. Soc. Am., 49:1762-1769.

Liberman, M. C., 1978, Auditory nerve response from cats raised in a low noise chamber, J. Acoust. Soc. Am., 63:442-455.

Liberman, M. C., 1980, Efferent synapses in the inner hair cell area of the cat cochlea: An electron-microscopic study of serial sections, Hearing Res., 3:189-204.

Liberman, M. C., 1982, Single-neuron labeling in the cat auditory nerve, Science, 216:1239-1241.

Liberman, M. C. and Brown, M. C., 1986, Physiology and anatomy of single olivocochlear neurons in the cat, Hearing Res., 24:17-36.

Mountain, D. C., 1980, Changes in endolymphatic potential and crossed olivocochlear bundle stimulation alter cochlear mechanics, Science, 210:71-72.

Neely, S. T. and Kim, D. O., 1983, An active cochlear model showing sharp tuning and high sensitivity, Hearing Res., 15:103-112.

Sewell, W. F., 1984, The relation between the endocochlear potential and spontaneous activity in auditory nerve fibres of the cat, J. Physiol., 347:685-696.

Siegel, J. H. and Dallos, P., 1986, Spike activity recorded from the organ of Corti, Hearing Res., 22:245-248.

Siegel, J. H. and Kim, D. O., 1982, Efferent neural control of cochlear mechanics? Olivocochlear bundle stimulation affects cochlear bio-mechanical nonlinearity, Hearing Res., 6:171-182.

Sohmer, H., 1966, A comparison of the efferent effects of the homolateral and contralateral olivo-cochlear bundles, Acta Oto-laryng., 62:74-87.

Spoendlin, H., 1966, The organization of the cochlear receptor, in: Adv. in Oto-Rhino-Laryng. 13, S. Karger, N. Y., 1-227.

Teas, D. C., Konishi, T., and Nielsen, D. W., 1972, Electrophysiological studies on the spatial distribution of the crossed olivo-cochlear bundle along the guinea pig cochlea, J. Acoust. Soc. Am., 51:1256-1264.

Warr, W. B. and Guinan, J. J., 1979, Efferent innervation of the organ of Corti:two separate systems, Brain Res., 173:152-155.

Warr, W. B., White, J. S. and Nyffeler, M. J., 1982, Olivocochlear neurons: Quantitative comparison of the lateral and medial efferent systems in adult and newborn cats, Soc. Neurosci. Abstr., 8:346.

Wiederhold, M. L., 1970, Variations in the effects of electric stimulation of the crossed olivocochlear bundle on cat single auditory-nerve fiber responses to tone bursts, J. Acoust. Soc. Am., 48:966-977.

Wiederhold, M. L. and Kiang, N. Y. S., 1970, Effects of electrical stimulation of the crossed olivocochlear bundle on single auditory nerve fibers in the cat, J. Acoust. Soc. Am., 48:950-965.

PHYSIOLOGY OF COCHLEAR EFFERENTS IN THE MAMMAL

Donald Robertson and Mark Gummer

Department of Physiology
University of Western Australia
Nedlands, Western Australia, 6009

INTRODUCTION

In the last few years there have been a number of new descriptions of the physiological properties of the efferent systems projecting to the mammalian auditory receptor organ (Robertson, 1984; Robertson and Gummer, 1985; Liberman and Brown, 1986). Without detracting in any way from the efforts of earlier investigators, it is fair to say that these recent studies have benefited from a number of advances since the early landmark recordings of single efferent fibre activity (Fex, 1962; Rupert et al., 1968).

The application of new surgical approaches to the efferent fibre bundles, together with reliable and accurate means of monitoring cochlear condition and inter-aural acoustic cross-talk have enabled quantitative classification of the different response categories of efferents to be made. Furthermore, the combination of microelectrode recordings and single-fibre intracellular injection of horseradish peroxidase have enabled direct correlations to be made between the physiology and anatomy of single efferent neurones.

At the same time, the advent of horseradish peroxidase and other neuronal labelling techniques has lead to a greater understanding of the organization of the brainstem centres of origin of this efferent projection in a number of species (Warr, 1975; Warr and Guinan, 1979; White and Warr, 1983;Robertson et al., 1987b).

It is the purpose of this review to present some of the more recent results from such studies together with some new data, with particular reference to the physiology of the efferent system projecting to the outer hair cells in guinea pig and cat.

PHYSIOLOGICAL PROPERTIES OF SINGLE EFFERENTS

General Properties

Under barbiturate anaesthesia, all efferents have low spontaneous firing rates. Efferent neurones are, however, easily distinguished by their very regular firing pattern in response to appropriate acoustic stimulation and by the long latency of their responses to sound compared to primary afferent neurones.

The extreme regularity of interspike intervals which is one of the most striking features of efferent discharge results in interspike interval histograms of distinctly symmetrical shape in contrast to the highly

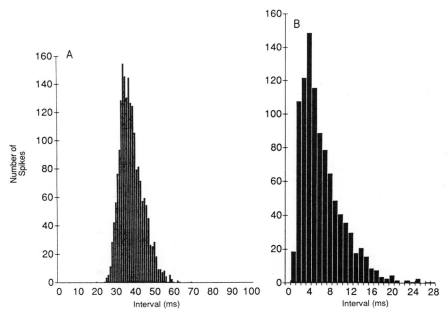

Fig. 1. A: Typical interspike interval histogram of spike activity to a CF tone 30 dB above threshold obtained from a single olivo-cochlear efferent in the guinea pig spiral ganglion. Note the roughly symmetric shape of the histograms and the long minimum modal interval. B: Typical interspike interval histogram of spontaneous discharge of primary afferent neurones showing characteristic asymmetric shape and short modal interval.

skewed distribution of interspike intervals found for all primary afferent neurones (Fig. 1). The symmetric distribution of interspike intervals is preserved over the full range of stimulus intensity, and a minimum modal interval of approximately 10-40 ms is observed in all efferents at their maximum driven discharge rate.

Efferent neurones fall into 3 distinct response categories. In a sample of 694 efferents obtained by microelectrode recordings from the intraganglionic spiral bundle in the guinea pig cochlea (Gummer, un-published data) 61.7% were excited by ipsilateral sound, and 35.2% by contralateral sound. In both these major response classes the neurones were only driven by sound in the non-preferred ear when this exceeded the cross-talk threshold calculated from either single afferent or N_1 measurements in the same animals. The third response category (3.7%) are neurones driven by sound presented to either ear. Inter-aural cross-talk cannot explain this binaural response property. In most cases, these neurones were approximately equally sensitive to either ear, though some showed lower thresholds to either the ipsi or contralateral ear stimula-tion. The quantitative aspects of these latest data in guinea pig differ slightly from an earlier report based on a smaller sample size (Robertson and Gummer, 1985). They can be compared with results in cat, using a different surgical approach, in which the above 3 response categories comprised 59, 29 and 11% respectively (Liberman and Brown, 1986). The binaurally-responsive group in cat has been further classified into binaural-ipsi (8%) and binaural-contra (3%), according to the ear in which sound stimulation drives the neurone most effectively. In both species, the characteristic frequencies are not always identical in each ear for binaural units, though generally they lie within 1/2 octave or less of each other.

Tuning Properties

In the guinea pig, efferent neurones are found with CF's ranging from 0.5 to 26 kHz, with the majority found between 4 and 12 kHz (Fig. 2). All 3 response categories appear to be represented across the full frequency range in guinea pig, though there is some evidence in cat that binaurally responsive efferents are found mainly with low frequency CF's (Liberman and Brown, 1986).

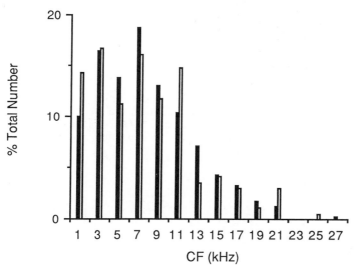

Fig. 2. Distribution of CF's for a large number of efferent neurones in barbiturate (solid bars) and Urethane (grey bars)-anaesthetized guinea pigs. There were no clear differences in the distribution of CF's in the 3 response categories.

Improvements in the understanding of cochlear physiology in general have enabled meaningful comparisons to be made between the tuning properties of cochlear efferent and afferent neurones. It is possible that early descriptions of the tuning properties of cochlear efferents were confounded by inadvertant deterioration in the tuning properties of the peripheral receptor organ which ultimately provides the acoustic drive to the efferents. We now know that provided the functional integrity of the peripheral receptor is assured, the majority of efferents show sensitivity and sharpness of threshold tuning curves which is equivalent to that exhibited by primary afferent neurones of similar CF and spontaneous firing rate. The most parsimonious, though clearly not the only explanation for this finding, is that efferent neurones within the brainstem nuclei receive tightly organized afferent drive with little convergence of input from neighbouring frequency domains of the peripheral receptor epithelium.

Binaural Interactions

   Though the vast majority of cochlear efferents are monaural when
tested with sound in one ear at a time, there is mounting evidence that
binaural interactions do exist for many supposedly monaural efferents
(Fig. 3).

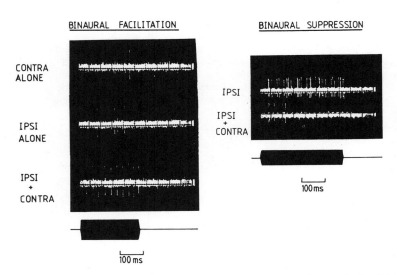

Fig. 3. Oscilloscope tracings showing examples of binaural facilitation
         and binaural suppression in two different efferent neurones.
         Responses to four consecutive sweeps are superimposed in the
         case of suppression.

   These interactions are usually only discernible when sound is
presented simultaneously in both ears. Such binaural effects can be either
facilitatory or suppressive (Robertson and Gummer, 1985; Liberman and
Brown, 1986). In a sample of 65 efferent neurones in the guinea pig,
78% of monaurally-classified fibres showed either binaural facilitation
or suppression when a CF tone was presented to the non-preferred ear
simultaneously with a CF tone in the preferred ear (Gummer and Robertson,
unpublished results). Approximately equal numbers of the monaural-ipsi
units showed binaural facilitation or suppression. For the monaural-contra
group, suppression was found twice as frequently as facilitation. The
effect is generally to shift the input-output curves for preferred-ear
CF stimulation either to the left or right, producing a change in threshold of 10-20 dB. All these effects are found at non-preferred ear sound
pressures at least 20 dB below cross-talk threshold, making it unlikely
that they are artifactual changes resulting from uncontrolled masking or
suppression of preferred-ear afferent drive.

   The quantitative aspects of these binaural interaction data must be
treated with caution, since the instability of recordings from single
efferents means that a complete study of binaural effects over the full
frequency range is not always realizable for each fibre. It is therefore
possible that a given fibre may show both facilitation and suppression in
particular regions of its response curve. Nonetheless, the results show
that the input to many "monaural" efferents is probably very complex. In

many cases efferent neurones in the brainstem must receive synaptic input from both ears, though in most cases the drive from only one cochlea is sufficiently secure to excite the neurone when sound is presented to one ear at a time.

In the case of suppression of firing by sound in the non-preferred ear it should be borne in mind that the very low spontaneous firing rate of most efferents would make it difficult for suppression to be demon-strated for one-ear stimulation in any case. The presentation of a CF tone in the preferred ear merely provides a background discharge against which these occult inhibitory effects can be clearly seen. In cases where spontaneous firing rate is appreciable, inhibitory effects of monaural stimulation have been seen (Liberman and Brown, 1986; Gummer, unpublished observations).

The cases of binaural facilitation are perhaps more interesting. It is clear that synaptic drive from the non-preferred ear is insufficient to affect the discharge when sound is presented to that ear alone. However, this input is able to modulate the excitability of the cell body when this is tested by its response to preferred ear stimulation. It might be postulated that weak excitatory synapses originating from the non-preferred ear pathways may be located on distal dendrites of many "monaural" efferent neurones in the brainstem nuclei.

Effects of Anaesthesia

The type of anaesthetic regime has been shown to have two clear effects on the properties of single efferent neurones in guinea pigs (Gummer, unpublished results; Robertson and Gummer, 1985).

Under Nembutal anaesthesia, the majority of efferents show no spontaneous activity. In a sample of 102 neurones in the guinea pig cochlea, 59% of efferents had no spontaneous firing rate, and the mean spontaneous rate for all fibres was 1.8 spike/s. Under Urethane anaesthesie (1.5g/kg) however, only 29% had no spontaneous activity and the mean spontaneous discharge rate rose to 5.3 spikes/s in a sample of 122 fibres (Gummer, unpublished data).

Furthermore, under both anaesthetic regimes, clearly non-stationary characteristics of the spontaneous discharge were found. Spontaneous discharge of single efferents under Nembutal tended to increase with a tendency to "bursting" patterns of discharge as the time for renewal of barbiturate anaesthetic approached.

Response latency also varies with anaesthetic regime. In a sample of 227 fibres under barbiturate anaesthesia, the mean minimum latency of the first spike to a suprathreshold CF tone burst was 22.4 ms. Under Urethane, the minimum latency was 16.3 ms (sample of 123 units). These differences were statistically significant (Gummer, unpublished data).

Although it is not possible to say with certainty whether Urethane or barbiturate anaesthesia more nearly represents the "normal" situation, these results strongly suggest that the physiological status of the brainstem synaptic pathways can affect some of the efferent response properties. It is therefore likely that the behaviour of efferents in the unanaesthetized animal is different from that seen in the anaesthetized animal.

Response to Amplitude-Modulated Sound

Efferent neurones in the guinea pig respond in a striking manner to amplitude-modulated sound (Gummer et al., 1986; Gummer, unpublished results). A typical modulation transfer function derived using a continous, 30% amplitude-modulated CF tone is shown in Fig. 4, compared to the modulation transfer function of a typical primary afferent neurone.

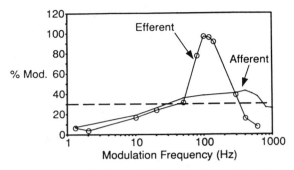

Fig. 4.   Typical examples of the modulation transfer function obtained for a single efferent and primary afferent neurone in the guinea pig cochlea. Stimulus was a 30% amplitude-modulated continuous CF tone.

The transfer function has a maximum gain of about 12 dB at a modulation frequency of about 100 Hz and falls off markedly above this frequency. This result was obtained for all 3 response categories. In a total of 18 fibres from which complete modulation data was obtained, 4 fibres showed atypical transfer functions with large gains at quite low frequency (10 Hz or less) suggesting that they may constitute a further functional subclass of efferent neurones with different adaptation or other characteristics from the majority of efferents.

The responses to amplitude-modulated sound enable precise estimates or response latency to be made from the measured group delays. In a study of 18 efferents in Urethane anaesthetized guinea pigs, group delays calculated from the modulation transfer function were compared to minimum latencies to the onset of a tone burst stimulus in the same fibres. In 17 of the 10 cases the calculated delay was less than the minimum onset latency, sometimes by as much as 35 ms. Furthermore, the range of calculated delays was very small (8.2 ± 1.0 ms) compared with the onset latencies in the same fibres (24.2 ± 12.5 ms).

These results suggest a fundamental difference between the onset and steady-state characteristics of the efferent system. They imply that during continuous stimulation there is an improvement in the synaptic efficiency at some point within the efferent feedback loop which reduces latency and jitter and enables these neurones to respond precisely to rapid change in the input amplitude of an ongoing sound. Further evidence for short-term plasticity in synaptic transmission within this pathway comes from observations of a build-up in response to repeated brief tone bursts in some fibres and the apparently critical dependence of discharge rate on stimulus repetition rate in some cases (Liberman and Brown, 1986). In addition, it has been shown that brief exposure to loud sound can produce a dramatic decrease in onset latency in many efferents, again suggesting some form of synaptic facilitation (Gummer, unpublished results).

Correlation with Anatomy

Both direct and indirect experimental approaches have been used in an attempt to relate the physiological properties of efferent neurones to their nuclei of origin in the brainstem, and their site of termination within the cochlea.

There is a very suggestive correlation between the relative numbers of the different response categories found in microelectrode recordings and the distribution of efferent neurones found within the brainstem. For example, in the guinea pig after injection of fluorescent retrograde label in the cochlea, approximately 69% of large medial system neurones are located in a variety of nuclei contralateral to the injected cochlea, and 31% are local ipsilaterally (Robertson et al., 1978b). Of all these large neurones about 3-9% have been shown by double-labelling techniques to project to both cochleas (Robertson et al., 1978b, c). It is therefore tempting to assign the ipsilaterally responsive monaural group of efferents to the contralateral nuclei, the contralaterally responsive monaural group to the ipsilateral nuclei and the binaurally responsive neurones to the bilaterally projecting cells found in the brainstem. A similar picture emerges in the cat, except that bilaterally projecting cells have not yet been identified histologically in this species.

Direct evidence consistent with these correlations has been obtained in the cat, in which a small number of single physiologically-characterized efferent fibres has been labelled with horseradish peroxidase and traced to their cell bodies of origin in the brainstem (Liberman and Brown, 1986). The results are consistent with the notion that efferent cell bodies are located in medial nuclei on the opposite side of the brain to the cochlea from which their excitatory drive originates.

The most important results obtained from intracellular microinjection of horseradish peroxidase pertain to the question of the site of termination of efferent fibres within the cochlea. In all instances in both cat and guinea pig responsive efferents are found to terminate on the outer hair cells (Fig. 5). There is some evidence from electronmicroscopic examination of labelled fibres that some branches may make en passant synaptic connections with radial afferent fibres beneath the inner hair cells (White et al., 1986). However, all the available data suggest that the acoustically-responsive efferents from which recordings have so far been obtained belong to the medial system of efferent neurons whose ultimate targets within the organ of Corti are the outer hair cells.

The single fibre labelling data also provide direct evidence for a tonotopic organization of the efferent projection to the peripheral receptor. Efferent neurones in the guinea pig project to a restricted array of outer hair cells at a location along the length of the cochlea which is in excellent agreement with their characteristic frequency and the known tonotopic place-frequency map derived from afferent recordings (Robertson and Gummer, 1986). There appears to be some difference between the guinea pig and the cat in the extent of the terminal arborizations, with efferents in the cat often showing extensive branching within the modiolus and the ganglionic spiral bundle, and innervating much larger lengths of the receptor (Liberman and Brown, 1986). Such extensive branching has not been seen routinely in the guinea pig, and injected efferents usually show a single fibre which can course for several millimetres within the ganglion before crossing the osseous spiral lamina to ramify within a restricted cochlear domain. It is possible that the site of HRP injection, or some other sampling bias may influence the picture obtained when cat and guinea pig are compared, but overall these data, together with the tuning properties, suggest that the medial efferents form a precise feedback loop, apparently designed to affect the properties of outer hair cells close to the frequency region of the cochlea either ipsilateral or contralateral to that from which the neurones receive their input.

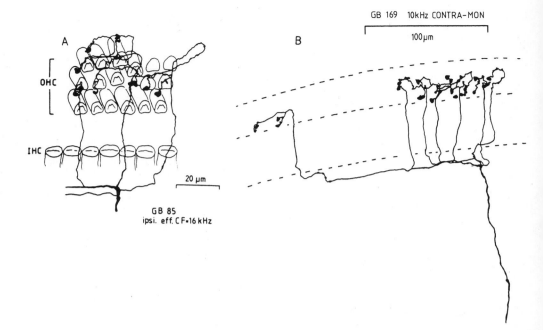

Fig. 5. A:Detailed camera lucida drawing of part of the terminal arborization of a single olivocochlear efferent injected intracellularly
with HRP in the intraganglionic spiral bundle in the guinea
pig. Note large terminals on bases of outer hair cells. B: Low
power camera lucida drawing of a single guinea pig olivocochlear
efferent, (contralateral-monaural response type), illustrating
typical distribution of endings on outer hair cells.

CONCLUSIONS

What have we learnt from these new experiments?

1.   In response to acoustic stimulation, efferent neurones are sensitive
and sharply tuned.

2.   There are at least 3 major classes of efferents based on their
acoustic response categories. Within these classes, however, there are
further complex response properties revealed by simultaneous binaural
stimulation and amplitude modulation.

3.   The projection of individual efferents to the cochlea is tonotopic,
and more or less precise. In the guinea pig, the efferents terminate on
a restricted region of outer hair cells close to the origin of their
input.

It should be stressed that these are not properties which would be
normally associated with a coarse, imprecise control system. The more
data that become available, the more complex does the olivocochlear
projection appear to be, both from a physiological and anatomical point
of view. A satisfying explanation for the function of the efferents
should encompass this complexity.

4.   Some evidence has been presented that the conditions of anaesthesia
and stimulation can influence spontaneous activity and speed of synaptic
transmission within this pathway. It is quite possible that under

traditional conditions of recording and stimulation, we are looking of a system responding well below its capacity.

5. All responsive efferents injected with horseradish peroxidase terminate on outer hair cells and therefore belong to the medial system. Thus, all the physiological data described apply to the medial system neurones, which comprise up to 50% of the total olivocochlear projection are also organized in cochleotopic manner (Robertson et al., 1987a; Guinan et al., 1984). However, data on the physiological response properties of this second major efferent system have not, so far, been obtainable.

ACKNOWLEDGEMENTS

This work was supported by grants from the National Health and Medical Research Council (Australia), the Australian University Grants Scheme and the University of Western Australia. The authors gratefully acknowledge the advice and encouragement of B. M. Johnstone, G. K. Yates and R. Patuzzi, and helpful discussions with J. J. Guinan Jr. and D. C. Mountain.

REFERENCES

Fex, J., 1965, Auditory activity in uncrossed centrifugal cochlear fibres in cat. A study of a feedback system, II. Acta Physiol. Scand., 64:43-57.

Fex, J., 1962, Auditory activity in centrifugal and centripetal cochlear fibres in cat. A study of a feedback system, Acta Physiol. Scand., 55, Suppl. 189:1-68.

Guinan, J. J. Jr., Warr, W. B. and Norris, B. A., 1984, Topographic organization of the olivocochlear projections from the lateral and medial zones of the superior olivary complex, J. Comp. Neurol., 226:21-27.

Gummer, M., Robertson, D., Johnstone, B. M. and Yates, G. K., 1986, Coding of amplitude modulation by efferent neurones in the guinea pig cochlea, Proc. Austr. Physiol. Pharmacol. Soc., 17:15.

Liberman, M. C. and Brown, M. C., 1986, Physiology and anatomy of single olivocochlear neurons in the cat, Hearing Res., 24:17-36.

Robertson, D., Anderson, C.-J. and Cole, K. S., 1987a, Segregation of efferent projections to different turns of the guinea pig cochlea, Hearing Res., 25:69-76.

Robertson, D., Cole, K. S. and Harvey, A. R., 1987b, Brainstem organiza- tion of efferent projections to the guinea pig cochlea studied using the fluorescent tracers fast blue and diamidino yellow, Exp. Brain Res., 66:449-457.

Robertson, D., Cole, S. and Corbett, K., 1987c, Quantitative estimation of bilaterally projecting medial olivocochlear neurones in the guinea pig brainstem, Hearing Res., 27:177-181.

Robertson, D. and Gummer, M., 1985, Physiological and morphological characterization of efferent neurones in the guinea pig cochlea, Hearing Res., 20:63-78.

Robertson, D., 1984, Horseradish peroxidase injection of physiologically characterized afferent and efferent neurones in the guinea pig spiral ganglion, Hearing Res., 15:113-122.

Warr, W. B., 1975, Olivo cochlear and vestibular efferent neurons of feline brainstem:their location morphology and number determined by retrograde axonal transport and acetylcholinesterase histochemistry, J. Comp. Neurol., 16:159-182.

Warr, W. B. and Guinan, J. J. Jr., 1979, Efferent innervation of the organ of Corti: Two separate systems, Brain Res., 173:153-155.

White, J. S. and Warr, W. B., 1983, The dual origins of the olivocochlear bundle in the albino rat, J. Comp. Neurol., 219:201-214.

White, J. S., Robertson, D. and Warr, W. B., 1986, Electron-microscopic observations on an HRP-filled, physiologically-characterized medial olivocochlear axon in the guinea pig cochlea, Abstr. 16th. Ann. Meeting Soc. Neurosci. (U.S.A.), 12:1264.

DESCENDING CENTRAL AUDITORY PATHWAY - STRUCTURE AND FUNCTION

J. Syka, J. Popelář, R. Druga* and A. Vlková

Institute of Experimental Medicine
Czechoslovak Academy of Sciences
*Institute of Anatomy
Medical School, Charles University
Prague, Czechoslovakia

The presence of many descending fibers in the auditory pathway has been known for many years and the complexity of the descending auditory pathway is well recognized. Recent findings revealing the role of the olivocochlear bundle in the control of cochlear micromechanics have opened a new era in the investigation of the role of descending fibres in the auditory pathway. The advent of axonal transport techniques for tracing neuronal connections stimulated reinvestigation of the connections of the olivocochlear bundle originally described by Rasmussen (1946, 1953). New concepts have emerged from the work of Warr and Guinan (1979), who defined two populations of the olivocochlear efferents: lateral and medial olivocochlear neurons. These two populations may be distinguished with respect to their origin in different parts of the superior olivary complex (SOC) and differing projections to the cochlea. The mediator of the olivocochlear bundle is undoubtedly acetylcholine, as was originally proposed by Rasmussen (1960). However, recently also other possible mediators were found with immunohistochemical methods such as enkephalins (Fex and Altschuler, 1981; Eybalin et al., 1984), aspartate and GABA (Fex et al., 1986).

The olivocochlear bundle seems to be the terminal part of the complex descending auditory pathway, which originates in pyramidal cells of the cortical layer V. Fig. 1 shows the most important parts of the descending auditory pathway in the rat, compiled from our findings (Syka et al., 1980; Druga and Syka, 1984 a, b; Druga et al., 1985) and reports of other investigators (White and Warr, 1983; Faye-Lund, 1985, 1986). Although some differences exist between species, the essential features of the descending auditory pathway in the rat are similar to that found in the cat (Diamond et al., 1969; Anderson et al., 1980; Hashikawa, 1983; Warr, 1975; Adams, 1983) and in the guinea-pig (Strutz and Spatz, 1980; Robertson, 1985; Syka et al., 1988).

In contrast to the mass of accumulated anatomical data, the physiology of the descending auditory pathway, especially of its higher parts, is essentialy unknown. In principle it has been found that electrical stimulation of higher centers inhibits neuronal activity in lower centers of the auditory pathway. Watanabe et al. (1966) and Andersen et al. (1972) reported that electrical stimulation of the auditory cortex produced both

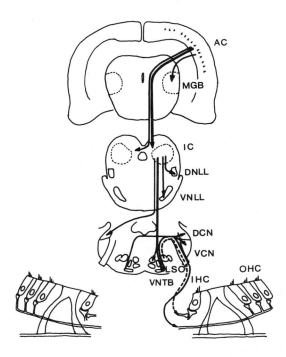

Fig. 1. Schematic drawing of the descending auditory pathway in the rat.
AC-auditory cortex, MGB-medial geniculate body, IC-inferior
colliculus, DNLL-dorsal nucleus of the lateral lemniscus, VNLL-
ventral nucleus of the lateral lemniscus, LSO-lateral superior
olive, VNTB-ventral nucleus of the trapezoid body, DCN-dorsal
cochlear nucleus, VCN-ventral cochlear nucleus, IHC-inner hair
cells, OHC-outer hair cells.

excitation and inhibition of single units in the medial geniculate body.
Comis and Whitfield (1968) observed inhibition of neurons in the cochlear
nucleus after stimulation of the lateral lemniscus. Similar inhibition
was obtained when they applied adrenaline to cochlear nucleus neurons.
On the contrary, electrical stimulation of the SOC resulted in excitation of the cochlear nucleus neurons, which could be mimicked by local
application of acetylcholine. Excitatory neurons were localized mostly
in the more medial parts of the SOC, whereas stimulation of more lateral
parts of the SOC resulted in inhibition of neurons in the cochlear nucleus
(Comis, 1970).

This paper summarizes our recent anatomical and physiological data
on the descending auditory pathway in the rat and guinea-pig. Morphological data on the corticocollicular pathway and the pathway descending
from the inferior colliculus to the SOC are presented as well as the
responses of inferior colliculus neurons to electrical stimulation of
the auditory cortex. In addition changes of auditory evoked responses in
the inferior colliculus are described when they were preceded by electrical
stimulation of the auditory cortex.

Structure of the Corticotectal and Tectoolivary Pathway

Corticotectal descending pathway originates in pyramidal cells of
the cortical layer V. In the rat, pyramidal cells which project to the
inferior colliculus are localized in the temporal cortex and partially
also in the parietal cortex (Syka et al., 1981; Druga and Syka, 1984b;

Druga et al., 1988), whereas in the guinea pig the neurons projecting to the IC are predominantly in the temporal cortex and partially in the posterior area of the claustrocortex isocorticalis (Druga et al., 1988). In the inferior colliculus of the rat descending fibers from the cortex terminate mainly in two subdivisions of the ipsilateral IC: in the dorsal cortex and in the external cortex, although some scattered terminals were also found in the central IC nucleus (Druga and Syka, 1984b). Descending fibers also terminate in the contralateral IC, however, only in its dorsal cortex. Fibers terminating in two different projection areas (dorsal and external cortices) originate in two different parts of the auditory cortex of the rat (Faye-Lund, 1985): those projecting to the dorsal cortex in the ventral part of the cortex (area 41 and 36) and those projecting to the external cortex in the dorsal and rostral parts of the AC (area 22).

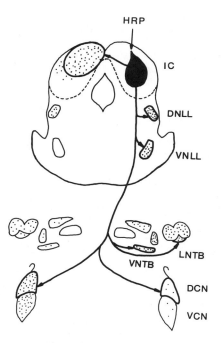

Fig. 2.   Anterograde and retrograde labeling in the rat auditory pathway after injection of HRP into the inferior colliculus. Arrows indicate the structures with anterograde labeling, dots indicate schematically the retrogradely labeled cells. For abbreviations- see text to Fig. 1.

Numerous fibers also descend from the inferior colliculus in the rat (Druga et al., 1985) and guinea-pig (Syka et al., 1988). Fig. 2 shows projection of fibers descending from the inferior colliculus in the rat traced with HRP anterograde transport. Fibers descending from the IC terminate in the ipsilateral dorsal and ventral nuclei of the lateral lemniscus, in the ipsilateral superior olivary complex (SOC) and in the dorsal cochlear nuclei mainly contralaterally. The main target in the SOC is the ventral nucleus of the trapezoid body, however, also other nuclei of the SOC (as periolivary nuclei and the lateral nucleus of the trapezoid body) obtain scarce projections from the IC. Similar projec- tions were observed in the rat by Faye-Lund (1986) who described the targets in the ipsilateral SOC as rostral and medioventral zones of the periolivary region. Fibers projecting from the IC to the SOC are particularly interesting because of their possible link with the olivocochlear bundle neurons. White and Warr (1983) reported that the origin of medial olivocochlear bundle neurons in the rat is the ventral nucleus of the trapezoid body (VTNB).

The origin of fibers projecting from the IC to the SOC was studied in the rat by Faye-Lund (1986) and in the guinea-pig by Syka et al., (1988), who injected HRP into different parts of the SOC. In both species HRP labeled neurons were found in the external cortex and few of them also in the central nucleus. Labeled neurons were of various sizes, some of them were large multipolar neurons. Labeled cells were never observed in the dorsal cortex and in the dorsomedial part of the central IC nucleus. Heavy labeling in the IC was obtained after HRP application in the medial part of the SOC (particularly when the tracer spread to the VNTB), whereas injections into the lateral part of the SOC were almost never combined with the labeling of IC neurons. The results suggest that the lateral olivocochlear system is not in direct contact with the IC efferents. In addition to the already described descending auditory fibers there exist further fibers descending from the nuclei of the lateral lemniscus to cochlear nuclei and to the SOC (Whitley and Henkel, 1984). Also many fibers descend from the medial group of olivary nuclei to cochlear nuclei bilaterally (Covey et al., 1984).

Influence of Electrical Stimulation of the Auditory Cortex on Neuronal Activity in the Inferior Colliculus

Responses of IC neurons to electrical stimulation of the AC were recorded at first in experiments aimed to analyze the role of the pathway descending from the AC to the IC (Syka and Popelář, 1984). In 17 Sprague-Dawley rats, anaesthetized with pentobarbital, the AC was stimulated with electrical pulses (duration 0.2 ms, repetition rate 1 Hz and intensity ranging from 0.2 to 1.5 mA). Eighty-four neurons out of the 162 tested (i.e. 52.4%) were influenced by electrical stimulation (ES) of the AC. In the majority of cases, the ES resulted in a brief burst of excitation which occurred after a latency of 3-15 ms (Figs. 3, 4, 5). The burst consisted of 1-10 spikes and the number of spikes increased with the ES intensity. The excitation threshold in some neurons was around 0.2 mA, but in some units it was necessary to increase the intensity up to 0.8-1.0 mA until the response was obtained. Fig. 3 (left side) shows a neuron (RCI 72 I) which is weakly excited after ES stimulation of the AC. The excitation is followed by a period of inhibition, demonstrated by suppression of the spontaneous activity for approximately 100 ms. The suppression of spontaneous activity by ES is similar to that evoked by acoustic stimulation at the best frequency (4.3 kHz). Similarly the neuron RCI 72 F (Fig. 3, right side) displays inhibition of spontaneous activity evoked both by electrical and acoustic stimulations. Inhibition resulting from the AC stimulation may also suppress the excitatory response evoked by acoustic stimulation. Fig. 4 shows a neuron (RCI 43 H) which discharges one spike after AC stimulation. This excitatory response is followed by inhibition lasting at least 40 ms, because the acoustically evoked response is inhibited when preceded by AC stimulation. A decrease in the ES current from 0.5 mA to 0.3 mA again results in excitation evoked by acoustic stimulation. Similarly an increase in the delay of acoustic stimulation after ES from 10 ms to 20 ms is accompanied by the reappearence of the excitatory response to an acoustic stimulus (Fig. 4, neuron RCI 72 P). Combination of the short-latency excitation with subsequent inhibition was the most frequent type of response in the IC (Fig. 5). In 25% neurons which reacted to auditory cortex stimulation only an inhibitory effect was found. The inhibition which in this case suppressed either spontaneous or acoustically evoked activity occurred after 3-15 ms latency and lasted 30-300 ms.

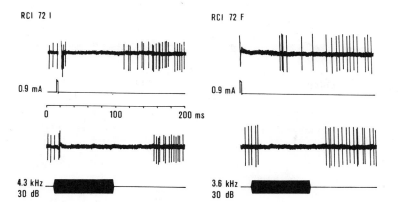

Fig. 3. Responses of two inferior colliculus neurons to electrical
stimulation of the AC and to acoustical stimulation. All
responses represent 5 consecutive summated individual responses,
photographed from the screen of a memory oscilloscope.

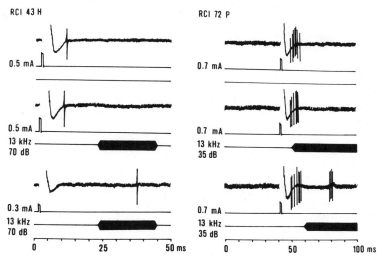

Fig. 4. Responses of two inferior colliculus neurons to electrical
stimulation of the AC and to acoustical stimulation. For details
see text to Fig. 3.

The data have shown that a significant number of neurons in the rat
IC react to electrical stimulation of the AC. While the excitation
occurring after a brief latency is probably a direct orthodromic response
of fibers descending from the layer V of the AC, the origin of the later
inhibition is less certain. It cannot be excluded that the inhibitory
response results from the activation of inhibitory interneurons within
the IC.

In another series of experiments we recorded depth profiles of gross
electrical responses in the IC evoked either by electrical stimulation
of the AC or by acoustical stimulation. The recording electrode was
introduced into the IC either dorsoventrally through the overlying cortex
or caudorostrally through the cerebellum. Acoustically evoked responses
dominated in the central part of the IC, whereas the electrical stimulation

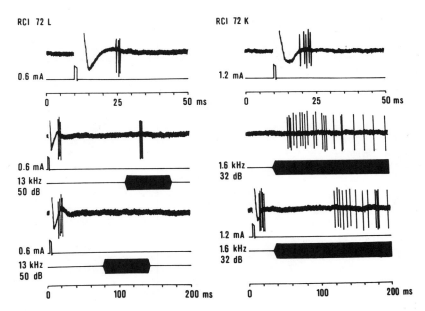

Fig. 5.   Responses of two inferior colliculus neurons to electrical
          stimulation of the AC and to acoustical stimulation. For details
          see text to Fig. 3.

of the AC resulted in evoked responses in marginal parts of the IC (Fig.
6). This finding is consistent with the projection of afferent fibers
to the IC: the central IC nucleus represents the major target of ascending
fibers whereas dorsal and external cortices are targets of descending
cortical fibers.

<u>Interaction of Auditory Evoked Responses with Electrical
Stimulation of the Auditory Cortex</u>

     Analysis of neuronal responses evoked by stimulation of the AC led
to the conclusion that the acoustically evoked activity in the IC can
be significantly suppressed by preceding electrical stimulation of the
AC. It was expected that this suppression observed at neuronal level can
be observed also in auditory evoked responses when the cortical electrical
stimulation will precede the acoustical stimulus. In 7 rats we recorded
click evoked responses in the IC. Stainless steel electrodes were
introduced into the IC several times in each animal; the electrode
tracks were histologically controlled. The cortical auditory field was
mapped with the aid of auditory evoked responses; bipolar silver-ball
electrodes were then placed into three different parts of the AC: the
dorsal, central and ventral.

     Fig. 7 shows the depth profile of changes in the IC click-evoked
responses preceded by electrical stimulation of the auditory cortex.
Maximal suppression of the evoked response in comparison with the control
was observed in superficial layers of the IC. In this region even the
N_1 wave of the evoked response was suppressed. Suppression is dependent
on the time which elapses between the electrical stimulation and the
occurrence of the click (Fig. 8). The effect is maximal when the ES
precedes the click by about 10 ms, it is less expressed at larger delays.
With a constant delay the suppression may be significantly enhanced by
increasing the ES current from 0.15 mA to 1.5 mA. Another factor which
influences the effectiveness of the AC stimulation is the site of stimula-

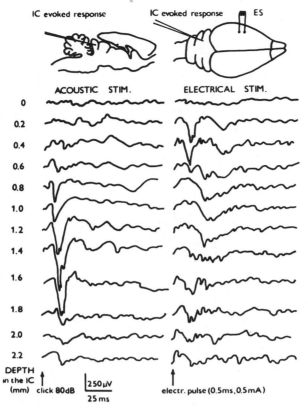

Fig. 6. Depth profiles of evoked responses to acoustical stimulation
(left side) and electrical stimulation of the AC (right side)
in the rat inferior colliculus.

tion. Maximum suppression was usually obtained when the stimulating
electrodes were placed in the most ventral part of the auditory cortex
(Fig. 9). When the stimulating electrodes were placed ventrally, the P_2
wave was suppressed maximally at all locations in the IC and the N_1
wave was significantly suppressed at least in some places. The central
and particularly dorsal locations of stimulating electrodes only
infrequently suppressed evoked responses in the IC. The relationship
between the site of AC stimulation and the site of recording in the IC
was investigated in detail in a further series of experiments.

Distribution of Neuronal Responses to Auditory Cortex Stimulation in the Inferior Colliculus

     Responses of 236 neurons to auditory cortex stimulation were recorded
in the inferior colliculus of anaesthetized rats. Single electrical
pulses with duration 0.2 ms served as stimuli. Microelectrode tracks were
reconstructed from histological sections and the localization of individual
neurons was estimated on the basis of the indicated microdrive position.
Fig. 10 summarizes the data obtained. Units which responded to electrical
stimulation of the AC are represented by filled circles, those not
responding by open circles. In principle few neurons in the central nucleus
of the IC responded to AC stimulation whereas many responding neurons
were observed in the external and particularly in the dorsal cortices of
the IC. The distribution of individual types of responses to AC stimula-

IC evoked responses

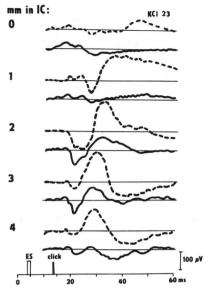

Fig. 7. Suppression of acoustical evoked responses in the IC by preceding electrical stimulation of the AC. Dashed line - control response to acoustical stimulus only, full line - auditory evoked response after preceding electrical stimulation of the AC. Averaged values for 32 responses. Depth of the electrode tip in the IC indicated on the left side.

IC evoked responses

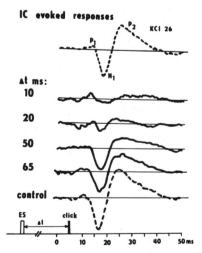

Fig. 8. Influence of the time interval between the electrical stimulation of the AC and the acoustical stimulation (click) on the suppression of the auditory evoked response in the IC. Averaged values for 32 responses.

tion within the IC is indicated in Fig. 11. Three classes of responses could be distinguished: excitation only, inhibition only and combination

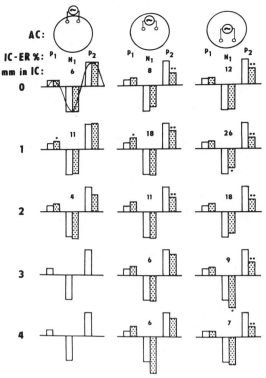

Fig. 9.   Suppression of P_1-N_1-P_2 waves of the auditory evoked response
in the IC after preceding electrical stimulation of different
sites in the AC. Depth of the electrode tip in the IC indicated
on the left side. White columns - average values of individual
wave amplitudes in control stimulations as 100% (click only);
stippled columns - average values of individual wave amplitudes
(in % relative to control stimulation) for stimulations with
preceding electrical stimulation of the AC. Number of averaged
cases indicated above each complex of columns.

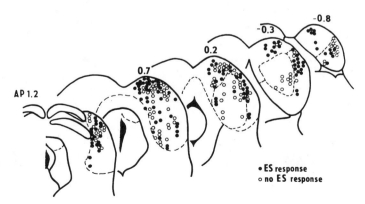

Fig. 10.   Distribution of neurons in the rat IC which respond to
electrical stimulation of the AC (full circles) and which do
not respond (empty circles).

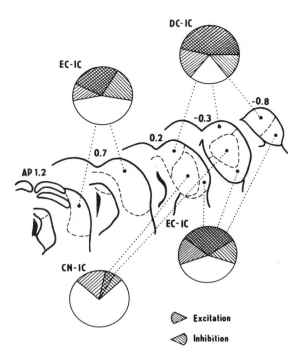

Fig. 11. Distribution of excitatory and inhibitory responses to electrical stimulation of the AC in different parts of the IC. DC-IC: dorsal cortex, EC-IC: external cortex, CN-IC: central nucleus of the inferior colliculus.

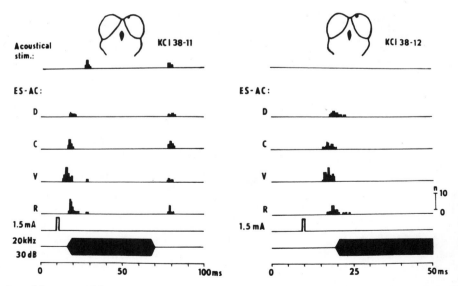

Fig. 12. PST histograms of two neurons in the IC to acoustical stimulation only and to acoustical stimulation preceded by electrical stimulation of the AC. Four sites of stimulation on the AC: D-dorsal, C-caudal, V-ventral, R-rostral. Summed responses of 64 stimulations.

of excitation and inhibition. Individual types of responses were
practically equally distributed in three structures of the IC: the external
and dorsal cortices and the central nucleus of the IC.

The degree of suppression of IC evoked responses depends on the position of stimulating electrodes in the auditory cortex as demonstrated
in Fig. 9. We therefore investigated the influence of stimulation of
individual parts of the auditory cortex on neuronal responses in the IC.
Four stimulating ball electrodes were placed on the surface of the auditory
cortex with a common central electrode. Thus four pairs of bipolar
electrodes were obtained: dorsal, caudal, ventral and rostral (Fig. 13,
above). In addition to responses to acoustical stimulation four types of
responses to AC stimulation were registered for each neuron in the IC.
Fig. 12 shows characteristic responses of two neurons of the external
cortex. Neuron KCI 38-11 (left side) reacted to a tone pip at its
characteristic frequency with an on-off reaction. Preceding single
electrical pulse stimulation of the auditory cortex resulted in excitation
which is followed by inhibition. The on-part of the response to the tone
pip is suppressed whereas the off-response remained intact. The shortest
latency was present when the ventral part of the AC was stimulated. Neuron
KC1 38-12 did not react to acoustical stimulation, the excitation elicited
by AC stimulation had the shortest latency when the ventral part of the
AC was stimulated. Whilst in both these representative neurons all stimulation sites evoked a reaction, in many other neurons stimulation of only
some of the four sites had an effect. Fig. 13 summarizes the results of
experiments with respect to the localization of neurons in three main
parts of the IC. The efficacy of cortical stimulation is demonstrated by
the stippled area within each quadrant representing the four parts of
the AC. The effects of stimulation of all four parts of the AC were
similar for neurons in the external IC cortex, whereas stimulation of the
ventral AC part was significantly more effective than stimulation of any
other AC part for neurons in the dorsal IC cortex.

The results agree in principle with the known anatomical data for
the projection of cortical descending fibers in the rat (Druga et al.,
1988; Faye-Lund, 1985). For example, Faye-Lund (1985) described projections to the external cortex of the IC to originate in the rostrodorsal
parts of the AC and projections to the dorsal cortex in the ventrocaudal
parts of the AC. Druga et al. (1988) found the origin of descending
projections to the external cortex in area Te 1 (i.e. rostrodorsally in
the AC) whereas fibers to the dorsal cortex originated in more extensive
temporal areas (i.e. also in ventral and caudal parts of the AC). The
high efficacy of stimulation of the ventral part observed in the present
study suggests that this part of the AC could play a dominant role
in influencing inferior colliculus activity. We can assume that the
function of the dorsal cortex of the IC is mainly to modulate the
processing of information in the IC by feedback loops through the
auditory cortex. The dorsal cortex receives a very limited ascending
afferent input and also does not send descending fibers to lower auditory
nuclei.

In contrast to this the role of the external cortex of the IC in
the efferent descending activity is less clear. Neurons of the external
cortex are multimodal (e.g. Aitkin et al., 1978) and they may subserve
integrative multimodal activities similarly as neurons in the superior
colliculus. However, the external cortex represents also the origin of
descending projections to lower auditory nuclei (Faye-Lund, 1986; Syka
et al., 1988) and thus a possible link with the efferent olivocochlear
bundle. If any direct influence of the auditory cortex on cochlear hair
cell activity exists, the three neuronal pathways from the auditory cortex
through the external nucleus of the inferior colliculus and the superior

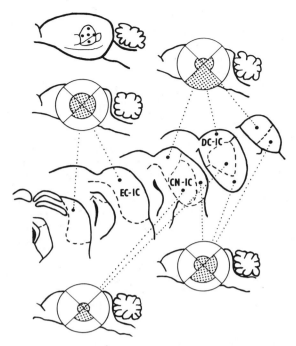

Fig. 13. Efficacy of electrical stimulation of four different sites in
the AC (represented by individual quadrants of the circle) on
neuronal activity in different parts of the IC. The extent of
the stippled area represents the relative efficacy of electrical
stimulation upon neurons which responded to stimulation of
the indicated part of the AC. Localizations of electrodes
on the AC surface depicted in the uppermost drawing. For
abbreviations see Fig. 11.

olivary complex represents the shortest route. In addition to the well
recognized role of the olivocochlear bundle we must consider seriously the
role of the complex descending auditory pathway in the processing of
auditory information. It cannot be excluded that the complicated pattern
of responses of auditory neurons to sound could, in some cases, be
explained on the basis of interaction with nuclei localized at higher levels
of the auditory pathway.

REFERENCES

Adams, J. C., 1983, Cytology of periolivary cells and the organization of
their projections in the cat, J. Comp. Neurol., 215:275-289.
Aitkin, L. M., Dickhaus, H., Schult, W. and Zimmerman, M., 1978, External
nucleus of inferior colliculus: auditory and spinal somatosensory
afferents and their interactions, J. Neurophysiol., 41:837-847.
Andersen, P., Junge, K. and Sveen, O., 1972, Corticofugal facilitation of
thalamic transmission, Brain Behav. Evol., 6:170-184.
Anderson, R. A., Snyder, R. L. and Merzenich, M. M., 1980, The topographic
organization of corticocollicular projections from physiologically
identified loci in the AI, AII and anterior auditory cortical fields
in the cat. J. Comp. Neurol., 191:479-494.
Comis, S. D. and Whitfield, I. C., 1968, Influence of centrifugal pathways
on unit activity in the cochlear nucleus. J. Neurophysiol., 31:62-68.
Comis, S. D., 1970, Centrifugal inhibitory processes affecting neurones
in the cat cochlear nucleus. J. Physiol. (Lond.), 210:751-760.

Covey, E., Jones, D. R. and Casseday, J. H., 1984, Projections from the superior olivary complex to the cochlear nucleus in the tree shrew, J. Comp. Neurol., 226:289-305.

Diamond, I. T., Jones, E. G. and Powell, T. P. S., 1969, The projection of the auditory cortex upon the diencephalon and brainstem in the cat. Brain Res., 15:305-340.

Druga, R. and Syka, J., 1984a, Ascending and descending projections to the inferior colliculus in the rat, Physiol. Bohemosl., 33:31-42.

Druga, R. and Syka, J., 1984b, Neocortical projections to the inferior colliculus in the rat, Physiol. Bohemosl., 33:251-253.

Druga, R., Syka, J. and Markow, G., 1985, Descending part of the auditory pathway- projections from the inferior colliculus in the rat, Physiol. Bohemosl., 34:414.

Druga, R., Syka, J. and Rajkowska-Markow, G., 1988, Localization of cortical neurons projecting to the inferior colliculus in the rat and guinea pig, this volume.

Eybalin, M., Cupo, A. A. and Pujol, R., 1984, Met-enkephalin characterization in the cochlea: high performance liquid chromatography and immunoelectronmicroscopy, Brain Res., 305:313-322.

Faye-Lund, H., 1985, The neocortical projection to the inferior colliculus in the albino rat, Anat. Embryol., 173:53-70.

Faye-Lund, H., 1986, Projection from the inferior colliculus to the superior olivary complex in the albino rat, Anat. Embryol., 175:35-52.

Fex, J. and Altschuler, R. A., 1981, Enkephalin-like immunoreactivity of olivocochlear nerve fibers in cochlea of guinea pig and cat, Proc. Natl. Acad. Sci. USA. 78:1255-1259.

Fex, J., Altschuler, R. A., Kachar, B., Wenthold, R. J. and Zempel, J. M., 1986, GABA visualized by immunocytochemistry in the guinea pig cochlea in axons and endings of efferent neurons, Brain Res., 366: 106-117.

Hashikawa, T., 1983, The inferior colliculopontine neurons of the cat in relation to other collicular descending neurons, J. Comp. Neurol., 219:241-249.

Rasmussen, G. L., 1946, The olivary peduncle and other fiber projections of the superior olivary complex, J. Comp. Neurol., 84:141-219.

Rasmussen, G. L., 1953, Further observations of the efferent olivocochlear bundle, J. Comp. Neurol., 99:61-74.

Rasmussen, G. L., 1960, Efferent fibres of the cochlear nerve and cochlear nucleus, in:" Neural Mechanisms of Auditory and Vestibular Systems", G. L. Rasmussen and W. F. Windle, eds., Charles C. Thomas, Springfield, 105-115.

Robertson, D., 1985, Brainstem location of efferent neurones projecting to the guinea pig cochlea, Hearing Res., 20:79-84.

Strutz, J. and Spatz, W. B., 1980, Superior olivary and extraolivary origin of centrifugal innervation of the cochlea in the guinea pig. A horseradish peroxidase study. Neurosci. Letters, 17:227-230.

Syka, J., Druga, R., Popelář, J. and Kalinová, B., 1980, Functional organization of the inferior colliculus, in: "Neuronal Mechanisms of Hearing", J. Syka and L. Aitkin, eds., Plenum Press, New York, 137-155.

Syka, J. and Popelář, J., 1984, Inferior colliculus in the rat:neuronal responses to stimulation of the auditory cortex, Neurosci. Letters., 51:235-240.

Syka, J., Robertson, D., and Johnstone, B. M., 1988, Efferent descending projections from the inferior colliculus in guinea pig, this volume.

Warr, W. B., 1975, Olivocochlear and vestibular efferent neurons of the feline brain-stem: their location, morphology and number determined by retrograde axonal transport and acetylcholinesterase histochemistry, J. Comp. Neurol., 161:159-182.

Warr, W. B. and Guinan, J. J., 1979, Efferent innervation of the organ of

Corti:Two separate systems, <u>Brain Res.</u>, 173:152-155.

Watanabe, T., Yanagisawa, K., Kanzaki, Y. and Katsuki, Y., 1966, Cortical efferent flow influencing unit responses of medial geniculate body to sound stimulation, <u>Exp. Brain Res.</u>, 2:302-317.

White, J. S. and Warr, B. W., 1983, The dual origins of the olivocochlear bundle in the albino rat. <u>J. Comp. Neurol.</u>, 219:203-214.

Whitley, J. M. and Henkel, C. K., 1984, Topographical organization of the inferior collicular projection and other connections of the ventral nucleus of the lateral lemniscus in the cat, <u>J. Comp. Neurol.</u>, 229:257-270.

LOCALIZATION OF CORTICAL NEURONS PROJECTING TO THE INFERIOR

COLLICULUS IN THE RAT AND GUINEA PIG

R. Druga, J. Syka* ang G. Rajkowska-Markow**

Inst. of Anatomy, Medical Faculty
Charles University
*Inst. of Experimental Medicine
Czechoslov. Acad. Sciences
Prague, Czechoslovakia
**Inst. Exp. Biology, Warsaw, Poland

The descendent part of the acoustic pathway is a complicated multi-synaptic projection system originating in the cerebral cortex and terminating on the receptors of the Corti organ. One of the important components of this system are cortico- collicular projections. Investigations in several mammalian species have demonstrated a direct projection from the auditory cortex to the inferior colliculus (Kuypers and Lawrence, 1967; Diamond et al., 1969; Rockel and Jones, 1973, a, b; Forbes and Moskowitz, 1974; Martin et al., 1975; Casseday et al., 1976; Cooper and Young, 1976; FitzPatrick and Imig, 1978; Adams, 1980; Willard and Martin, 1983, 1984).

In most of these studies neocortical projection to the inferior colliculus is described as a predominantly ipsilateral system (with a weak contralateral component) terminating in the external and dorsal nucleus of the inferior colliculus (IC). According to the prevailing opinion, the central nucleus of IC does not receive a direct neocortical input.

Several authors attempted to define, in the rat, the cortical field which gives rise to fibers terminating in the IC (Beyerl, 1978; Syka et al., 1981; Druga and Syka, 1984). Quite recently, Faye-Lund (1985) published a study of neocortical projections to the rat IC by anterograde fiber degeneration following large lesions of the cerebral cortex and by anterograde transport of WGA - HRP. According Faye-Lund's findings dorsal and external nucleus of the IC receive fibers from rather discrete loci of the temporal neocortical field.

Our approach to solving problems concerning the organization of neo-cortical projections to IC utilized the results from HRP injections into the dorsal and external nucleus of IC in 12 rats and 6 guinea pigs.

MATERIAL AND METHOD

The distribution of cortical neurons projecting to the IC was studied in 12 rats (weight 250-300 g) and in 6 guinea pigs (weight 300-350 g) using the horseradish peroxidase labeling technique. Small amounts of HRP (Sigma VI, 30% w/v, 0.1-0.2 μl) were injected stereotaxically with a Hamilton 1 μl syringe under pressure. After survival times of 48-72 h, the animals

were anaesthetized with pentobarbital and perfused through the left
ventricle of the heart with normal saline followed by a mixture of 0.4%
paraformaldehyde and 1.25% glutaraldehyde with 0.05 mol/l phosphate
buffer added (pH 7.4). The brains were removed from the skull, divided
into blocks and stored in the same fixative for 24 h at 4°C. Frozen section (40 μm thick) were cut in the coronal plane and treated with diaminobenzidine (Graham and Karnovsky, 1966). One series of sections was mounted
unstained, the other was lightly stained with cresyl violet. The sections were subsequently examined using both bright-field and dark-field
microscopy.

For the division of the IC in both rodent representatives we used
criteria proposed by Faye-Lund and Osen (1985). Localization of the HRP-
positive neurons was indicated by means of neocortical maps for the rat
and guinea pig. The maps are based on quantitative cytoarchitectonic
procedure using a computer controlled image analyser according to Wree
et al. (1981) and Zilles (1985).

RESULTS

Localization of Cortical Neurons Projecting to the IC in the Rat

After injections of small amounts of HRP into dorsal cortex of the
IC, retrogradely labeled neurons were found in all temporal neocortical
fields which Zilles (1985) designated as Te 1, Te 2 and Te 3. The
majority of labeled cells was located in area Te 3 and Te 1. Besides some
HRP-positive neurons occurred rostrally to area Te 1, in the caudal part
of parietal area Par 2.

Injections of HRP into external cortex of IC only labeled neocortical
neurons located in the temporal area Te 1. (Figs. 1 a and b).

Cortical neurons projecting to the external and dorsal cortex of IC
were localized exclusively in the Vth layer of the respective fields and
all the labeled cells were large or medium-sized pyramids. Ipsilateral
labeling prevailed but some of the HRP-positive cells were found in the
contralateral temporal region.

Localization of Cortical Neurons Projecting to the IC in the Guinea Pig

After injections of HRP into the dorsal cortex of IC in the guinea
pig, HRP - positive neurons were found in the anterior and basal parts
of the temporal area designated by Wree et al. (1981) as Te 2, almost in
the whole area Te 1.3 and in area Te 1. Smaller numbers of labeled cells
were also found in the close vicinity of the rhinal sulcus, in the
posterior part of area designated by Wree et al. as" claustrocortex isocorticalis" (Cli). (Fig. 2 a, b).

Injections of HRP into the external cortex of IC resulted in labeling
limited to the anterior part of area Te 1.3 and to area Te 1. No labeled
cells were found in area Te 2 and in the posterior part "claustrocortex
isocorticalis". Similarly as in rats, labeled neurons were confined to
the Vth layer of respective areas and all of them were large and medium-
sized pyramidal cells. Furthermore, a small number of retrogradely
filled neurons was found in the contralateral temporal areas.

DISCUSSION

If we compare our findings in rats with the results of Faye-Lund
(1985) it must be taken into consideration that we have used different
cytoarchitectonic parcellation of temporal neocortex based on Zilles (1985)

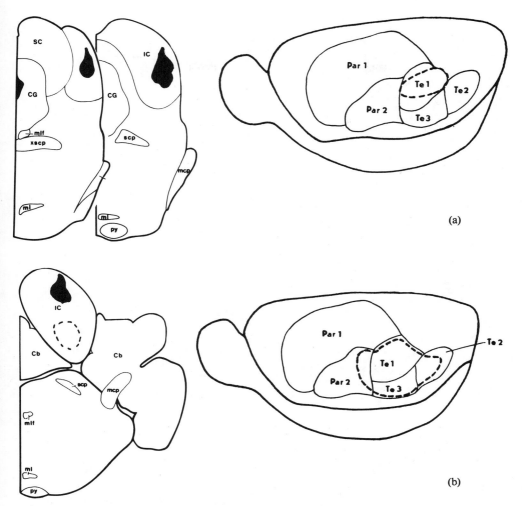

Fig. 1. a) Frontal sections of the inferior colliculus in the rat. The black area depicts the deposit of injected HRP into the external nucleus of the IC. Broken line indicates the distribution of labeled neurons in temporal area Te 1.
b) The black area depicts the deposit of HRP in dorsal nucleus of IC. Broken line indicates labeling in all temporal areas and in the posterior part of area Par 2.

quantitative analysis, while Faye-Lund used Krieg's (1946) map of the rat neocortex.

Apart from the differences between the cytoarchitectonic maps used by Faye-Lund and by us our results concerning the organization of cortico-collicular projections in the rat are not in contradiction with their findings. According to Faye-Lund (1985), ejection sites projecting exclusively to the external cortex of IC cover the rostrodorsal part of the cortical field. In the central position and caudoventrally ejection sites project to the dorsal collicular cortex.

According to the present findings, the temporal field projecting to the external collicular cortex covers the dorsal and dorsorostral part of the respective neocortical region and corresponds approximately to

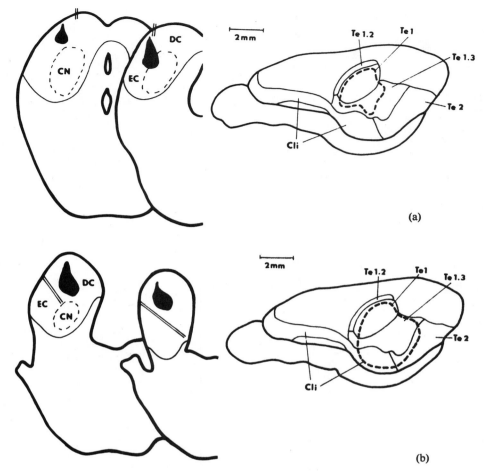

Fig. 2. a) The extent of injected HRP (black area) into the external
nucleus of IC in the guinea pig. Labeling was limited to temporal
area Te 1 and to rostral part of area Te 1.3. (broken line).
b) After injections of HRP into the dorsal nucleus of the IC,
HRP-positive neurons were found in all temporal areas and
additionally in the posterior part of area Cli.

Zilles' area Te 1. The cortical field projecting to the dorsal collicular
cortex is more extensive and covers, besides area Te 1, also area Te 3,
the anterior part of area Te 2 and the posterior part of parietal Par 2.

A similar organization of neocortical projections to the IC was found
in the guinea pig. After injections of HRP limited to the external IC
nucleus, retrogradely labeled cells were confined to a relatively small
area that includes area Te 1 and the anterior part of area Te 1.3. After
injections involving the dorsal cortex of IC, the labeled field was more
extensive and also covered the anterior part of area Te 2, the posterior
part of claustrocortex isocorticalis (Cli) and almost the whole of area
Te 1.3. Labeling in the posterior part of Cli indicates that this area
is probably a part of the acoustic cortical field. In the context of these
findings it is of interest to note that Hellweg et al. (1977) distinguished
two auditory cortical fields in the guinea pig adjoining the rhinal
fissure in their basal part. If we compare the location of Hellweg's
auditory area with the cytoarchitectonic map used (Wree et al. 1981),

Hellweg's area probably corresponds to area Te 1, Te 1.3, and partly to the posterior part of claustrocortex isocorticalis.

Redies (1987) distinguished two large tonotopic areas (anterior field "A" and dorsocaudal field "DC") in the guinea pig's auditory cortex. A third tonotopic area, a small field "S", is located anterior to field "A". Fields "A" and "DC" are located posterior to Sylvian fissure and probably correspond to Wree's area Te 1 and at least to part of area Te 1.3.

SUMMARY

The retrograde transport of HRP was used to study the cells of origin of the neocortical projection to the dorsal and external nucleus of the IC in the rat and guinea pig. In both these rodent representatives the organization of neocortical projections to the IC is similar. Cortico-collicular projections are predominantly ipsilateral with a weak contra-lateral component. Injections of HRP into the external IC nucleus resulted in the labeling of area Te 1 (rat) and of areas Te 1 and Te 1.3 (guinea pig). Injections of HRP into the dorsal IC nucleus resulted in labeling of all temporal areas in the rat and in the guinea pig. All labeled neurons were located in the Vth layer of the respective cortical areas.

REFERENCES

Adams, J. C., 1980, Crossed and descending projections to the inferior colliculus, Neurosci. Lett., 19:1-5.

Beyerl, B. D., 1978, Afferent projections to the central nucleus of the inferior colliculus in the rat, Brain Res., 145:209-223.

Casseday, J. H., Diamond, I. T. and Harting, J. K., 1976, Auditory pathways to the cortex in Tupaia glis, J. Comp. Neurol., 166:303-340.

Cooper, M. H. and Young, P. A., 1976, Cortical projections to the inferior colliculus of the cat, Exp. Neurol., 51:488-502.

Diamond, I. T., Jones, E. G. and Powell, T. P. S., 1969, The projection of the auditory cortex upon the diencephalon and brain stem in the cat, Brain Res., 15:305-340.

Druga, R. and Syka, J., 1984, Ascending and descending projections to the inferior colliculus in the rat, Physiol. Bohemoslov., 33:31-42.

Faye-Lund, H., 1985, The neocortical projection to the inferior colliculus in the albino rat, Anat. Embryol., 173:53-70.

Faye-Lund, H. and Osen, K., 1985, Anatomy of the inferior colliculus in rat, Anat. Embryol., 171:1-20.

FitzPatrick, K. A. and Imig, T. J., 1978, Projections of auditory cortex upon the thalamus and midbrain in the owl monkey, J. Comp. Neurol., 177:537-556.

Forbes, B. F. and Moskowitz, N., 1974, Projections of auditory responsive cortex in the squirrel monkey, Brain Res., 67:239-254.

Graham, R. C. and Karnovsky, M. J., 1966, The early stages of absorption of injected horseradish peroxidase in the proximal tubules of mouse kidney. Ultrastructural histochemistry by a new technique, J. Histo-chem. Cytochem., 14:291-302.

Hellweg, F. C., Koch, R. and Vollrath, M., 1977, Representation of the cochlea in the neocortex of guinea pig, Exp. Brain Res., 29:467-474.

Krieg, W. J. S., 1946, Connections of the cerebral cortex. I. The albino rat. A. Topography of the cortical areas, J. Comp. Neurol., 84:221-276.

Kuypers, H. G. J. M. and Lawrence, D. G., 1967, Cortical projections to the red nucleus and the brain stem in the rhesus monkey, Brain Res., 4:151-188.

Martin, G. F., Bresnahan, J. C., Henkel, C. K. and Megirian, D., 1975, Corticobulbar fibers in the North American opossum (Didelphis marsupialis virginiana) with notes on the Tasmanian brush-tailed possum (Trichosurus valpecula) and other marsupials, J. Anat., 120: 439-484.

Redies, H., 1987, Tonotope Strukturen im auditorischen Thalamus und Cortex des Meerschweinchens. Dissertation. Georg - August - Universität zu Göttingen, 1-124.

Paxinos, G. and Watson, C., 1982, The rat brain in stereotaxic coordinates. Academic Press, Sydney, New York, London.

Rockel, A. J. and Jones, E. G., 1973a, The neuronal organization of the inferior colliculus of the adult cat. I. The central nucleus, J. Comp. Neurol., 147:11-60.

Rockel, A. J. and Jones, E. G., 1973b, The neuronal organization of the inferior colliculus of the adult cat. II. The pericentral nucleus, J. Comp. Neurol., 149:301-334.

Syka, J., Druga, R., Popelář, J. and Kalinová, B., 1981, Functional organization of the inferior colliculus, in: "Neuronal Mechanisms of Hearing", J. Syka and L. Aitkin eds., Plenum Press, New York - London, 137-153.

Willard, F. H. and Martin, G. F., 1983, The auditory brainstem nuclei and some of their projections to the inferior colliculus in the North American opossum, Neuroscience, 10:1203-1232.

Willard, F. H. and Martin, G. F., 1984, Collateral inervation of the inferior colliculus in the North American opossum: A study using fluorescent markers in a double- labeling paradigm. Brain Res., 303:171-182.

Wree, A., Zilles, K. and Schleicher, A., 1981, A quantitative approach to cytoarchitectonics. VII: The areal pattern of the cortex of the guinea pig, Anat. Embryol., 162:81-103.

Zilles, K., 1985, The cortex of the rat. A stereotaxic atlas. Springer Verlag, Berlin - Heidelberg - New York - Tokyo, 1-121.

EFFERENT DESCENDING PROJECTIONS FROM THE INFERIOR COLLICULUS IN

GUINEA PIG

J. Syka, D. Robertson* and B. M. Johnstone*

Institute of Experimental Medicine
Czechoslovak Academy of Sciences
Prague 2, Czechoslovakia
* Department of Physiology
University of Western Australia
Nedlands, 6009, Australia

Efferent olivocochlear neurons have been studied in the guinea pig both morphologically (Strutz and Spatz, 1980; Strutz and Beilenberg, 1984; Robertson, 1985) and electrophysiologically (Robertson, 1984; Robertson and Gummer, 1985). They originate in the lateral superior olive (LSO) ipsilateral to the injected cochlea and bilaterally in the dorsomedial periolivary region (DPMO), in the medial, ventral and lateral nuclei of the trapezoid body (MNTB, VNTB, LNTB) and in the vicinity of the ventral nucleus of the lateral lemniscus. Descending fibers are also known to project from the auditory cortex to the inferior colliculus (in guinea pig see Druga et al., this volume). The aim of our work was to investigate the origin and termination of fibers descending from the inferior colliculus (IC) in guinea pig. We were particularly interested in fibers projecting from the inferior colliculus to the superior olivary complex (SOC) because of possible contact with neurons projecting to the cochlea via the olivo-cochlear bundle.

METHODS

17 pigmented guinea pigs (250 - 300 g) were used in this study. They were anaesthetised by pentobarbital and the skull was opened above the occipital cortex and the cerebellum. The animal's head was then fixed in a stereotaxic apparatus and a microelectrode was inserted into the IC or the SOC. Tone pips delivered binaurally through hollow ear bars served as acoustic stimuli. Characteristic evoked potentials or single unit activity were recorded in the IC or the SOC; they were used as indicators of the position of the electrode in the structure. 1% solution of horseradish peroxidase conjugated to wheat germ (WGA-HRP) was injected iontophoretically into the structure (current approximately 2 μA, positive 1 s pulses, 3 times for 5 min). The electrode was left in the brain for 15 min after last application. After survival times of 24 h or 48 h the animals were anaesthetized with pentobarbital and perfused transcardially with 0.9% saline followed by 2.5% glutaraldehyde in 0.1 M phosphate buffer. After 30 min fixation the animals were perfused with 30% sucrose in 0.1M phosphate buffer and the brains removed and stored overnight in the same solution at 4° C. Serial transverse frozen sections were cut at thickness ranging from 35 to 50 μm. Sections were reacted for the presence of HRP

with the tetramethylbenzidine (TMB) method. In several animals alternate sections were reacted with diaminobenzidine (DAB) method.

RESULTS

In 9 animals WGA-HRP was injected into the IC and anterograde labelling was observed in different auditory nuclei (Table 1). Three main targets of descending fibers were the ipsilateral dorsal nucleus of the lateral lemniscus (DNLL), ipsilateral ventral nucleus of the lateral lemniscus (VNLL) and ipsilateral ventral nucleus of the trapezoid body (VNTB). Anterograde labelling was found also in the contralateral IC and in the ipsilateral medial geniculate body. Cochlear nuclei of both sides were practically without anterograde labelling. In most cases the area of injection was small and limited to different parts of the IC. Anterograde labelling in the VNTB did not depend on the location of injection in the IC, i.e. all injections to the IC resulted in anterograde labelling in the VNTB.

Table 1. Distribution of anterograde labelling in auditory nuclei after WGA-HRP injection to inferior colliculus.

IC INJECTION - ANTEROGRADE TRANSPORT WGA HRP

ANIMAL No.	DNLL i	DNLL c	VNLL i	VNLL c	LSO i	LSO c	MSO i	MSO c	VNTB i	VNTB c	LNTB i	LNTB c	MNTB i	MNTB c	DPMO i	DPMO c	CN i	CN c	IC c
8	X		X																X
11	X		X																
12	X								X										X
19	X		X						X										
22	X		X						X										X
48	X								X										
50	X		X						X	X			X		X				
53	X								X		X								
55			X						X										X

Retrogradely labelled cells after IC injections were observed in large quantities in the contralateral cochlear nuclei, ipsilateral dorsal nucleus of the lateral lemniscus (DNLL), ipsilateral ventral nucleus of the lateral lemniscus (VNLL) and in the contralateral DNLL. Less frequently occurred retrogradely labelled cells in the SOC, where the largest distribution was found in the ipsilateral MSO, VNTB and DPMO. Small numbers of retrogradely labelled cells were found in the LSO on both sides. Ipsilateral cochlear nuclei were practically devoid of retrogradely labelled neurons. The distribution of labelled cells in subcollicular nuclei was dependent on the location of HRP injection in the IC; some sites in the IC were associated with heavy labelling in the cochlear nuclei, whereas injection of HRP in other sites resulted in the occurrence of many labelled cells in the lemniscal nuclei or in the SOC. We did not analyse these relationships in details, though the results were in principle similar to those described in cat by Kudo and Nakamura (this volume).

In 8 animals the WGA-HRP was injected in different parts of the SOC. Insertion of the electrode tip to the SOC was accompanied by a characteristic evoked potential with latency about 3 ms. Excitation from the ipsilateral ear and inhibition from the contralateral ear were constantly found when the electrode tip was in the LSO; reversed effectiveness, i.e. ipsilateral inhibition and contralateral excitation was observed at electrode locations within medial parts of the SOC. Injection of the WGA-HRP to medial and lateral parts of the SOC resulted in different pattern of retrograde labelling. Table 2 shows distribution of retrogradely labelled cells in three animals with different sites of the HRP injection in the SOC. In GP 57 the application site comprised MSO, VNTB, MNTB and DMPO with only the two last mentioned structures being involved in the heavily stained center. Retrogradely labelled cells were found predominantly in the contralateral cochlear nuclei, with few cells in the ipsilateral

Table 2. Distribution of retrograde and anterograde labelling in auditory nuclei after WGA-HRP injection to different parts of the superior olivary complex.

GP 57 INJECTION WGA HRP MEDIAL PART SOC

	VCN		DCN		VNLL		DNLL		IC	
	i	c	i	c	i	c	i	c	i	c
RETROGR.	4	1124	0	13	6	0	34	0	32	0
ANTEROGR.	+ +		+	+	+ +		+		+ +	+

GP 58 INJECTION WGA HRP PREDOMINANT MEDIAL PART SOC

	VCN		DCN		VNLL		DNLL		IC	
	i	c	i	c	i	c	i	c	i	c
RETROGR.	347	2550	0	1	0	3	1	10	225	5
ANTEROGR.	+ +	+	+	+	+	+	+ +		+ +	+

GP 34 INJECTION WGA HRP LATERAL PART SOC

	VCN		DCN		VNLL		DNLL		IC	
	i	c	i	c	i	c	i	c	i	c
RETROGR.	192	0	0	0	0	0	0	0	3	0
ANTEROGR.					+	+	+		+	

VNLL, DNLL and IC. Anterograde labelling was present mainly on the ipsilateral side, in the VCN, DCN, VNLL, DNLL and IC. Similarly in the animal GP 58 the injection comprised predominantly the medial part of the SOC. Large numbers of retrogradely labelled neurons were present in the contralateral VCN with less retrogradely labelled neurons in the ipsilateral VCN. Practically no labelled cells occurred in the DCN, VNLL and DNLL, however, significant numbers of labelled cells were found in the ipsilateral IC. The heavily stained center of the HRP injection comprised MSO, DPMO, VNTB and also ventral parts of the VNLL. Anterograde labelling was widely spread, practically in all auditory mesencephalic and brainstem nuclei.

In contrast with medial SOC injections the LSO injections resulted in very limited labelling of neurons in other auditory nuclei. Significant numbers of labelled cells were present only in the ipsilateral VCN. Anterograde labelling was also very scarce and present only in the ascending pathways. In all animals with SOC injections labelled fibers of the olivo-cochlear bundle were observed, especially those fibers which cross under the floor of the fourth ventricle. In two animals with medial SOC injections the labelling of fibers in the organ of Corti has been investigated. In both cases labelling occurred under OHC in the contralateral cochlea. Scattered but definite labelling was observed also in the nuclei of the commissure of the IC. Neurons of the dorsal cortex of the IC were never labelled after SOC injections.

Fig. 1 shows characteristic retrograde labelling of neurons in the

ipsilateral IC after WGA-HRP injections in the medial part of the SOC.
Medium and large size labelled neurons are present mainly in the ventromedial parts of the external IC cortex. Few scattered neurons are
labelled also in the central IC nucleus in parts adjacent to the external
cortex.

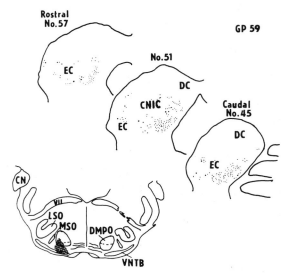

Fig. 1. Distribution of retrogradely labelled neurons in the inferior
colliculus after WGA-HRP injection to medial part of the superior
olivary complex. CNIC - central nucleus of the inferior colliculus, EC - external cortex of the IC, DC - dorsal cortex of
the IC, CN - cochlear nucleus, LSO - lateral superior olive,
MSO - medial superior olive, DMPO - dorsomedial periolivary
complex, VNTB - ventral nucleus of the trapezoid body.

DISCUSSION

As in the rat (Druga et al., 1985; Faye-Lund, 1986) neurons of the
IC in guinea pig project to subcollicular nuclei, particularly to the
ipsilateral DNLL, VNLL and SOC. Fibers descending to the SOC originate
in the external cortex of the IC and terminate mainly in the VNTB.
Although anterograde labelling was observed in the VNTB also after WGA-
HRP injections to the dorsal cortex of the IC the absence of retrograde
labelling after injections in SOC suggests that in this case the transsynaptic transport of the WGA-HRP may be involved. The termination of
fibers descending from the IC predominantly in the VNTB is of interest
because Robertson (1985) has shown that the VNTB belongs to those SOC
nuclei which project to the cochlea via the olivocochlear bundle. In the
rat VNTB represents the main nucleus of origin of medial olivocochlear
bundle fibers (White and Warr, 1983). The lack of projection of descending
fibers from the IC to the LSO is surprising. This contrasts with the large
numbers of the small LSO neurons which send their axons to the cochlea in
guinea pig (Robertson, 1985).

On the basis of the findings described it may be proposed that in guinea pig as in the rat, there exists a three neuronal descending link between the auditory cortex and the cochlea with interpolations in the external cortex of the IC and in the VNTB. Fibers descending from the auditory cortex to the dorsal cortex of the IC apparently do not have direct contact with neurons of the external IC cortex. They can, however, modulate their activity through interpolated IC neurons. Whether any direct influence of the auditory cortex on the cochlear activity exists remains to be investigated. So far only the effects of cortical stimulation on the neuronal activity in the IC (Syka and Popelář, 1984; Syka et al., this volume) and the influence of the olivocochlear bundle on the cochlear activity (e.g., Robertson and Gummer, this volume, Guinan, this volume) have been studied.

ACKNOWLEDGEMENTS

Supported by a grant from the N.H.M.R.C. (Australia) and the University of Western Australia.

REFERENCES

Druga, R., Syka, J. and Markow, G., 1985, Descending part of the auditory pathway- projections from the inferior colliculus in the rat, Physiol. Bohemosl., 34:414.

Druga, R., Syka, J. and Rajkowska-Markow, G., 1988, Localization of cortical neurons projecting to the inferior colliculus in the rat and guinea pig, this volume.

Faye-Lund, H., 1986, Projection from the inferior colliculus to the superior olivary complex in the albino rat, Anat. Embryol., 175:35-52.

Guinan, Jr. J. J., 1988, Physiology of the olivocochlear efferents, this volume.

Kudo, N. and Nakamura, Y., 1988, Organization of lateral lemniscal fibers converging onto the inferior colliculus in the cat; an anatomical review, this volume.

Robertson, D., 1984, Horseradish peroxidase injection of physiologically characterized afferent and efferent neurones in the guinea pig spiral ganglion, Hearing Res., 15:113-122.

Robertson, D., 1985, Brainstem location of efferent neurones projecting to the guinea pig cochlea, Hearing Res., 20:79-84.

Robertson, D. and Gummer, M., 1985, Physiological and morphological characterization of efferent neurones in the guinea pig cochlea, Hearing Res., 20:63-78.

Robertson, D. and Gummer, M., 1988, Physiology of cochlear efferents in mammals, this volume.

Strutz, J. and Spatz, W. B., 1980, Superior olivary and extraolivary origin of centrifugal inervation of the cochlea in guinea pig: a horseradish peroxidase study, Neurosci. Lett., 17:227-230.

Strutz, J. and Beilenberg, K., 1984, Efferent acoustic neurons within the lateral superior olivary nucleus of the guinea pig, Brain Res., 299:174-177.

Syka, J. and Popelář, J., 1984, Inferior colliculus in the rat: neuronal responses to stimulation of the auditory cortex, Neurosci. Lett., 51:235-240.

Syka, J., Popelář, J., Druga, R. and Vlková, A., 1988, Descending central auditory pathway-structure and function, this volume.

SOME SPECULATIONS ON THE FUNCTION OF THE DESCENDING AUDITORY PATHWAYS

H. Faye-Lund

Anatomical Institute
University of Oslo
Karl Johansgt. 47, Oslo I, Norway

The descending auditory pathways are still less known than the ascending, and the central parts of the descending pathways are less familiar than the peripheral parts.

The present investigation concerns the more central descending pathways in rat, an increasingly used experimental animal, partly because of economical reasons.

The inferior colliculus, IC, is a near obligatory relee-station for ascending as well as descending auditory pathways. In line with this IC has the highest spontaneous activity in the rat brain (Clerici and Coleman, 1986). IC in rat was found to consist of a central nucleus, a dorsal cortex and an external cortex (Faye-Lund and Osen, 1985) (Fig. 1). This is also the case in other species, but the relative size of the parts vary. Thus the central nucleus in rat IC is smaller and the cortices larger than in cat. The central nucleus, defined as the laminated, tonotopically arranged part of IC, is located in the caudal and medial part of the structure, and is surrounded caudally and dorsally by the dorsal cortex, and laterally, medially, ventrally and rostrally by the external cortex. The last mentioned part constitutes about half the rostrocaudal and one third of the mediolateral IC. Both the external and the dorsal cortices consist of three layers based on cyto- and myelo-architecture.

The descending fibers from neocortex to IC are mainly known to terminate in the peripheral parts of IC (among others, see: Noort 1969, Rockel and Jones 1973 a, b, Adams 1980, Andersen et al., 1980), while the ascending fibers mostly terminate in the central nucleus (among others, see: Noort 1969, Elverland 1978, Adams 1979). By degeneration studies we found that in rat only the auditory neocortical fields project to IC, and that their termination areas were the external and dorsal cortices (Faye-Lund, 1985). The central nucleus, however, did not receive any direct neocortical input. After discrete applications of horseradish peroxidase in auditory neocortical fields the results spoke in favour of three separate projections from three origins in neocortex. Thus area 41 was found to project to layer 2-3 of the dorsal cortex bilaterally, area 22 to layer 2 and the superficial part of layer 3 of the external cortex ipsilaterally while area 36 most probably project to layer 1 of the dorsal cortex ipsilaterally (Fig. 1).

The colliculoolivary projection in rat was found by degeneration and antero- and retrograde horseradish peroxidase studies (Faye-Lund, 1986). The fibers were shown to originate in layer 3 in the external cortex and the adjacent part of the central nucleus. The fibers terminated in the rostral and medioventral zone of the ipsilateral periolivary region (Fig. 1). The terminal area carefully overlapped the large olivo-cochlear cells, demonstrated by acetylcholinesterase (Osen et al., 1984). Degeneration in areas known to contain the motoneurons of the stapedial and tensor tympani muscles was also seen (Lyon, 1975, 1978, Joseph et al., 1985). In this periolivary area are also found smaller periolivary cells projecting to the cochlear nuclei (Adams, 1983). Both cell types mainly have a crossed projection in rat. Judged from these findings both cell types could constitute the next link in a descending projection to the outer hair cells, and/or cells in the cochlear nuclei on the opposite side.

As a conclusion, the present findings make possible a 3-neuronal chain from the auditory cortex, via the external cortex of the ipsilateral IC, and the periolivary region of the ipsilateral superior olivary complex to the outer haircells in cochlea and cells in the cochlear nuclei. An electronmicroscopical demonstration of these synapses, however, remains.

The external cortex might because of its electrophysiological qualities and connections be a center for auditory reflexes (Aitkin et al., 1984; Willard and Ryugo, 1983). It is therefore interesting that the main part of the colliculoolivary fibers originate from here. The external cortex has connections with the dorsal column nuclei (Bjorkland and Boivie, 1984; Wiberg and Blomqvist, 1984) and the trigeminal nuclei (Willard and Ryugo, 1979). An interaction between auditory and somato-sensory systems might therefore take place here, and reflexes originating in this part of IC could be influenced by multisensory impulses.

Auditory reflexes might from their effect be of several kinds. One kind can through its efferent part affect structures in cochlea itself. The crossed olivocochlear system and the middle ear reflexes are among these. The two systems at the same time overlap and supply each other functionally. The two mechanisms work in the extreme ends of the spectrum of frequencies, the middle ear reflexes mostly in the low frequency, and the crossed olivocochlear system mainly in the high frequency area (Guinan et al., 1984). Both protect the inner hair cells against noise damage (Borg et al., 1979; Engstrom, 1983).

The large olivocochlear cells are compared to motoneurons which by influencing the form of the outer hair cells and the stiffness of their stereocilia, reduce the cochlear response to sound and modulate the sensitivity and "tuning" of the hair cells (Brownell, 1983; Strelioff and Flock, 1982). If the 3-neuronal chain, mentioned earlier exists, the outer hair cells may be under direct neocortical control via IC and the periolivary region. The reason why the inner hair cells are more easily damaged by noise after removal of the outer hair cells, may be the lack of such a centrally dependent inhibition.

The middle ear reflexes by regulating the sound transmission through the ossicles reduce the stimulation of cochlea and increase the dynamic area of the ear (Borg and Zakrisson, 1973). The reflex arches are only partly known. A simple connection between the cochlear nuclei and the motor neurons of the tensor tympany muscle have been demonstrated (Fig. 2) (Itoh et al., 1985). There is reason to believe that the middle ear reflexes are under neocortical controle, among other things the fact that the reflexes are activated before an expected stimulus (Djupesland,

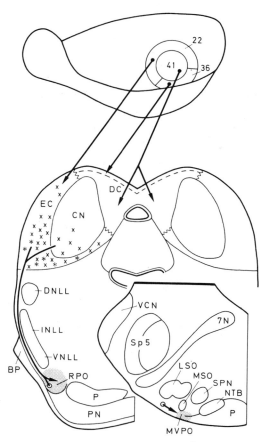

Fig. 1.   This semidiagrammatic drawing shows part of the descending
auditory pathway in rat. It includes the auditory cortex,
inferior colliculus, pons, and the superior olivary complex
and summarizes the results of the present and preceding studies.
The crosses (cells < 20 um) and asterisks (cells > 20 um)
indicate the actual mean number of colliculoolivary cells in
a 50 μm thick section. The auditory cortical areas 41, 22 and
36 have different targets in the colliculus. Only area 22
supplies the superficial part of layer 3 external cortex (EC),
which gives rise to fibers descending further to subcollicular
levels. The field of origin of the colliculoolivary fibers in
addition includes the most deepest part of EC and the adjacent
part of the central nucleus (CN). The colliculoolivary fibers
terminate in the rostral (RPO) and medioventral (MVPO) zones
of the ipsilateral periolivary region (terminal field indicated
by shading). The direct bilateral projection from IC to the
cochlear nuclei is not indicated.

1967). This neocortical control is made possible via several pathways.
One being the direct projection via IC to the trigeminal and facial
motor nuclei, the other via neurons in the superior olivary complex
(Borg, 1973).

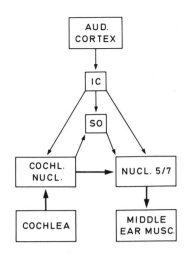

Fig. 2. This diagram shows the possible connections between neocortex, IC, superior olivary complex (SO), cochlea, cochlear nuclei, the motor nuclei of the stapedius and tensor tympani muscles and the middle ear muscles themselves.

REFERENCES

Adams, J. C., 1979, Ascending projections to the inferior colliculus, J. Comp. Neur., 183:519-538.

Adams, J. C., 1980, Crossed and descending projections to the inferior colliculus, Neurosci. Lett. 19:1-5.

Adams, J. C., 1983, Cytology of periolivary cells and the organization of their projections in the cat, J. Comp. Neur., 215:275-289.

Aitkin, L. M., Irving, D. R. and Webster, W. R., 1984, Central neural mechanisms of hearing, in: "Handbook of Physiology - The Nervous System", Brochart, J. M., Mountcastle, V. B., Darian-Smith, I., Geiger, S. R., eds., Vol 3, part 2., Waverly Press Inc., Baltimore MD.

Andersen, R. A., Snyder, R. L. and Merzenich, M. M., 1980, The topographic organization of corticocollicular projections from physiologically identified loci in the AI, AII and anterior auditory cortical fields of the cat, J. Comp. Neur., 191:479-494.

Bjorkeland, M. and Boivie, J-, 1984, An anatomical study of the projections from the dorsal column nuclei to the midbrain in cat, Anat. Embryol., 170:29-43.

Borg, E., 1973, On the neuronal organization of the acoustic middle ear reflex, A physiological and anatomical study, Brain Res., 49:101-123.

Borg, E. and Zakrisson, J-E., 1973, Stapedius reflex and speech features, J. Acoust. Soc. Am., 54:525-527.

Borg, E., Nilsson, R. and Liden, G., 1979, Fatigue and recovery of the human acoustic stapedius reflex in industrial noise, J. Acoust. Soc. Am., 65:846-848.

Brownell, W. E., 1983, Observations on a mobile response in isolated outer hair cells, in: "Mechanisms of Hearing", Webster, W. R., Aitkin, L. M., eds., Monash University Press, Clayton, Victoria, Australia 3168, 5.

Clerici, W. J. and Coleman, J. R., 1986, Resting and high-frequency evoked 2-deoxyglucose uptake in the rat inferior colliculus: developmental changes and effect of short-term conduction blockade, Dev. Brain Res., 27:127-137.

Djupesland, G., 1967, Contractions of the tympanic muscles in man, Universitetsforlaget, Oslo. Norwegian monographs on Medical Science.

Elverland, H. H., 1978, Ascending and intrinsic projections of the superior olivary complex in the cat, Exp. Brain Res., 32:117-134.

Engström, B., 1983, Stereocilia in sensory cells in normal and hearing impaired ears. A morphological, physiological and behavioural study, Scand. Audiol. Suppl., 19:1-34.

Faye-Lund, H., 1985, The neocortical projection to the inferior colliculus in the albino rat, Anat. Embryol., 173:53-70.

Faye-Lund, H., 1986, Projection from the inferior colliculus to the superior olivary complex in the albino rat, Anat. Embryol., 175: 35-52.

Faye-Lund, H. and Osen, K. K., 1985, Anatomy of the inferior colliculus in rat, Anat. Embryol., 171:1-20.

Guinan, J. J., Warr, W. B. and Norris, B. E., 1984, Topographic organization of the olivocochlear projections from the lateral and medial zones of the superior olivary complex, J. Comp. Neur., 226:21-27.

Itoh, K., Nomura, S., Konishi, A., Yasui, Y., Sugimoto, T. and Muzino, N., 1986, A morphological evidence of direct connections from the cochlear nuclei to tensor tympani motoneurons in the cat: a possible afferent limb of the acoustic middle ear reflex pathways, Brain Res., 375:214-219.

Joseph, M. P., Guinan, J. J., Fullerton, B. C., Norris, B. E. and Kiang, N. Y. S., 1985, Number and distribution of stapedius motoneurons in cats, J. Comp. Neur., 232:43-54.

Lyon, M. J., 1975, Localization of the efferent neurons of the tensor tympany muscle of the newborn kitten using horseradish peroxidase, Exp. Neur., 49:439-455.

Lyon, M. J., 1978, The central location of the motor neurons to the stapedius muscle in the cat, Brain Res., 143:437-444.

Noort, J. van, 1969, The structure and connections of the inferior colliculus. An investigation of the lower auditory system, Proefschrift, van Gorcum and Comp, NV Leiden, 1-118.

Osen, K. K., Mugnaini, E., Dahl, A-L. and Christiansen, A. H., 1984, Histochemical localization of acetylcholinesterase in the cochlear and the superior olivary nuclei. A reappraisal with emphasis on the cochlear granule cell system, Arch. Ital. Biol., 122:169-212.

Rockel, A. J. and Jones, E. G., 1973a, The neuronal organization of the inferior colliculus of the adult cat. I The central nucleus, J. Comp. Neur., 147:11-60.

Rockel, A. J. and Jones, E. G., 1973b, The neuronal organization of the inferior colliculus of the adult cat. II The pericentral nucleus, J. Comp. Neur., 149:301-334.

Strelioff, D. and Flock, A., 1982, Mechanical properties of hair bundles of receptor cells in the guinea pig cochlea, Soc. Neurosci. Abstr., 5:40.

Wiberg, M. and Blomqvist, A., 1984, The projection to the mesencefalon from the dorsal column nuclei. An anatomical study in the cat, Brain Res., 311:225-244.

Willard, F. H. and Ryugo, D. K., 1979, External nucleus of the inferior colliculus:A site of overlap for ascending auditory and somatosensory projections in the mouse, Soc. Neurosci. Abstr., 5:33.

Willard, F. H. and Ryugo, D. K., 1983, Anatomy of the central auditory system, in: "The auditory psychobiology of the mouse", Willot, J. F., ed., Thomas, Springfield, 201-304.

PROCESSING OF COMPLEX ACOUSTIC STIMULI

HUMAN AUDITORY PHYSIOLOGY STUDIED WITH POSITRON EMISSION TOMOGRAPHY

Judith L. Lauter*, Peter Herscovitch**, Marcus E. Raichle**

*Speech and Hearing Sciences
University of Arizona
Tucson AZ 85721 USA
**Edward Mallinckrodt Institute of Radiology
Washington University School of Medicine
St. Louis, MO 63110, USA

Although the past 40 years have seen significant progress in our understanding of the organization of sensory nervous systems in a number of animals, access to the details of human sensory CNS structure and function has been hampered by the lack of noninvasive, high-resolution technology. However, within the last decade, a number of new devices have appeared that provide relatively noninvasive access to the human brain: e.g. CT and MRI for anatomical imaging, and MEG, BEAM, and PET for topographic physiological studies.

METHODS

PET

Positron emission tomography represents a modification of tissue autoradiography techniques, and depends on the ability of radiation products of positron-emitting isotopes to penetrate the human skull, and thus become externally detectable. When a bolus of water labelled with oxygen-fifteen is injected into a subject's arm vein, a ring of detectors surrounding the subject's head can generate a data array that can be used to reconstruct an image showing a topography of greater and lesser concentrations of isotope. The resulting images represent the brain as a series of slices ranging from the top of the brain down into cerebellum. A color scale indicates regions of greater and lesser isotope concentration, or, if blood samples are taken during scanning to monitor actual isotope levels, regions of greater and lesser blood flow. Using appropriate software, such images can also be combined to produce difference images, showing derived maps of areas undergoing greater or lesser change in blood flow/isotope concentration from control scan to a scan taken under stimulation conditions.

For these studies, positron emission tomography was performed using a PETT VI system (Ter-Pogossian et al., 1982; Yamamoto et al., 1982). Data are recorded simultaneously for 7 slices with a center-to-center separation of 14.4 mm; the in-plane (i.e., transverse) reconstructed resolution is about 12.4 mm in the center of the field of view, and slice (axial) thickness is about 13.9 mm at the center. Each scan is 40 sec in

313

length, and is performed following the intravenous bolus injection of about 10 ml of saline containing 55-80 mCi of 0-fifteen-labelled water (half life: 123 sec). Cerebral blood flow (CBF: ml/(min x 100 g)) is calculated using a PET adaptation of the Kety tissue autoradiographic technique previously described and validated in our laboratory (Herscovitch et al., 1983; Raichle et al., 1983).

For auditory studies, we have designed a sound-delivery system based on insert receivers set in plastic tubing that snaps into standard earmolds. This fits underneath the face mask (see below) , and allows not only the shielding of text stimuli from ambient noise, but also the isolation of sounds to the two ears, to distinguish monaural, binaural, and dichotic presentations. The frequency response of this system has been shaped to mimic the filter characteristics of the outer ear, so that the signal presented at the eardrum is "ecologically valid" in its acoustical makeup (see Lauter et al., 1985 for a complete description).

Subjects

Normal young adults with no history of neurological or hearing disorders served as subjects; each was paid for his/her participation. Prior to testing, each subject received an orientation visit to the laboratory, when all procedures were explained, and a consent form was read and signed.

Subject preparation preceding each session included the percutaneous insertion of a radial arterial catheter, under local anesthesia, to permit frequent sampling of arterial blood during scans, and the insertion of an intravenous catheter in the opposite arm for isotope injection. The head was positioned with a special head holder which utilized an individual molded plastic face mask to prevent movement during the study. A laser permanently attached to the wall projected a line onto the mask that corresponded to the position of the lowest PET slice. A lateral skull radiograph with this line marked by a radiopaque wire provided a record of the subject's exact position in relation to the PET slices. The overlapping position of radiopaque markers placed in the external auditory canals (the earmold rings) confirmed that the head was not rotated about the anterior-posterior or vertical axes. After the head was in place, a transmission scan used for individual attenuation correction was performed with a ring source of activity containing germanium-68/gallium -68. During scans the room was darkened and the subject's eyes were covered with gauze pads. Ambient noise during each scan was limited to the sound of cooling fans for the electronic equipment.

Stimuli

A variety of sounds has been used in our test series. Results to be reviewed here will focus on experiments using pure tones and synthetic syllables.

Pure tones. Pure tones were generated using a General Radio 1310A oscillator, an electronic switch and pulse generator built at Central Institute for the Deaf in St. Louis, and a Hewlett-Packard 350D attenuator. Tones were monitored using a Monsanto 113A counter, Telequipment S54A oscilloscope and Hewlett-Packard 400GL voltmeter.

Tones of 500 Hz and 4 kHz were used for testing. Tone were pulsed with a duty cycle of 50%, approximately 500 msec on/off, with a rise/fall time of 50 msec. The subject's threshold for each frequency tested was determined just prior to scanning for that frequency. All tones were presented at 50 dB SL, monaurally to the right ear. For each experimental

314

scan, the sound was turned on approximately 1 min prior to isotope injection, and was presented throughout the scan; thus total presentation time was approximately 2 min.

Synthetic syllables. A tape recording of a set of synthetic nonsense stop-consonant-vowel (stop CV) syllables used in our dichotic listening experiments (e.g., Lauter, 1982) was presented to subjects via a Nagra tape recorder. In preparing this recording, the original 250-msec version of each syllable was edited to leave only the first 50 msec, including acoustical information regarding both consonant and vowel. The tape recording consisted of a constant cycling of the syllable string (ba-da-ga-pa-ta-ka-ba-da-ga...etc.). The rate of syllable repetition, overall level of the recording, and ear of presentation were manipulated in separate experiments (see below). As with the tones, the subject's threshold for the tape recording was obtained just prior to the text scan.

Anatomical Localization

In order to determined where in the three-dimensional data complex to look for auditory responses, we used an anatomical localization scheme developed in our laboratory (cf Fox et al., 1984) that is independent of the appearance of the CBF images. This method yields both slice number and transverse-plane coordinates for a predicted region of interest (ROI) selected from a standard stereotaxic atlas of the human brain (Talairach et al., 1967).

We identified two ROIs for the scans involving pure tones and syllables: primary auditory cortex, and a region surrounding the angular gyrus, often designated as "language cortex." Tomographic images from each subject were then used to create "percent-difference images" (Fox and Raichle, 1984), comparing control and experimental conditions. These images are based on blood-flow values normalized to control for global changes in blood flow occurring between scans, and to highlight areas of maximum change from control to stimulated condition that occur independent of any global changes in CBF.

RESULTS

Pure tones. Examination of activity changes within the estimated region of primary auditory cortex for each hemisphere of each subject revealed systematic shifts of the area of maximum change from condition to condition. In each subject, maximum change always occurred in the left-hemisphere A1 region (i.e., contralateral to stimulation). Also, for each subject, the contralateral region of greatest activity change during stimulation with the 500 Hz tone was more lateral and anterior, and the region that responded best to the 4 kHz tone was more medial and posterior. The orientation of these regions for the five subjects tested in six sessions agree well with those reported for tonotopic responses in monkey auditory cortex using electrophysiological methods (e.g., Brugge and Merzenich, 1973). (See Lauter et al., 1985 for a complete description of these results.)

Synthetic syllables. Results are available to date for single-subject examples of the effects of manipulating rate, level, and ear of presentation of the recorded syllables. Clear qualitative changes were observed in the rCBF images in response to the syllables, occurring in the angular-gyrus "language cortex" region previously defined for each subject. Analysis of the quantitative changes in rCBF as a function of the dimensional manipulations indicate that as rate and level are increased, there is a corresponding increase in rCBF; as ear of presentation is changed, related shifts in activation seem to reflect the predominance of

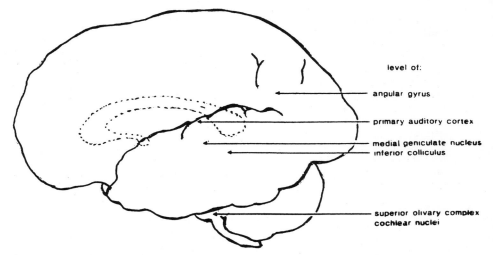

Fig. 1. Schematic of nuclear levels within the human auditory CNS.
in these results, but since they are based on data from single subjects,
the suggestions cannot be taken as conclusive.

contralateral response. There are suggestions of asymmetrical responses

   Multi-level activation of auditory nuclei. The human auditory CNS is
advantageously arranged for study with a tomographic device such as the
PETT VI (Fig. 1). This is in contrast, for example, with the visual
system, which lies essentially within the dimensions of a single PET
slice. As a result, it might be possible to view, in a single 40-sec scan,
responses in more than one auditory center to a particular stimulus.

   Fig. 2 presents a series of difference images taken in a single scan
of one of our test subjects, representing comparisons between a control
scan and a stimulation scan in which synthetic syllables were presented
binaurally at a rate of 20 per second at a level of 50 dB SL. The regions
represented in the two most rostral slices (Panel A, upper two images)
contain no known auditory centers. The slice shown in the lower left of
Panel A, however, at the level of the angular gyrus for this subject,
shows clear bilateral activation, and there is an apparent asymmetry of
substantial proportions--a 16% "left hemisphere advantage" in terms of
rCBF change. Further analysis will be required to determine whether this
difference is statistically significant. The lower-right slice is at the
level of primary auditory cortex; note again bilateral activation, more
symmetrical at this level. The top left slice of Panel B represents the
level of the thalamus: the striking bilateral, symmetrical activation
seen here may be interpretable as response in posterior thalamus, perhaps
representing a combination of MGN and pulvinar. The top right slice is at
the level of the midbrain; the midline activation could indicate inferior
colliculus response, with separation of the two halves of the IC beyond
the resolution of PETT VI. The last slice may be through the cerebellum;
significance of the small response seen here is unknown.

DISCUSSION

   This library of PET auditory activation studies in normal human brains
will be used to answer a variety of additional questions. However, results

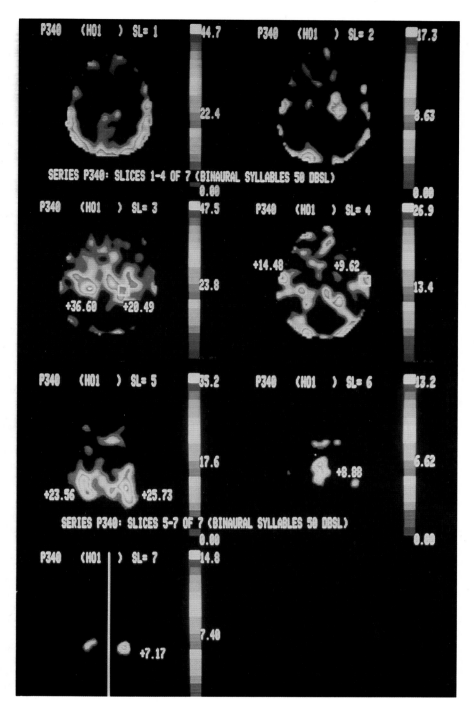

Fig. 2. Simultaneous multi-level activation of human auditory CNS. Series
of 7 slices (Panel A, SL 1 most rostral, Panel B, SL 7 most
caudal) from a single 40-sec scan on one subject. Slices are shown
as "difference images", comparing control with stimulation by
synthetic syllables presented binaurally at a rate of 20 per
second at a level of 50 dB SL. (See text for details.)

to date already suggest that positron emission tomography holds enormous potential as a revolutionary tool for the study of normal human sensory physiology. In the auditory system, responses to both simple and complex sounds can be observed, at a number of auditory CNS levels, with brain activity integrated over as little as 40 sec. The new generation of PET machines (e.g., "SuperPet") will provide improved spatial resolution and much better temporal resolution, sufficient for "evoked rCBF response" studies.

We believe that our results with auditory stimulation, combined with parallel findings in other modalities, point to the possibility of a new physiology, pursued via noninvasive techniques, and designed to emphasize human nervous systems and the complex interactions between stimulus, presentation, and subject variables that are the hallmark of everyday behavior.

Preparation of this paper supported by U.S. Air Force Office of Scientific Research, Life Sciences Directorate.

REFERENCES

Brugge, J. F. and Merzenich, M. M., 1973, Responses of neurons in auditory cortex of the macaque monkey to monaural and binaural stimulation, J. Neurophysiol., 36:1138-1158.

Fox, P. T. and Raichle, M. E., 1984, Stimulus rate dependence of regional cerebral blood flow in human striate cortex, demonstrated by positron emission tomography, J. Neurophysiol., 51:1109-1120.

Fox, P. T., Perlmutter, J. S., and Raichle, M. E., 1984, A stereotactic method of anatomical localization for positron emission tomography, J. Comp. Assist. Tomogr., 9:141-153.

Herscovitch, P., Markham, J., and Raichle, M. E., 1983, Brain blood flow measured with intravenous H2-O15. I. Theory and error analysis, J. Nucl. Med., 24:782-789.

Lauter, J. L., 1982, Dichotic identification of complex sounds:Absolute and relative ear advantages, J. Acoust. Soc. Amer., 71:701-707.

Lauter, J. L, Herscovitch, P., Formby, C., and Raichle, M. E., 1985, Tonotopic organization in human auditory cortex revealed by positron emission tomography, Hearing Res., 20:199-205.

Reichle, M. E., Martin, W. R. W., Herscovitch, P., Mintun, M. A., and Markham, J., 1983, Brain blood flow measured with intravenous H2-O15. II. Implementation and validation, J. Nucl. Med., 24:790-798.

Talairach, J., Szikla, O., and Tournous, P., 1967, Atlas d'Anatomie stereotaxique du Telencepahle, Masson, Paris.

Ter-Pogossian, M. M., Ficke, D. C., Hood, J. T., Yamamoto, M. and Mullani, N. A., 1982, PETT VI: Positron emission tomography utilizing cesium fluoride scintillation detectors, J. Computer Assist. Tomogr., 6: 125-133.

Yamamoto, M., Ficke, D. C., Ter-Pogossian, M. M., 1982, Performance study of PETT VI, a positron computed tomograph with 288 cesium fluoride detectors, IEEE Trans. Nucl. Sci. NS, 29:529-533.

MEDIAL GENICULATE BODY UNIT RESPONSES TO CAT CRIES

J. Buchwald, L. Dickerson, J. Harrison and C. Hinman

Department of Physiology
Brain Research Institute and
Mental Retardation Research Center
UCLA School of Medicine
Los Angeles, California 90024 USA

Participation of different auditory pathways for various kinds of
auditory information processing is suggested by a variety of behavioral
and anatomical data. Behavioral testing procedures indicate that bilateral lesions of the primary auditory cortex cause no loss in discrimination
of temporal sequencing of auditory stimuli (summarized in Colavita, 1979)
nor in discriminations between complex vocal stimuli (cat: Dewson, 1964;
dog: Heffner, 1977), whereas bilateral lesions of the insular and temporal
cortical areas produce marked disruption in these functions (cat: Goldberg et al., 1957; Cornwall, 1967; Kelly, 1973; Colavita et al., 1974;
Dewson, 1964). Insular and temporal cortical regions have been further separated from each other insofar as bilateral lesions confined to insular
cortex produce loss of temporal pattern discrimination, not produced by
temporal cortex lesions (cat: Colavita et al., 1979), while lesions largely confined to temporal cortex result in loss of speech sound discriminations (e.g., "u" versus "i", cat: Dewson, 1964).

Such functional differentiation is supported by anatomical data which
indicate that, while the major thalamic projections to areas AI, II, and
EP originate in the principal division of the medial geniculate body
(MCB), the major projection to temporal cortex originates in the caudal
division of the MGB, and the major projection to insular cortex originates
in the medial division of the posterior nuclear group of the thalamus
(Winer et al., 1977; Diamond, 1978). Furthermore, brainstem input to each
of these thalamic subdivisions appears to originate differentially so
that the parallel but separate anatomical identity of the forebrain pathways is extended caudally to the region of the lateral lemniscus and inferior colliculus (Diamond, 1978).

We have been interested in examining these regions electrophysiologically to test the hypothesis that single units in a particular thalamo-
cortical subsystem will preferentially respond to the same stimulus class
as that class which no longer can be discriminated following lesions of
the subsystem. In our work thus far we have focused upon the thalamic level of these systems in the cat and have studied the major, i.e., principal, caudal, and magnocellular, divisions of the MGB.

Pilot experiments indicated marked changes in evoked potential and unit responses for some of the thalamic regions of interest when the animal was anesthetized with pentobarbital. Thus, all recordings were made from awake unanesthetized adult cats restrained by head-cap fittings which secured the skull in the stereotaxic plane but induced no discomfort. One or two final recording sessions were carried out under barbiturate anesthesia (35 mg/kg) prior to terminating the animal.

The cat was placed at a calibrated site in a sound isolation chamber, and stimuli were delivered through a speaker located 15 cm in front of the nose. All stimuli were 70 ± 6 db SPL, calibrated as acoustic output measured at the external meatus. Continuous online EEG recordings were used to monitor arousal levels, as were concurrent observations of the animal through a one-way mirror. Only data obtained during periods of EEG desynchrony and behavioral arousal have been included in this study.

Initially evoked potential responses to click stimuli were mapped in the subdivisions of the MGB. Typical differences were noted both in the awake cat and during barbiturate anesthesia. In the ventral and dorsal portion of the principal division of the MGB the classical 6-8 ms click evoked response (EP) was regularly recorded. This response decreased moderately at click rates faster than 20/s and was resistant to barbiturate anesthesia. In contrast, in the caudal division of the MGB, click EPs were rarely recorded and, if present, were always very small in amplitude. In the magnocellular division a complex bimodal response appeared, the first component of which resembled the EPs of the principal division while the second component had a 12-15 ms peak latency, decremented markedly to click rates of 20/s and was depressed by barbiturate.

Single unit recordings were carried out with a stimulus protocol which included a set of vocal stimuli, pure tones, and clicks. An estimate of best frequency was provided by 125 ms duration pure tones delivered with a 5 ms rise-fall time over a 0.4 to 10 KHz range. Click responsiveness was tested with a 0.1 ms square wave stimulus. The vocal stimuli were derived from one class of cat calls, the kitten isolation call, known to be of behavioral significance in terms of maternal retrieval (Buchwald, 1981). The taped stimulus protocol of approximately 20 min duration presented 15 trials of each stimulus at a rate of 1/2 sec, with a series of isolation calls both at the beginning and end of the protocol.

Histological confirmation of stereotaxic locations was accomplished by passing current through a stainless steel microelectrode at a number of MGB sites just before sacrifice. Histological sections of the brain were processed for the Gomori reaction to reveal these marker sites as blue dots. Reconstruction of the marks on brain atlas diagrams of the MGB provided a stereotaxic basis for localizing the electrode tracks and units of the preceding recording sessions.

Of the 555 units studied in the MGB, approximately one-third was in each of the major subdivisions. In all divisions, responses to the "call" stimulus class were elicited. However, in the principal divisions essentially none of the units was responsive only to the call stimuli. In contrast, in the caudal division approximately 30 % of the acoustically responsive units responded only to the call stimuli, while relatively few responded to tone or click. In the magnocellular division, responses occurred about equally to click, tone and call stimuli with relatively little selectivity to any class.

A variety of response patterns characterized units throughout the MGB, none of which appeared to be regionally restricted. Responses of sustained increase or decrease in discharge, with or without rebound, were found in all MGB divisions, as were "on", "off", and "on-off" responses and a number of other complex patterns. Moreover, a unit response pattern to the complete "call" stimulus was not necessarily predictable from its responses to call components or to tones.

In general, unit and EP data obtained in these experiments have been consistent. Both indicate a marked responsivity to click stimuli in the principal division which is largely absent in the caudal division of the MGB. In the magnocellular division the degree of EP and unit sensitivity to clicks was intermediate between that of the principal and caudal MGB. The EPs of the principal division further contrasted with those of the magnocellular MGB in that rapid click rates and barbiturate anesthesia had relatively mild effects on the former, but caused marked changes in the latter. Thus, on the basis of the click responsivity, rate effects and barbiturate sensitivity, a clear functional differentiation exists at the thalamic level between the principal, caudal and magnocellular divisions of MGB.

A major objective of the present experiments was to test the hypothesis that units in MGB subdivisions might respond selectively to stimuli which can no longer be discriminated following lesions of specific cortical MGB subdivisions. In previous studies it has been shown that similar discrimination deficits occur with lesions at either the thalamic or cortical level of a particular auditory subsystem (Colavita and Weisberg, 1979). We were particularly interested in comparing responsivity of units in the caudal division of the MGB with those in other divisions to a species/specific vocal signal with a vowel-like harmonic structure. The caudal division of the MGB provides the major input to temporal cortex (Winer et al., 1977; Diamond, 1978) and lesions in this thalamo-cortical subsystem result in a loss of discrimination of vowel sounds (Dewson, 1964). Although responses to the kitten "isolation call" occurred in all MGB divisions, only in the caudal division did a large proportion of the units, approximately 30 %, respond selectively to this stimulus class.

Taken together these electrophysiological data support previous behavioral and anatomical data which suggest different functional roles for the various auditory forebrain systems represented by the divisions of the MGB and their cortical projections.

REFERENCES

Brown, K. A., Buchwald, J. S., Johnson J. R. and Mikolich D. J., 1978, Vocalization in the cat and kitten, Developmental Psychobiol., 11: 559-570.
Buchwald, J. S., 1981, Development of acoustic communication processes in an experimental model, in: "Pre-term Birth and Psychological Development", Friedman,S. L. and Sigman, M. eds., Academic Press, New York, 107-126.
Colavita, F. B., Szeliga, F. V. and Zimmer, S. D., 1974, Temporal pattern discrimination in cats with insular-temporal lesions, Brain Res., 79:153-156.
Colavita, F. B. and Weisberg, D. H., 1979, Insular cortex and perception of temporal pattern , Physiol. and Behav., 22:827-83o.
Cornwall, P., 1967, Loss of auditory pattern discrimination following insular temporal lesions in cats, J. Comp. Physiol. Psychol., 63:165-168.
Dewson, J. H., 1964, Speech sound discrimination by cats, Science, 144: 555-556.
Diamond, I. T., 1978, The auditory cortex, in: "Evoked Electrical Activity in the Auditory Nervous System", Academic Press, New York, 463-485.

Goldberg, J. M., Diamond, I. T. and Neff, W. D., 1957, Auditory discrimination after ablation of temporal and insular cortex in the cat, Fed. Proc., 16:47.

Heffner, H. H., 1977, Effect of auditory cortex ablation on the perception of meaningful sounds, Soc. Neurosci. Abst., 3:6.

Kelly, J. B., 1973, The effects of insular and temporal lesions in cats on two types of auditory pattern discrimination, Brain Res., 62:71-87.

Winer, J. A., Diamond, I. T. and Raczkowski, D., 1977, Subdivisions of the auditory cortex of the cat: The retrograde transport of horseradish peroxidase to the medial geniculate body and posterior thalamic nuclei, J. Comp. Neurol., 176:387-418.

PROJECTIONS FROM SUPERIOR TEMPORAL GYRUS:CONVERGENT CONNECTIONS

BETWEEN STRUCTURES INVOLVED IN BOTH AUDITION AND PHONATION

A. Bieser and P. Müller-Preuss

Max-Planck-Institut für Psychiatrie

8000 München 40

INTRODUCTION

Acoustic communication in primates is basically controlled by two distinguishable systems in the brain. The anatomically well known auditory pathway receives and handles acoustical input, the production of vocalizations is generated by the vocalization areas. The latter ones have been identified by electrical brain stimulation and by brain lesion studies (Jürgens, 1979). So far only a few studies about anatomical and physiological interactions between structures in primates involved in both audition and phonation have been published. In squirrel monkeys Müller-Preuss et al. (1980) reported on anatomical and physiological evidence of a relationship between the limbic vocalization area and the auditory cortex. The authors found a partially inhibitory influence on auditory responsive cells in the auditory cortex during stimulation of cingulate cortex. Additional anatomical studies with the aid of 3H-Leu autoradiographic technique confirmed connections between cingular cortex and superior temporal gyrus (STG). An other inhibitory influence on auditory cortex neurons was revealed during stimulation by self produced vocalizations (Müller-Preuss and Ploog, 1981).

It was the aim of this study to look for projections from the auditory responsive STG to structures which also receive connections from vocalization areas. Such convergent connections between structures involved in both audition and vocalization may work as a relay station for interactions of phonation and audition.

METHODS

In 17 squirrel monkeys (Saimiri sciureus) the efferent projections from STG were investigated with the aid of the autoradiographic tracing technique. With a microsyringe 0.75 μl of tritium-labeled leucine (33 μl Ci/μl) was injected in STG with a microdrive in each animal. After a survival time of 15 days the monkeys were deeply anaesthetized and perfused with 4% formaldehyde. After one week of storage in formaldehyde the brains were dehydrated, paraffin embedded and cut serially at 7 μm in the coronal plane. The deparaffinized sections were coated with photographic emulsion (Kodak NTB 3) and kept in darkness for 7 weeks.

Then they were developed, fixed and stained with cresylviolet.

Identification of the diverse brain structures was done according to the stereotactic atlas of Emmers and Akert (1963).

RESULTS

Convergent connections between structures involved in both audition and vocalization were only found in the caudatum and the thalamus.

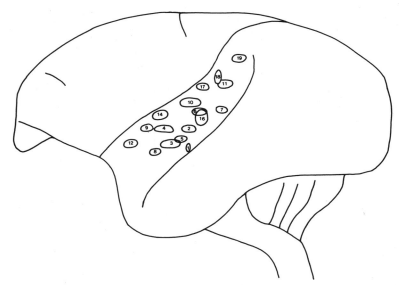

Fig. 1. Schematic lateral view of squirrel monkey's brain showing the injections and their range marked by circles. The numbers indicate different animals.

Projections from STG to the Caudatum

In 14 of 17 3H-Leu treated animals projections from STG to the caudate nucleus were identified. Fig. 1 shows the injection sites and their range in a schematic lateral view of the monkey brain. The numbers indicate the 17 different animals used in this study. The association cortex lateral and rostral to primary cortex (AI) has connections to the head and tail of the caudate nucleus. Labeled cells in the head were always found on the border to the ventriculus lateralis, whereas the tail was labeled ventrally or on the border to the tractus opticus. The distribution of silver grains in the caudate head is different when looking from rostral to caudal:the injections in the rostral and lateral area near AI (4, 9, 12, 2, 6) have connections to a small area of the head. The anterolateral and an area posterolateral from AI (1, 3, 5, 8, 16) projects to almost the whole part of the head. Rostral from the posterolateral association field close to the superior temporal sulcus there is only a connection to the tail (7). The injections caudomedial from AI have efferents to the middle part of the caudate nucleus. In this part of the caudatum the labeling changed from the border of the ventriculus lateralis to the middle part of caudate nucleus. No connections were found from injections in spot 10 and 14 (probably AI).

Projections to the Thalamus

The projections from AI (10, 14) and the adjacent rostral area near the silvian fissure (9, 12) are limited to only two thalamical nuclei: medial nucleus of pulvinar (PuM) and medial geniculate body (GM). The anterolateral association area has connections to GM, PuM and peripeduncular nucleus (Pd) (1, 3, 5, 8 and to some extent 4). A posterolateral group of injection sites (2, 6) has connections to PuM, GM and posterior nucleus of thalamus (P). The medial dorsal nucleus of thalamus (MD) receives connections from spot 4 and 16. Caudo-dorsal to primary auditory field a region with auditory and /or visual aspects can be distinguished. This area has more thalamic terminal fields than the above cited areas of STG.

Table 1. Summary of connections from STG to the thalamus

	VPL	MD	LP	R	GL	GM	PuM	Pd	P	Pul	PuL	PTc
1						+	+	+				
2						+	+		+			
3						+	+	+				
4		+				+	+	+				
5						+	+	+				
6						+	+		+			
7							+					
8						+	+	+				
9						+	+					
10						+	+					
11		+					+			+	+	+
12						+	+					
14						+	+					
16		+	+			+	+					
17		+	+			+	+					+
18	+		+	+	+		+			+	+	+
19			+		+					+	+	

Table 1 summarizes all connections from STG to the thalamus. It is evident that most areas of STG have efferent projections to the PuM (except injection No. 19) and only 3 of 17 injection sites have no connections to GM. These exceptions are the most caudo-dorsal visual areas (11, 18, 19). One further exception is No. 7 which has the simplest connectivity: only the PuM is labeled.

CONCLUSIONS

Regarding our particular interest in such terminal fields which receive connections not only from auditory responsive STG but also from vocalization structures, the following brain structures have to be taken into account:

1. The head and tail of caudate nucleus. It was shown by Müller-Preuss and Jürgens (1976) that these parts of the caudate nucleus receive connections from the 'cingular' vocalization area. The rostral, anterolateral and posterolateral field of STG projects to the same caudate region.

2. There are two thalamic nuclei which receive projections from STG and from different vocalization areas (Jürgens and Müller-Preuss, 1977 and unpublished data):MD and PuM. The heavy labelling in PuM is only absent in the most dorsal injection (19). The MD receives its connections from a small area in STG around the primary auditory cortex.

ACKNOWLEDGMENTS

Supported by Deutsche Forschungsgemeinschaft SFB 204

REFERENCES

Emmers, R. and Akert, K., 1963, A stereotaxic atlas of the brain of the squirrel monkey (Saimiri sciureus). Madison: University of Wisconsin Press

Jürgens, U., 1979, Neural control of vocalization in nonhuman primates. in: "Neurobiology of social communication in primates", H. D. Stecklis and M. J. Raleigh, eds., New York, Academic Press, 11-44.

Jürgens, U. and Müller-Preuss, P., 1977, Convergent projections of different limbic vocalization areas in the squirrel monkey, Exp. Brain Res., 29:75-83.

Müller-Preuss, P. and Jürgens, U., 1976, Projections from the 'cingular' vocalization area in the squirrel monkey, Brain Res., 103:29-43.

Müller-Preuss, P., Newman, J. D. and Jürgens, U., 1980, Anatomical and physiological evidence for a relationship between the 'cingular' vocalization area and the auditory cortex in the squirrel monkey, Brain Res., 202:307-315.

Müller-Preuss, P. and Ploog, D., 1981, Inhibition of auditory cortical neurons during phonation, Brain Res., 215:61-76.

NEURAL PROCESSING OF AM-SOUNDS WITHIN CENTRAL AUDITORY PATHWAY

P. Müller-Preuss, A. Bieser, A. Preuss and H. Fastl*

Max-Planck Institut für Psychiatrie
*Institut für Elektroakustik
München, FRG

INTRODUCTION

Studies concerned with the neural mechanisms of processing biologically relevant sounds are heavily burdened by the choice of an appropriate stimulus test repertoire. From the physical point of view, acoustic signals are determined by frequency and intensity parameters and their temporal courses. Various combinations of these parameters are the components also of bioacoustic signals and therefore have to be considered in the search for those parts of the signals which transmit the biologically relevant information. Our approach in solving such problems considered the parameter intensity or course of a sound's amplitude. This parameter is the subject of research mostly because it obviously bears information about the distance and the spatial location of a sound source within the acoustic environment of individuals. However, regarding the communicative behavior of species, the amplitude of a sound often signify emotional states and, beyond that, it's course can also deliver intentional information used for intraspecific acoustic communication between the members of a species. The information bearing parameters can be single or multiple as well as periodically repeated AM-elements.

One part of our approach is now concerned with the study of the neural reactivity to those sounds whose amplitude changes periodically. Sounds having such temporal properties are, firstly, part of the specie's vocal repertoire (Schott, 1975) and, secondly, have also been shown by psychoacoustic experiments (Terhardt, 1968; Fastl, 1983) to be of relevance in acoustic communication of man. So we designed experiments to evaluate the activity of neurons of central stations of the auditory pathway when subjected to acoustic stimuli modulated sinusoidally in amplitude.

METHODS

The experiments were performed on awake, adult squirrel monkeys, whose neural activity was recorded extracellularly with the aid of tungsten microelectrodes. The electrodes were implanted in the auditory cortex (superior temporal gyrus), thalamus (medial geniculate body) and midbrain (inferior colliculus), single and multi-unit activity was explored by using a hydraulic microelectrode holder. Neural activity was fed into a LSI 11/73 computer for peri-stimulus-time histogram (PSTH) calculations. Data obtained from 15 consecutive stimulus presentations were used for calculating one PSTH, which in turn served for further data analysis. Digitizing rate was 1000 points per second.

Acoustic stimulation consisted of tones of a particular frequency (mostly corresponding with the neurons characteristic frequency) and noise bursts of a length of 1 sec. Acoustic stimulation occurred in a sound shielded chamber through a loudspeaker placed 1.5 m in front of the animals head (free field, binaural situation). To cover a modulation frequency range as large as possible, frequency of modulation was selected in octave steps from 1 Hz to 256 Hz. An electronic switch determined stimulus length (1 sec) and gaps (1 sec) between consecutive stimuli, whereas another circuit allowed variation of the modulation depth (0-100%, in 10% steps) with constant peak amplitudes. Sound pressure level was controlled electronically or manually by an attenuator.

Electrode penetrations were histologically verified either by electro-lytic microlesions (some) or reconstructed together with the lesion-verified tracks on the basis of stereotactic data (most). For histological preparation, paraffin sections of the brain, after formaldehyde sections, were stained with cresylviolet.

The quantitative analysis of the transfer of the modulation frequency to neural activity was carried out by calculating the content of the particular modulation frequency within the PSTH spectra through a Fast Fourier Algorithm and/or a Cross Correlation of the PSTH with the amplitude envelope of the stimulating signal. Thus the power calculated at the particular frequency as well as the factor of the correlation served as "magnitude" or "value" of modulation transfer. Additionally, the total sum of spikes of a particular PSTH have been considered by the analysis.

RESULTS

The activity of about 450 auditory units within the pathway has been evaluated; 215 of them came from the midbrain, 115 from the thalamus and the remaining units from the auditory cortex. For the inferior colliculus, most of the neurons have been localized in the central nucleus, whereas units recorded in the medial geniculate body have been identified as coming from the whole nucleus, i.e. from all its sub-structures. Within the auditory cortex most units were from its primary area, but some have been localized rostrally to it, thus belonging to an area described in a similar species (Imig et al., 1977) as the rostral auditory field.

Almost all units recorded within the three auditory areas were influenced by amplitude-modulated stimuli. Concerning the transfer of a signal's modulation frequency, the most impressive result is that most of the units are sensitive within a particular band of AM-frequencies. There are only a few units which display a low pass response charac-teristic or have complex response patterns (i.e. multiple peaked). The center frequency, i.e. the MF a particular cell is most sensitive to (=best modulation frequency or BMF), varies between 2 and 256 Hz. It is also a general observation that, if a cell is not very sharply MF tuned, the high frequency slope of a bandpass is much steeper than its lower part. Concerning the particular regions, a comparison of the neural response characteristic shows a gradual decrease in the MF transfer while ascending the pathway: In the midbrain the greater number of units are sensitive to signals modulated with 64/32 Hz, whereas neurons of the thalamus transfer mainly MFs of the range between 32 and 16 Hz. In the auditory cortex the MFs decline to 16 and 4 Hz. The histograms on the right hand side of the figure show schematically the response properties of the particular structures. It can also be seen that there is a difference regarding the structures between the transfer of modulation,

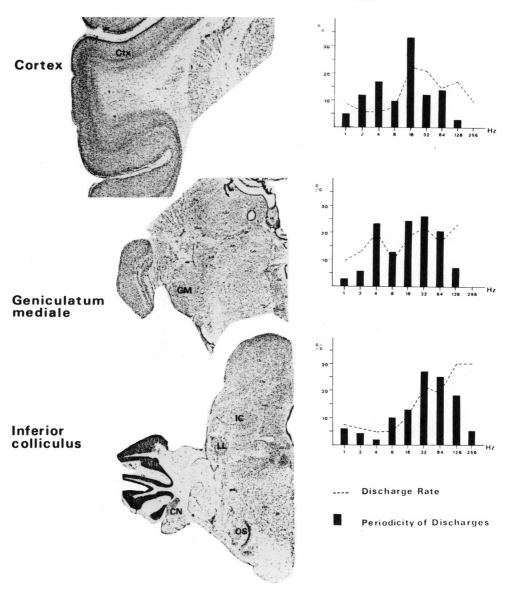

Cortex

Geniculatum mediale

Inferior colliculus

---- Discharge Rate

■ Periodicity of Discharges

Fig. 1. On the Left in frontal Nissl-stained sections of the squirrel
monkeys brain the particular structures of the auditory pathway
are shown. On the Right: in histograms the MF transfer properties
of the structures are indicated (black columns). Broken lines
show the corresponding discharge rates. For comparison, values
of the ordinate have been normalized. Abbrev.: Ctx.: Auditory
cortex; GM: Medial geniculate body; IC: Inferior colliculus;
LL: Dorsal nucleus of lateral lemniscus; OS: superior olive;
CN:cochlear nucleus.

i.e. the periodicity of a signal and the total spike activity: High MFs
evoke high activities in collicular neurons, which must not be necessarily accompanied by strong modulation transfer. In the lower most histo-

gram (IC), the discharge rate increases at MFs higher than 64 Hz, whereas the value of the periodicity decreases. In contrast, at the cortex the amount of the discharge rate parallels the value of periodicity.

If the modulation depth was changed from 100% in steps of 10% down to 0 modulation, a monotonic relation between depth and value of modulation transfer has been mostly observed: The deeper a signal is modulated, the stronger is the neural modulation. Only a few neurons have been detected which displayed a nonmonotonic response characteristic, i.e. had their strongest modulation at depths lower than 100%. Also there is only a weak difference between the particular regions insofar as more neurons responding nonmonotonically are found within the medial geniculate. In this respect, data of the cortex are up to now not sufficient to be added to the analysis.

A lack of nonmonotonic response properties is also registered when the intensity of a stimulus is varied: strongest modulation of a neurons activity occurred mostly at the greatest intensities tested (for this purpose the attenuator has been switched within the range from 90 dB to 30 dB in 10 dB steps).

As already indicated above, noise as well as tones were used as stimulus carriers, the latter if a neuron clearly showed a best response at a particular frequency, i.e. displayed tuning properties. An analysis of the best modulation frequency distribution of the neurons and the carrier (i.e. noise vs. tones) revealed no significant differences within a structure or between the structures. Furthermore, also no clear relationship between transfer of modulation frequency and characteristic frequency has been detected. These observations are supported by the histological evaluations: Within the central nucleus of the inferior colliculus the BMF distribution is quite uniform thus signifying no correlation with the tonotopic organization demonstrated for this auditory station (Fitzpatrick, 1975). Only a tendency of low BMFs occurring more often in the center of the IC rather than at its margins has been observed. The data obtained from the thalamus and the cortex also lack indications for a correlation between BMF and CF, again the number of units may be too small to allow such conclusions also for these auditory stations. On some occasions it was possible to hold neurons long enough for repetition of the stimulus repertoire. Even if the number of units tested over a longer period of time is small, a considerable amount of variability concerning a units BMF became obvious. In some cases the BMF varied over a range of 3 octave steps and, furthermore, on the single unit level the carrier could influence the BMF. However, this variability does not affect the overall response properties of the particular structures, as indicated in the figure.

CONCLUSIONS

The data presented here support the results from other studies, carried out on various stations of the auditory pathway of different species, insofar, as a decline of the frequency of amplitude modulation from the periphery to the system's central parts is a common observation (for example Møller, 1974; Rees and Møller, 1983; Schreiner and Urbas, 1986). Due to these results, cortical neurons will transfer preferably signals which have a slow repetition rate, whereas quickly modulated sounds will evoke greater activity in lower parts of the pathway. Those species-specific vocalizations which have periodic AM-components cover a range from 12 - 75 Hz, a range which coincides quite well with the AM-sensitivity of the structures shown here. The faster modulated calls are, from a behavioral point of view, used for the expression of more emotional situation, whereas the slower modulated one serve more for social contact within a group. The differential sensitivity of the particular stations to different AM-sounds could indicate a "hierarchical" organization of the processing of calls within the pathway, depending on the complexity of the behavioral situation a particular vocalization is used for. It is suggested, that periodical AM-modulations serve as a carrier for particular components within the sequence which contain the relevant information. Such components are processed preferably by neurons showing phasic response properties, a characteristic detected in an increasing amount at higher levels of the pathway (for example Newman, 1979; Creutzfeldt et al., 1980).

A comparison of the data presented here with the psychoacoustic evaluations show some noteworthy similarities: Firstly, also in man AM modulations are perceived in a bandpass like characteristic, a phenomenon described as fluctuation strength and roughness (Fastl, 1982), secondly, variations of the parameter intensity and depth cause motonic response characteristics. And last, not least, these response properties are quite carrier independent. Such a congruence between data gathered, on the one side, by neurophysiological methods on a nonhuman primate and, on the other side, obtained by psychoacoustic experiments in man, suggests that the processes of transmitting AM-sounds from acoustic energy to neural activities are controlled by similar mechanisms, at least within primate species.

ACKNOWLEDGEMENTS

Supported by "Deutsche Forschungsgemeinschaft SFB 204"

REFERENCES

Creutzfeldt, O., Hellweg, F.-C. and Schreiner, Chr., 1980, Thalamo-cortical transformation of responses to complex auditory stimuli, Exp. Brain Res., 39:87-104.
Fastl, H., 1982, Fluctuation strength and temporal masking patterns of amplitude-modulated broadband noise, Hearing Res.,59:59-69.
Fastl, H., 1983, Fluctuation strength of modulated tones and broadband noise, in: "Hearing - Physiological Bases and Psychophysics", Klinke, R. and Hartmann, R., eds., Springer, Berlin, 282-288.
FitzPatrick, A. K., 1975, Cellular architecture and topographic organization of the inferior colliculus of the squirrel monkeys, J. Comp. Neurol., 164:185-208.
Imig, T. J., Ruggero, M. A., Kitzes, L. M., Javel, E. and Brugge, J. F., 1977, Organization of auditory cortex in the owl monkey (Aotus trivirgatus), J. Comp. Neurol., 171:111-128.
Møller, A. R., 1974, Responses of units in the cochlear nucleus to sinusoidally amplitude-modulated tones, Exp. Neurol., 45:104-117.
Newman, J. D., 1979, Central nervous system processing of sounds in primates, in: "Neurobiology of social communication in primates: an evolutionary perspective", Steklis, H. D. and Raleigh, M. J., eds., New York, Academic Press, 60-109.
Rees, A. and Møller, A. R., 1983, Responses of neurons in the inferior colliculus of the rat to AM and FM tones, Hearing Res., 10:301-330.
Schott, D., 1975, Quantitative analysis of the vocal repertoire of Squirrel monkeys (Saimiri sciureus), Z. für Tierpsychol., 38:225-250.
Schreiner, C. E. and Urbas, J. V., 1986, Representation of amplitude modulation in the auditory cortex of the cat. I. The anterior auditory field (AAF), Hearing Res., 21:227-241.
Terhardt, E., 1968, Ueber akustische Rauhigkeit und Schwankungsstaerke, Acustica, 20:215-224.

AUDITORY LOCALIZATION

PROPERTIES OF CENTRAL AUDITORY NEURONES OF CATS RESPONDING

TO FREE-FIELD ACOUSTIC STIMULI

Lindsay Aitkin

Department of Physiology
Monash University
Melbourne, Australia

We can distinguish at least two advantages conferred by binaural, compared with monaural, hearing. First, the input from each ear can be summed, and thus enhanced, for purposes of spectral analysis - "what is it"? The resultant effects of the various peripheral and central auditory mechanisms enabling these analyses are most easily studied using free-field stimuli. This review summarizes studies carried out on cats during the last seven years in my laboratory that address the question: how do neurones in the brain's auditory pathway respond to sounds emitted from different points around the head? Can we recognise functional classes of auditory neurones with free-field stimulation? From the point of view of spectral analysis, we might expect to find neurones maximising the input from each ear, perhaps responding best to sounds coming from straight ahead, or in line with each pinna. For purposes of localisation there may be neurones that have sharp preferences for particular loca- tions in space, not just in the "maximum input" position. We might expect "maps" of acoustic spatial locations in a brain region, just es we find visual maps in the visual pathway. Some of the material to be reviewed has appeared in three journal articles (Aitkin et al., 1984, 1985; Aitkin and Martin, 1987).

The response of a given neurone to changes in the azimuthal position of a sound source around the head will be determined by both the changes in the physical cues that occur, and the nature of the connectivity of that neurone with the periphery.

Changes in Physical Parameters that Result from a Sound Emitter Being Moved around the Head

First, the difference in time between the arrival of the sound at each ear changes, from zero straight ahead to maximum values of approx. 300-400 μsec for cats (Roth et al, 1980) and 600-700 μsec for humans (Mills, 1958). Secondly, there are changes in the relative intensity of the sound at each ear, due to diffraction and reflection by the head.

Diffraction is related to sound frequency, being dependent on the relation between interaural width and the wavelength of the sound (Rayleigh, 1876). Measurements of the sound field around the cat's head made either with a microphone in one ear canal or using cochlear micro- phonic recordings indicate highest pressures when the speaker is roughly

in line with the geometric axis of the ipsilateral pinna if it were
considered as a horn (Calford et al., 1986; Calford and Pettigrew, 1984;
Irvine, 1987; Phillips et al., 1982). If the ipsilateral pinna is removed
the sound pressure drops on that side, so one needs to consider the pinna
as a source of pressure amplification as well as the head being a site
of attenuation. Together they produce an interaural intensity difference,
IID. Measurements of the maximum IID's in cats show that they are below
5dB at frequencies low for a cat (less than 2kHz) and rise to high values,
around 30 dB, at frequencies between 10 and 30 kHz.

The ridges and grooves of the pinna also reflect and diffract sounds
of high frequency to an extent that depends on the angle of incidence
of the sound wave. The spectral composition of vibrations entering the
ear canal may be different to those emanating from a sound source, and
will depend on both the angular location of the sound and its distance
from the ear. Animals that are sensitive to high frequencies, such as
cats, can often move their pinnae both conjunctively and disjunctively.
These movements will alter the spectral composition and IID of a sound
from a fixed sound source, and make the apparent sound field very complex
with moving stimuli.

The Gross Design of the Brain Auditory Pathway

One can recognise in cats a number of "parallel" brain stem pathways
that converge in the auditory midbrain, whose principal nucleus is the
inferior colliculus. These pathways have been defined in dichotic stimulation and tracing experiment (see the article by Kudo and Nakamura in this
volume); for simplicity they can be abbreviated to the following:

(i) contralateral excitatory, running directly from the dorsal or
ventral cochlear nucleus; (ii) binaural excitatory, relaying in the
medial superior olive; (iii) contralateral excitatory - ipsilateral
inhibitory (binaural inhibitory), relaying in the lateral or medial
superior olive and, less well documented (or smaller?) (iv) ipsilateral
excitatory, via the lateral superior olive.

Neurons in the inferior colliculus receive to a varying degree a
convergent input from these pathways (Semple and Aitkin, 1980, 1981).
Some neurones or groups of neurones receive input predominantly from one
nucleus - e.g. the ipsilateral medial superior olive (Aitkin and Schuck,
1985) - while others receive input from a number of sources (Maffi and
Aitkin, 1987). Thus, the processing performed in these pathways is
integrated in the midbrain. We should therefore expect to find neurones
in the inferior colliculus whose properties reflect the synaptic interactions that have occurred along the input pathways to this nucleus.

Responses of Central Auditory Neurones to Free-Field Stimuli

I will describe how neurones in the midbrain - the inferior colliculus - respond to sounds emitted from different speaker positions along
a near-horizontal plane. It should be noted that while these results
certainly pertain to sound localisation, they concern the coding of free
field stimuli generally. These experiments, carried out in collaboration
with Drs. Martin, Phillips, Calford and Pettigrew, involved recording
the discharges of single neurones in the inferior colliculus of anaesthetised cats to pure tone or noise stimuli presented at different positions on a plane passing through the centres of the acoustical axes of
the pinna, a little above the horizontal plane (Fig. 1). These positions
are referred to as stimuls azimuths.

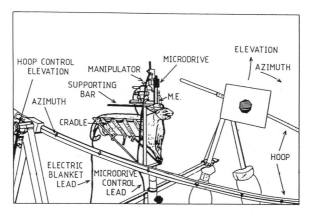

Fig. 1. Apparatus for studying neural responses to free-field stimuli in anechoic space. A curved hoop carries a lightweight speaker. The speaker is moved along the hoop, and the hoop rotated up and down by activation of stepping motor located outside the anechoic room. The anaesthetised animal is positioned with its head in the centre of the imaginary sphere described by hoop movements. A microelectrode (ME) may be advanced with a microdrive controlled from outside the anechoic room.

The sound frequency most effective in activating the neurone (the best frequency) is first determined with the speaker located near the contralateral pinna axis. This is an optimal starting point since it is well known that most neurones are excited by contralateral stimuli in the central auditory pathway beyond the medulla. The best frequency stimulus, or noise (which is usually very effective in exciting auditory neurones) is presented at an intensity 20 to 50 dB above threshold from different speaker positions. The number of spikes evoked by 20 stimuli is plotted against speaker position (azimuth), and the resultant azimuth function repeated at different stimulus intensities.

Thus two parameters are varied - intensity and position - keeping spectral composition constant. In most cases stimulus intensities are such that sound reaches both cochleas at all stimulus positions. The term "azimuth function" should thus be distinguished from the term "spatial receptive field" which is a measure of the effective sound field close to threshold (Middlebrooks and Pettigrew, 1981; Moore et al., 1984; Semple et al., 1983). Many spatial receptive fields relate only to contralateral monaural input.

Omnidirectional Units

These units are insensitive to stimulus location but may be influenced by stimulus intensity. The unit illustrated in Fig. 2 exhibits an increase in discharge rate that is monotonically related to stimulus intensity for both contralateral (filled circles) and ipsilateral (open circles) speaker positions. The two functions are nearly parallel and are separated on the intensity axis by about 15dB, approximately the maximum IID at 17.5 kHz (Irvine, 1987). This unit is thus likely to receive mainly monaural contralateral input; other omnidirectional units appear to be binaurally excitatory.

About one-third of our sample are of this type. Very few (less than 1%) of these behave as though only or predominantly driven through the

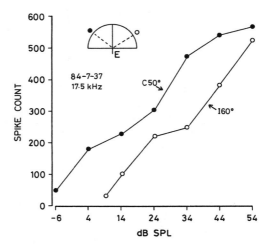

Fig. 2. Spike count in 20 trials (ordinate) of unit 84-7-37 plotted
against intensity in dB of 17.5 kHz tone pips, presented from an
azimuthal position 50 degrees from the midline on the contralateral side (black dots, inset and C50 degrees) or 60 degrees
ipsilateral to the midline (open circles, inset and I60 degrees).
Inset: The "hoop" viewed from above, intersecting with the
midsagittal plane and the interaural line. E: location of recording electrode contralateral to black dots.

ipsilateral ear. One explanation for this is that the ipsilateral excitatory pathway may converge with another pathway or pathways on to single
cells in the midbrain.

Frontal Units

     These could be considered a type of binaurally excitatory unit the
responses of which are determined by the mutual facilitation of input
from each ear in a balanced fashion. While they are found in the inferior
colliculus, they are more common in the auditory cortex and become the
most common type in certain premotor areas such as the cerebellum. They
are also more prevalent when noise stimuli, rather than tones, are used.

     Two examples are shown in Fig. 3, in which azimuth functions (spike
count plotted against stimulus azimuth) are derived at three different
intensities for one unit excited by noise (upper) and another by tones
(lower). For both units lowest threshold and peak firing rates occur
near the midline, and the functions broaden at high intensities.

Units with Intensity-Dependant Directionality

     These cells, acting as though receiving a strong monaural (contra-
lateral) input in addition to a binaural input, are very common in the
auditory midbrain. In the example of Fig. 4 the intensity functions
(upper) reveal that stimuli delivered from a contralateral location of
the speaker elicit a net excitatory response, whereas those from an
ipsilateral position evoke suppression of firing below spontaneous
activity levels. Such a unit is likely to receive input from the binaural
inhibitory channel. Inspection of the azimuth functions (lower) confirms
this but shows that the contralateral excitatory input is dominant and

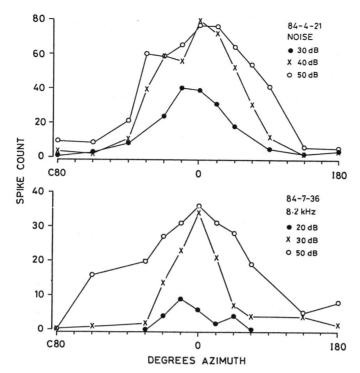

Fig. 3. Spike count in 20 trials (ordinate) plotted against speaker
location (abscissa) (azimuth functions) for two units stimulated
with noise (unit 84-4-21) and tone pips (unit 84-7-36), at three
different intensities. C80, I80:80 degrees contralateral and
ipsilateral azimuth.

excitation encroaches well into the ipsilateral hemifield at higher
intensities. At any given intensity the azimuth function is directional
but the directional feature (peak or border location) shifts in location
with changes in intensity.

Azimuth Selective Units

These directional units have discharge features that are largely
independent of intensity. They are interesting because they can be
"tagged" with a potential function - that of sound localisation. For this
reason they will be examined in more depth. We shall consider first units
with best frequencies below 2 kHz, where the IID's are small and binaural
interactions depend predominantly on interaural time differences.

More than 50% of low best frequency units of the central nucleus
of the inferior colliculus are azimuth selective, and some of these units
have very sharply peaked functions. (Fig. 5).

The main criterion for azimuth selectivity is that an important
feature (a peak or border) in the azimuth function has constancy (10
degrees or less variation) in azimuthal position over a significant range
of intensities. Although this is an arbitrary criterion it does differentiate these units from the more common type as seen in Fig. 4, particularly those with high best frequencies. It should be noted, however, that

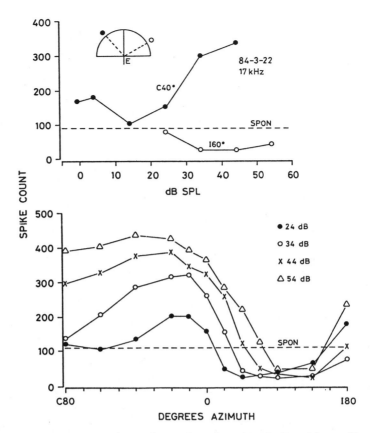

Fig. 4.   Intensity functions (upper) and azimuth functions (lower) for
unit 84-3-22 at 17 kHz. SPON = spontaneous activity level.

some units defined as azimuth selective over the range of intensities
studied here (20 to 50 dB above threshold) could have exhibited shifts
in directionality at higher intensities. Similarly, some units intensity-
dependent over the range studied may have become selective at higher
intensities.

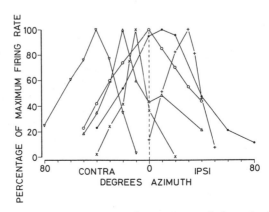

Fig. 5.   Azimuth functions for 6 units selected for their sharpness
and range of peaks (best azimuths). Modified from Aitkin et al.
(1985).

340

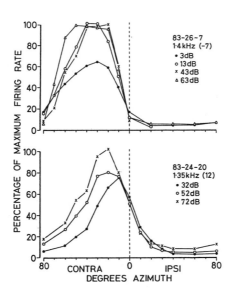

Fig. 6. Azimuth functions at different intensities for two units. Numbers
in brackets are threshold sound pressure levels. Modified from
Aitkin et al. (1985).

The two examples of Fig. 6 show how intensity may affect the shape
of the azimuth function. The upper functions show an increase in broadness
but retain a constant centre best azimuth of approximately 30 degrees
contralateral. The lower functions have a common border and a shift in
best azimuth from contralateral 10 to contralateral 20 degrees.

These sharply-peaked patterns are the products of binaural interactions, presumably in the time domain. If one or the other ear canal is
occluded, as in Fig. 7, the spike count is greatly reduced, sometimes to
zero (indicating the operation of binaural facilitation) and the function
becomes flat or non-existent.

The sharpest of these functions occur in the best frequency range
1.1 - 1.4 kHz. When the best azimuths of these units are related to their
locations in the central nucleus of the inferior colliculus (Fig. 8) a
tendency is observed for medial azimuths (10 to -10 degrees) to be
represented caudally and peripheral contralateral azimuths rostrally in
the lateral part of the central nucleus. These data support a concept of
the topographic organisation in the inferior colliculus of the azimuthal
location of low frequency sounds.

A topographic representation of stimulus azimuth, in terms of peak
discharge for high sound frequencies seems unlikely in the inferior
colliculus because most high frequency neurones tend to have plateau
rather than peaked functions (Fig. 9). Peaks are usually only demonstrated
at low intensities when their presence may be due to the directional
amplification produced by the contralateral pinna (Fig. 9B, 27dB).

High best frequency azimuth selective units are, instead, characterized by a discharge border whose azimuthal position is relatively little
altered by changes in stimulus intensity (Fig. 9) but is often markedly
altered by changes in pinna position (Fig. 10).

For example, with the contralateral pinna in the alert, forward-

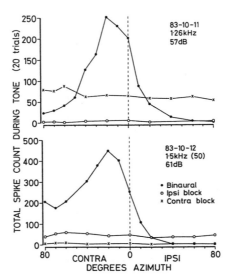

Fig. 7.   Azimuth functions of two units obtained under binaural conditions,
          or with the ipsi or contra ear canal blocked. Modified from
          Aitkin et al. (1985).

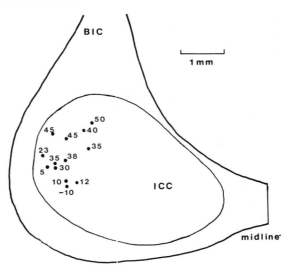

Fig. 8.   Inferior colliculus of a cat viewed from above, with the central
          nucleus (ICC) outlined. BIC:brachium of the inferior colliculus
          (rostral). Each dot is the point of entry of a microelectrode;
          each number the mean best azimuth for units of best frequency
          1.1 - 1.4 kHz encountered by the electrode. Positive numbers
          contralateral, negative value ipsilateral. Modified from Aitkin
          et al. (1985).

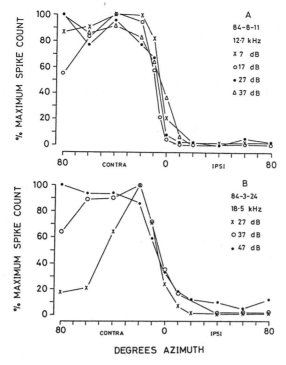

Fig. 9. Azimuth functions for two units stimulated by high frequency
tones. Modified from Aitkin and Martin (1987).

directed position the azimuth functions of unit 84-10-16 have borders
falling between approximately contralateral 20 and 0 degrees. With the
pinna in the flattened, backward position the functions shift by 20
degrees further peripherally, a shift of a similar order of degrees to
that of the angle of the pinna.

If azimuth functions are plotted in a normalised fashion it is
possible to calculate a mean function for each unit and derive a mean
border. Such borders obtained in animals with pinnas positioned in a
similar, alert fashion, show a tendency to be concentrated near the
midline (Fig. 11). This pattern prevails for both noise and tone functions. Although some borders occupy much of the contralateral or ipsilateral hemifield, most fall between contralateral and ipsilateral 20
degrees.

The question of azimuthal topography for noise- activated units
remains to be resolved. Similar proportions of noise-activated units have
fixed peaks and fixed borders (Fig. 11). The peaks are to some extent
scattered along the azimuth axis although they do concentrate near
contralateral 20 degrees.

CONCLUSION

There are units in the auditory midbrain whose properties reflect
clearly both the nature of on-going interaural cues and the monaural or
binaural properties of their dominant input pathways. Integration of input
processed separately by parallel brain stem pathways is important at
this level of the brain. A number of neuronal types have clear functional

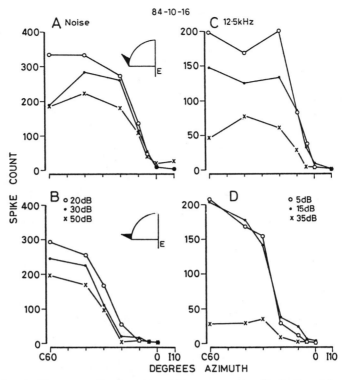

Fig. 10. The influence of pinna position on azimuthal functions of unit
84-10-6 generated either by noise (A,B) or 12.5 kHz tones (C,D)
The insets schematically depict a cat's head viewed from above
with the angle of the front of the pinna to the head depicted
by the black triangle, in relation to the contralateral recording site (E).

capacities - about one-third can be correlated with a function in sound
localisation at this level of the auditory pathway.

It is well known that lesions of the upper auditory system produce
profound contralateral acoustic localisation deficits in both humans and
animals (e.g. Jenkins and Masterton, 1982; Sanchez-Longo and Forster,
1958). Azimuth selective units in the inferior colliculus overwhelmingly
code the location of <u>contralateral stimuli</u> and may form a basis for
localisation. Some of these units (particularly low frequency and noise
responders) code a narrow azimuthal region, others (particularly high
frequency responders) code a broad contralateral range with a sharp
cut-off near the midline.

Frontal units have a peak response for sounds in the midline - they
may be special cases of azimuth selective units but behave as though
summing input from each ear in a facilitatory manner. Such units would
respond best if the acoustic stimulus was in the direction of frontal
gaze, and may have a role in stimulus identification rather than in
localisation.

One possible role for azimuth selective units is to indicate the
side from which a sound initially originates. In some way their activity
may initiate head, eye and pinna movements towards the sound source.

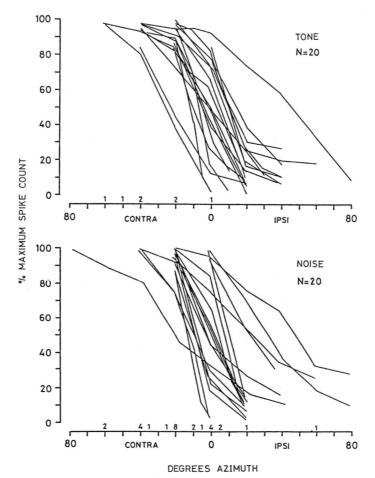

Fig. 11. Mean borders calculated for 20 units activated by noise and 20
by tones. The small numbers on the abscissa indicate the number
of units with peaked functions whose best azimuths lie at the
indicated azimuth. Note that very few (7) tone- activated units
but many (27) noise-activated units have peaked functions.

Binaural facilitatory ("frontal") units could indicate by their peak
discharge that the direction of frontal gaze is now towards the source of
the sound.

However, since accurate orienting movements can occur in the absence
of on-going acoustic stimulation in animals such as humans, cats, bats
and owls (e.g. Knudsen et al., 1979; Thompson and Masterton, 1978), there
must also be some internalised place of neuronal activity that can, when
activated, direct an appropriate orienting response. Such "maps" may take
various forms (e.g. Knudsen and Konishi, 1978; Wise and Irvine, 1985).
Whatever the form, it is likely that they will be composed of ensembles
of azimuth-selective units.

REFERENCES

Aitkin, L. M., Gates, G. R. and Phillips, S. C., 1984, Responses of
neurons in the inferior colliculus to variations in sound source
azimuth, J. Neurophysiol., 52:1-17.

Aitkin, L. M. and Martin, R. L., 1987, The representation of stimulus azimuth by high best-freqency azimuth-selective neurons in the central nucleus of the inferior colliculus of the cat, J. Neurophysiol., 57:1185-1200.

Aitkin, L. M., Pettigrew, J. D., Calford, M. B., Philips, S. C. and Wise, L. Z., 1985, The representation of stimulus azimuth by low frequency neurons in the inferior colliculus of the cat, J. Neurophysiol., 53: 43-59.

Aitkin, L. and Schuck, D., 1985, Low frequency neurons in the lateral central nucleus of the inferior colliculus receive their input predominantly from the medial superior olive, Hearing Res., 17:87-93.

Calford, M. B., Moore, D. R. and Hutchings, M. E., 1986, Central and peripheral contributions to coding of acoustic space by neurons in inferior colliculus of cat. J. Neurophysiol., 55:587-603.

Calford, M. B. and Pettigrew, J. D., 1984, Frequency dependence of directional amplification at the cat's pinna, Hearing Res., 14:13-19.

Irvine, D. R. F., 1987, Interaural intensity differences in the cat: changes in sound pressure level at the two ears associated with azimuthal displacements in the frontal horizontal plane, Hearing Res., 26:267-286.

Jenkins, W. M. and Masterton, R. B., 1982, Sound localization: effects of unilateral lesions in central auditory system, J. Neurophysiol., 47:987-1016.

Knudsen, E. I., Blasdel, G. G. and Konishi, M., 1979, Sound localization by the barn owl (Tyto alba) measured with the search coil technique, J. Comp. Physiol. A, 133:1-12.

Knudsen, E. I. and Konishi, M., 1978, Space and frequency are represented separately in the auditory midbrain of the owl, J. Neurophysiol., 41:870-884.

Maffi, C. L. and Aitkin, L. M., 1987, Differential projections to regions of the inferior colliculus of the cat responsive to high frequency sounds, Hearing Res., 26:211-219.

Middlebrooks, J. C. and Pettigrew, J. D., 1981, Functional classes of neurons in primary auditory cortex of the cat distinguished by sensitivity to sound location, J. Neurosci., 1:107-120.

Mills, A. W., 1958, On the minimum audible angle, J. Acoust. Soc. Am., 30:237-246.

Moore, D. R., Semple, M. N., Addison, P. and Aitkin, L. M., 1984, Properties of spatial receptive fields in the central nucleus of the cat inferior colliculus. I. Responses to tones of low intensity, Hearing Res., 13:159-174.

Phillips, D. P., Calford, M. B., Pettigrew, J. D., Aitkin, L. M. and Semple, M. N., 1982, Directionality of sound pressure transformation at the cat's pinna, Hearing Res., 8:13-28.

Rayleigh, Third Baron, 1876, Our perception of the direction of a source of sound, Nature, 14:32-33.

Roth, G. L., Kochhar, R. K. and Hind, J. E., 1980, Interaural time differences: implications regarding the neurophysiology of sound localization, J. Acoust. Soc. Am., 68:1643-1651.

Sanchez-Longo, L. P. and Forster, F. M., 1958, Clinical significance of impairment of sound localization, Neurology, 8:119-125.

Semple, M. N. and Aitkin, L. M., 1980, Physiology of pathway from dorsal cochlear nucleus to inferior colliculus revealed by electrical and auditory stimulation, Exp. Brain Res., 41:19-28.

Semple, M. N. and Aitkin, L. M., 1981, Integration and segregation of input to the cat inferior colliculus, in: "Neuronal Mechanisms of Hearing", Syka, J. and Aitkin, L., eds., Plenum, New York, 155-161.

Semple, M. N., Aitkin, L. M., Calford, M. B., Pettigrew, J. D. and Phillips, D. P., 1983, Spatial receptive fields in the cat inferior colliculus, Hearing Res., 10:203-215.

Thompson, G. C. and Masterton, R. B., 1978, Brain stem auditory pathways involved in reflexive head orientation to sound, J. Neurophysiol., 41:1183-1202.

Wise, L. Z. and Irvine, D. R. F., 1985, Topographic organization of interaural intensity difference sensitivity in deep layers of cat superior colliculus: implications for auditory spatial representation, J. Neurophysiol., 54:185-211.

INFORMATION PROCESSING CONCERNING MOVING SOUND SOURCES IN THE AUDITORY

CENTERS AND ITS UTILIZATION BY BRAIN INTEGRATIVE AND MOTOR STRUCTURES

J. A. Altman

I. P. Pavlov Institute of Physiology
Academy of Sciences of the USSR
Leningrad, USSR

It is noteworthy that although extensive literature has been devoted to psychoacoustic, behavioral (on animals) and electrophysiologic studies of sound source localization in the horizontal plane, until quite recently basic attention has been focused on the investigation of the localization of unmoved sound sources (Altman, 1972).

Needless to say that localization of moving sound sources and determination of various parameters of moving objects is as important (or even more important) for humans and animals as the localization of unmoved sound sources; in some cases it may even be of vital importance.

The purpose of this presentation is to describe the results on this subject which were obtained mainly in our laboratory with the use of electrophysiological method. Our studies were devoted to the processing of impulse activity observed in anesthetized cats at various levels of the auditory system above the level of binaural convergence of afferent pathways from the right and left ears. Extra- and intracellular recording of neuronal activity has been performed. As far as the stimulation is concerned, we have used a signal which with a dichotic paradigm simulates the movement of a sound source in space. It is well known from experiments in humans that when a sound signal is presented via two earphones the subject perceives it as a fused image (FI) localized within the head. At equal amplitude and simultaneous presentation of the signals to the right and left ears the FI is localized at the head midline. If the sound signals, e.g. clicks, are presented with an interaural time delay the subject perceives the FI shifted towards the side receiving the earlier stimulation (the lateralization effect).

It is this effect of the FI lateralization that was used to create a sound signal simulating the movement of a sound source. A click train of 1-2 s duration was presented binaurally with a time delay linearly changing with the repetition of the clicks in the train. At a particular click rate and particular parameters of interaural time delay such a signal produced a marked sensation of the FI movement in most of subjects. For instance, if at the onset of the click train the delay was long enough (e.g. 800 μs) and gradually diminished to reach zero level the subjects perceived a sensation of the FI movement from the leading ear to the midline. It is obvious that changing the value and the sign of the interaural delay, one can modify the direction, trajectory and velocity of the FI motion.

Let us now consider the electrophysiological findings obtained on cats. At the first level of binaural convergence of afferentation, i.e. in the superior olivary complex, neurons sensitive to the FI movement have been detected (Altman, 1972). The observed sample of 15 neurons responsive to the sound appeared to have 8 elements of this kind (53%). However, in responses of these neurons (determinated with the use of the method of poststimulus time (PST) histograms) only symmetric changes of impulse activity have been observed regardless of the direction of the FI movement (whether it was from the ear to the midline or in the reverse direction).

The observations of neurons of a higher, midbrain level of the auditory system - inferior colliculus - have shown that the character of the neuronal response to the FI movement could be quite different. The neuronal reaction had a certain specialization: in 13% of cases (and in 26% of cases when evoked potentials were estimated) the neurons responded asymmetrically as evident from the value of the response (and sometimes also from its pattern): e.g. the response to the sound simulating sound source movement from the midline to the ear was different from that with the movement in the reverse direction. In terms of the reaction of visual system neurons, these neurons could be designated as detectors of the direction of the sound source movement. Of course we do not assume that these neurons act exclusively toward the extraction of the mentioned parameter of sound signals - they can get involved in the response to sound stimuli with other parameters. However, it is now important to emphasize that they can distinguish the direction of the FI movement by means of a specialized response.

It is important that at the inferior colliculus level a certain topographic organization of neuronal systems responding to the "movement" of the sound source was revealed. This organization is related to a certain spatial distribution of the effects of predominance of contralateral stimulation over the ipsilateral one in the central nucleus of the inferior colliculus, as well as of the inhibitory effects associated with binaural stimulation of afferent impulses. It is suggested that a certain group on neurons, the so-called multipolar cells concentrated in the ventro-central area of the central nucleus of the inferior colliculus are the first to respond to the "movement" of the sound source (Radionova, 1987).

Further processing of the afferentation proceeds at the next level of the auditory system - the medial geniculate body (thalamic level). First, this center contains a higher amount of neurons specifically responding to the signals simulating the sound source movement of a particular direction. The tested sample of 50 neurons appeared to contain 11 detectors of movement direction (22%). Second, 73% of MGB neurons specifically responded only within a certain limited range of the sound source "movement" (i.e. within a restricted part of the signal), instead of giving an impulse response throughout the whole period of the "movement". In other words thalamic neurons of the auditory system respond to a certain sector of the space. This sector being different for different neurons, they can wholly cover the perimeter of the sound source movement in the horizontal plane. It is noteworthy that such neurons make-up only 7% at the superior olivary level and 20% at the level of the inferior colliculi. Finally, the third characteristic feature of MGB neuronal response is the appearance of an afterdischarge in 37% of the neurons (diffuse or rhythmical) which is maintained over several seconds after the sound signal ceases. This afterdischarge may reflect (as evident from the analysis of both the number of impulses and pattern of the afterdischarge) the physical characteristics of the sound source "movement", e.g. the velocity of its movement. Thus, the afterdischarge can maintain the information about the direction of the sound source movement for several seconds after the stimulus is withdrawn.

In this connection neuron selectivity to sound movement velocity should be noted: for 75% of thalamic neurons of the auditory system the velocity ranging from 30 to 90 degrees/s was most efficient with certain assumptions concerning its quantitative estimation.

As far as a sample of the neurons tested in the auditory cortex (area AI) is concerned, the neurons of this higher part of the auditory system retain the same properties as observed in the responses of thalamic auditory neurons. It should be noted that the responses of the cortical auditory neurons have been recorded not only extracellularly, but also intracellularly (Altman and Nikitin, 1985). It is obvious from these findings that specialization of cortical auditory neuronal responses (the neurons of layers III-IV of the auditory cortex have been mainly tested) is connected with marked inhibitory processes that are evident from prolonged postsynaptical hyperpolarization potentials. The inhibitory components of the response determine not only the time pattern and specialization of the responses in the course of the stimulus presentation, but also the afterdischarges of the cortical units.

Thus, it follows from the data presented above that the information of the parameters of sound source "movement" undergoes complex processing along the auditory pathway in the impulse activity of neurons at different levels of the auditory system. It should be also stressed that the values of the velocities of the FI movement which were efficient for the auditory neurons (generally from 30 to 1200 degrees/s) are rather similar to the parameters of moving sound source as revealed in discrimination (tests in psychoacoustical experiments in humans). This allows one to suggest that the above described processing may be likely as the neurophysiological basis of the localization performance.

Meanwhile, the above described results confirmed further in the investigations of other authors (Sovijärvi, 1973; Yin and Kuwada, 1983) have urged us to focus our attention on another important problem in the studies of the neurophysiological mechanisms of localization of moving sound sources. The problem is what is the further utilization of this auditory system activity? This problem seemed quite natural, as it has been unknown what form the information of spatial position of the sound source takes after it goes beyond the classic auditory pathway and how it is further processed. In this framework the study of sound source localization has certain advantages over the study of other properties of sound signals (e.g. pitch, intensity, duration, etc.). Localization of a sound source, including the tracing of its movement, is closely related to the motor activity of man and animals. This is well exemplified by a well known orienting response to the sound, which is a marked manifestation of localization behavior.

Turning to morphological data on the connections between the auditory system and motor cerebral structures we can trace the following common scheme of their projections: inferior colliculus and auditory cortex - pontine nuclei - auditory area of cerebellar vermis cortex - subcortical cerebellar nuclei - red nucleus. In addition, ascending afferent pathway branches at the thalamic level: some part of the afferent impulses pass through ventrolateral thalamic nucleus to the sensori-motor cortex and further to extrapyramidal centers, while the other part via motor midbrain center (red nucleus) enters the spinal cord. This morphological scheme has laid the basis for the investigations of further processing of information concerning sound source localization.

Even the first results obtained in this direction have been quite encouraging. It has been found that the neurons of the auditory cerebellar vermis cortex show a marked sensitivity to the "moving" sound source. The percentage of neurons specifically responsive to the signals simulating sound source movement in a particular direction appeared to be 1.5-3 times as high (31%) as in the centers of the classic auditory pathway (13-14% in the inferior colliculus, 22% in the MGB, 23% in the auditory cortex). After it has been established that the only output cells of cerebellar vermis cortex - Purkinje cells - can also exhibit a specialized response to the direction of sound source movement we have supposed that the cells of subcortical cerebellar nuclei may be also sensitive to those parameters of the sound signals that simulate the sound source movement. Electrophysiological study of neurons of the fastigial nucleus, to which the pathways from the auditory cerebellar cortex project, provide additional evidence for high sensitivity of these units to sound source "movement". 39% of these neurons displayed a response to sound source "movement", the neuronal responses being different in all cases from the responses to "unmoved" sound signals with the same interaural delays. This allowed us to consider these neurons as detectors of sound source movement (Bekhterev and Kudryavtseva, 1983). The neuronal activity of the ventrolateral nucleus has not been studied yet. However, the study of neurons of sensori-motor cortex area SII (Kulikov and Bekhterev, 1979) have detected an extremely high percent of neurons sensitive to the FI movement. Under the conditions of anesthesia this percentage amounted to 80%, and in nonanesthetized preparations - to 90% (Kulikov and Bekhterev, 1979). A high percentage of neurons (71%) responding to sound source movement has been also observed in the red nucleus, it was 89% in its magnocellular part, and 58% in the parvocellular part (Schinkarenko, 1984).

Thus, on the basis of our findings several statements concerning further processing of information about a moving sound source beyond classic auditory pathways can be put forward. First, localization parameters of a moving sound source are reflected in the activity of various integrative motor centers, which differ both functionally and hierarchically in the nervous system. Some of these centers are directly connected with spinal cord neurons, i.e. with the neurons that actually form the motor activity of the animal. Second, and most unexpected, the percentage of these neurons in the above-listed structures appeared in many cases to be several times as high as in classic auditory pathway centers.

How shall we explain this massive representation of the information about spatial characteristics of a sound signal in the mentioned integrative structures? The following explanation seems rather likely. For the localization of a sound source, first and foremost the analysis of outer (extrapersonal) acoustic space is needed. This analysis performed with the help of consecutive processings of afferent impulsation in the auditory system leads to formation of a model of this space in the brain. However, to discriminate the direction of the sound source movement (as well as for its localization in general) this model alone is not enough. The localization of the sound source also requires a certain reference level with which the model of outer acoustic space could be compared. We believe that the body scheme could be used as this reference level: the information constantly available in the brain can lay the basis for the decision about the relationship between the body (or its parts) and outer acoustic space (Altman, 1983). It is obvious that numerous inter connections among various brain structures (direct or indirect) can ensure the comparison of these two models in the brain (i.e.

of outer acoustic space and body scheme). However, the most rapid and efficient comparison of these models in the brain can be achieved if they are formed in the same brain structures. It is especially important now to emphasize that all the above-listed integrative motor centers are well known to have a pronounced somatotopical organization, or in other words, contain body scheme elements in their nervous network. Therefore, detailed information about the parameters of the sound source movement available in these brain structures allows to compare them to both the model of the acoustic space and the body scheme, and make a decision about the position of the sound source in space. It should be noted that the neurons of the integrative motor structures tested are also sensitive to stimuli of other modalities. We believe that this supports the proposed hypothesis, since the description of an outer stimulus provided not only via the auditory channel, but also via other sensory channels will undoubtedly facilitate its recognition by integrative and associative brain structures.

Processing of the information about sound source movement and its characteristics (the region of movement, its direction and speed) in classic divisions of the auditory system and integrative motor centers of the brain in anesthetized animal is largely evident from the changes of the percentage of neurons selectively responding to modifications of particular parameters of the "movement". It can be seen from these Figs. that the certification of the fact of the movement itself requires involvement of a large number of neurons (72%) at the level of the inferior colliculus, which is the key center for the localization of the sound source in the classic auditory system. To determine the region and speed of the movement it is required that higher centers of the auditory system – MGB and the cortex exert activity high enough (in terms of percentage of the neurons involved). The fulfillment of a more complicated task, – discrimination of the sound direction - is mediated by a relatively insignificant part of the auditory neurons, cortical regions included (14-23%).

The role of integrative motor centers with respect to the response of their neurons to different parameters of sound source movement is very significant. A large part of the sound-sensitive neurons of these brain regions, and often virtually all of them, are involved in the establishment of the fact of the movement itself, its region and speed. The activity of the neurons (in terms of percentage of the neurons involved) of the red nucleus and sensori-motor cortex is expecially high, even with respect to the selectivity of their responses to different directions of the sound source movement, which is the most complicated task of the localization of moving sound sources.

REFERENCES

Altman, J. A., 1972, Sound localization, Neurophysiological mechanisms, Nauka, Leningrad, (in Russian).
Altman, J. A., 1983, Localization of a moving sound source, Nauka, Leningrad, (in Russian).
Altman, J. A. and Nikitin, N. I., 1985, Inhibitory processes in the cat's auditory cortex neurons in dichotic stimulation, J. Evol. Physiol. Biochim., 21:463-469, (in Russian).
Bekhterev, N. N. and Kudryavtseva, I. N., 1983, Fastigial nuclei neurons reactions in cats on auditory stimuli, Physiol. J. USSR, 69:1143-1150, (in Russian).

Kotelenko, L. M. and Kudryavtseva, In.N., 1985, Characteristics of the afterdischarge in medial geniculate body neurons in cat, Physiol. J. USSR, 71:453-460 (in Russian).

Kotelenko, L. M., 1985, Pattern estimations of the afterdischarge of the medial geniculate body in cat, Physiol. J. USSR, 71:1540-1547, (in Russian).

Kulikov, G. A. and Bekhterev, N. N., 1979, Sensori-motor cortex neurons reactions to monaural and binaural stimulation in cat, Physiol. J. USSR, 65:801-811.

Radionova, E. A., 1987, Sound signal analysis in the auditory system. Neurophysiological mechanisms, Nauka, Leningrad.

Shinkarenko, S. A., 1984, Activity of n. ruber neurons in cat to interaural differences in auditory stimulation, Physiol. J. USSR, 70:291-298, (in Russian).

Sovijärvi, A. R. A., 1973, Single neurons responses to complex and moving sounds in the primary auditory cortex of the cat, Acad. Diss. Helsinki.

Yin, T. C. T. and Kuwada, S., 1983, Binaural interaction in low-frequency neurons in inferior colliculus of the cat. II. Effects of changing rate and direction in interaural phase, J. Neurophysiol., 50:1000-1019.

PARTICIPANTS

ADAMS, JOE C.
    Department of Otolaryngology, Medical University, Charleston,
    SC 29425, USA
AITKIN, LINDSAY M.
    Department of Physiology, Monash University, Clayton, Victoria,
    3168, Australia
ALTMAN, JACOB A.
    I. P. Pavlov Inst. of Physiol., Nab. Makarova 6, 199034, Leningrad,
    USSR
BAREŠ, KAREL
    Department of Otolaryngology, Medical Faculty, Charles University,
    U nemocnice 2, 12000 Prague 2
BERNDT, HARTMUT
    ENT Department, Humboldt University, Schumannstrasse 20/21,
    1040 Berlin, GDR
BIEDERMANN, MANFRED
    Institute of Physiology, Teichgraben 8, 6900 Jena, GDR
BIESER, ARMIN
    Department of Primate Behaviour, Max-Planck Inst. for Psychiatry,
    Kraepelinstr. 2, 8000 Munchen 40, FRG
BLACKBURN, CAROL C.
    Department of Biomedical Eng., John Hopkins University, 720 Rutland
    Ave, Baltimore, MD 21205, USA
BLAŽEKOVÁ, LUDMILA
    Research Institute of Occupational Medicine, Limbova 14,
    83301 Bratislava, ČSSR
BOETTCHER, FLINT A.
    University of Texas at Dallas, Callier Center, 1966 Inwood Road,
    Dallas, Texas 75235, USA
BRANIŠ, MARTIN
    Institute of Experimental Medicine, Czechoslovak Academy of Sciences,
    Lidových milicí 61, 12000 Prague 2, ČSSR
BUCHWALD, JENIFFER S.
    Brain Res. Institute UCLA, Los Angeles, CA 90024, USA
ČADA, KAREL
    Department of Otolaryngology, Medical Faculty, Pekařská 53,
    65691 Brno, ČSSR
CARNEY, LAUREL H.
    Department of Neurophysiology, University of Wisconsin, 1300
    University Avenue, Madison, Wisconsin, 53706 USA
COLLIA, FRANCISCO
    Department of Histology, Medical Faculty, University of Salamanca,
    37007 Salamanca, Spain

DALLOS, PETER
    Auditory Research Laboratory, Northwestern University, 2299 Sheridan
    Road, Evanston, Illinois, 60201, USA
DIDIER, ANNE
    Laboratory of Exp. Audiology, INSERM, U 229, Hospital Pellegrin,
    33076 Bordeaux, France
DRUGA, ROSTISLAV
    Department of Anatomy, Charles University, Medical Faculty,
    U nemocnice 5, 12000 Prague 2, ČSSR
EPPING, WILLEM J. M.
    Department of Medical Physics, University of Nijmegen, P.O.Box 9101,
    6500 HB Nijmegen, Netherland
ERNST, ARNE
    Department of Otolaryngology, Martin Luther University, Leninallee
    12, Halle 4020, GRD
EVANS, EDWARD F.
    Department of Communication and Neuroscience, University of Keele,
    Keele, Staffordshire, ST5 5BG, Great Britain
FAYE-LUND, HILDE
    Anatomical Institute, Karl Johans Gate 47, Oslo 1, Norway
FEX, JORGEN
    National Institute of Health, Building 36, Room 5D08, Bethesda,
    Maryland, 20892, USA
FISCHER, WOLFGANG H.
    Department of Otolaryngology, University of Ulm, P.O.Box 4066,
    D-7900 Ulm, FRG
FLOCK, ÅKE
    Department of Physiology II, Karolinska Institute, Solnavagen 1,
    Stockholm, S-10401, Sweden
FRANĚK, MAREK
    Institute of Theory and History of Art, Czechoslovak Academy of
    Sciences, Haštalská 6, 11000 Prague 1, ČSSR
FRIAUF, ECKHARD
    Department of Zoophysiology, University of Tübingen, Morgenstelle
    28, D-7400 Tübingen, FRG
FURNESS, DAVID N.
    Department of Neuroscience, University of Keele, Keele ST5 5BG,
    Staffordshire, Great Britain
FURUKAWA, TARO
    Department of Physiology, Tokyo Medical-Dental Univ., 1-5-45
    Yushima, Bunkyo-ku, Tokyo 113, Japan
GODFREY, DONALD A.
    Department of Physiology, Oral Roberts University, 7777 South Lewis,
    Tulsa, Oklahoma 74171, USA
GUINAN, JOHN J. JR.
    Eaton-Peabody Laboratory, 243 Charles Street, Boston, MA 02114,
    USA
HACKNEY, CAROLE M.
    Department of Communication and Neuroscience, University of Keele,
    Keele, Staffordshire ST5 5BG, Great Britain
HARRISON, ROBERT V.
    Department of Otolaryngology, Hospital for Sick Children,
    555 University Avenue, Toronto, Ontario M5G 1X8, Canada
HÄUSLER, UDO
    Institute of Zoology, Technical University Munich, Lichtenbergstr. 4,
    D-8046 Garching, FRG
HEIL, PETER
    Inst. of Zoology, Technical University, Schnittspahnstr. 3,
    D-6100 Darmstadt, FRG
HERRERA, MANUEL
    Department of Morphology, Medical Faculty, University of Alicante,
    Alicante, Spain

HOFMAN, VLADIMIR
    Department of Otolaryngology, Medical Faculty, Charles University,
    V úvalu 84, 15018 Prague 5, ČSSR
HOSE, BERND
    Institute of Zoology, Techn. University, Schnittspahnstr. 3,
    D-6100 Darmstadt, FRG
HRUBÝ, JAROSLAV
    Institute of Physiological Regulations, Czechoslovak Academy of
    Sciences, Nad Závěrkou 17, 16000 Prague 6, ČSSR
IKEDA, KAZUHIKO
    Institute of Virology and Immunobiology, Univ. of Würzburg,
    Versbacher str. 7, D-8700 Würzburg, FRG
IMIG, THOMAS
    Department of Physiology, Medical Center, University of Kansas,
    Kansas City, KS 66103, USA
IRVINE, DEXTER R. F.
    Department of Psychology, Monash University, Clayton, Victoria 3168,
    Australia
JILEK, MILAN
    Institute of Experimental Medicine, Czechoslovak Academy of Sciences,
    Lidových milicí 61, 12000 Prague 2, ČSSR
KLINKE, RAINER
    Zentrum der Physiologie, Theodor-Stern-Kai 7, D-6000 Frankfurt a. M.
    70, FRG
KÖPPL, CHRISTINE
    Institute of Zoology, Lichtenbergstr. 4, 8046 Garching, FRG
KUBEC, ZBYNĚK
    Institute of Experimental Medicine, Czechoslovak Academy of Sciences,
    Lidových milicí 61, 12000 Prague 2, ČSSR
KUDO, MOTOI
    Department of Anatomy, School of Medicine, Kanazawa University,
    Kanazawa 920, Japan
LANGNER, GERALD
    Institute of Zoology, Technical University, Schnittspahnstr. 3,
    Darmstadt D-6100, FRG
LAUTER, JUDITH L.
    Department of Speech and Hearing Sciences, University of Arizona,
    Tucson, AZ 85721, USA
MAFFI, CONSTANZA
    Department of Otolaryngology, University of Melbourne, Parkville,
    Victoria, 3052, Australia
MANLEY, GEOFFREY A.
    Institute of Zoology, Tech. University Munich, Lichtenbergstr. 4,
    8046 Garching, FRG
MASTERTON, BRUCE R.
    Department of Psychology, Florida State University, Tallahassee,
    Florida 32306, USA
MELICHAR, IVO
    Institute of Experimental Medicine, Czechoslovak Academy of
    Sciences, Lidovych milici 61, 12000 Prague 2, ČSSR
McMULLEN, TERESA A.
    Dept. of Biomedical, Engineering, Boston University, Boston, MA
    02215, USA
MELSSEN, WILLEM J.
    Dept. Medical and Biophysics, Greet Groote Plein Noord 21,
    Nijmegen NL - 6525 EZ, Netherland
MERCHAN, MIGUEL A.
    Dept. of Biology, Faculty of Medicine, University of Salamanca,
    37007 Salamanca, Spain
MEYER zum GOTTESBERGE, ANGELE-M.
    Research Inst. of ENT Department, University Düsseldorf,
    Moorenstrasse 5, 4000 Düsseldorf, FRG

MILLER, JOSEF J.
      Kresge Hearing Research Inst., 1301 E. Ann Street, Ann Arbor,
      MI 48109, USA
MØLLER, AAGE R.
      Department of Neurological Surgery, Presbyterian-University
      Hospital, 230 Lothrop Street, Pittsburgh, PA 15213, USA
MOORE, JEAN K.
      Dept. of Anatomical Sciences, SUNY at Stony Brook, Stony Brook,
      NY 11794, USA
MOREL, ANNE
      INSERM U-94, 16 Av. du Doyen Lepine, 69500 Bron, France
MOUNTAIN, DAVID
      Dept. of Biomedical Engineering, Boston University, 110
      Cummington Street, Boston, MA 02215, USA
MOUSHEGIAN, GEORGE
      Callier CTR for Com. Dis., University of Texas at Dallas, Dallas,
      TX 75235, USA
MÜLLER, CHRISTIAN M.
      Max-Planck-Institute for Brain Research, Deutschordenstr. 46,
      6000 Frankfurt a.M. 71, FRG
NAVARA, MICHAL
      Department of Otolaryngology, Medical Faculty, Charles University,
      U nemocnice 2, 12000 Prague 2, CSSR
NOVÁK, MILOŇ
      Department of Otolaryngology, Medical Faculty, Charles University,
      U nemocnice 2, 12000 Prague 2, ČSSR
NOVOTNÝ, ZDENĚK
      ENT Department, Medical Faculty, Charles Univ., Karlovo nám. 32,
      12000 Prague 2, ČSSR
OSEN, KIRSTEN K.
      Anatomical Institute, University of Oslo, 0162 Oslo 1, Norway
PARKER, DAVID J.
      Dept. Communication and Neuroscience, University of Keele,
      Keele, Staffs., ST5 5BG, Great Britain
PICKA, JIŘÍ
      Institute of Physiological Regulations, Czechoslovak Academy of
      Sciences, Nad Záverkou 17, 16000 Prague 6, ČSSR
POPELÁŘ, JIŘÍ
      Institute of Experimental Med. Czechoslovak Academy of Sciences,
      Lidových milicí 61, 12000 Prague 2, ČSSR
PUJOL, REMY
      INSERM U-254, Hospital St. Charles, 34059 Montpellier Cedex, France
RADIONOVA, ELENA A.
      I. P. Pavlov Inst. of Physiology, Nab. Makarova 6, Leningrad
      199034, USSR
RAJKOWSKA-MARKOW, GRAZYNA
      Nencki Inst. of Exper. Biology, Pasteura 3, 02-093 Warsaw, Poland
REDIES, HERMANN
      Max-Planck-Institute of Biophysical Chemistry, Am Fassberg,
      Postfach 2841, D-3400 Göttingen-Nikolausberg, FRG
REDMAN, J. R.
      Dept. of Psychology, Monash University, Clayton, Victoria 3168,
      Australia
RIBAUPIERRE de, FRANCOIS
      Institute of Physiology, Bugnon Street 7, 1005 Lausanne,
      Switzerland
RICHTER, FRANK
      Inst. of Physiology, Teichgraben 8, 6900 Jena, GRD
RIGHETTI de, FRANCESCO
      Institute of Physiology, Bugnon Street 7, 1005 Lausanne,
      Switzerland

ROBERTSON, DONALD
     Department of Physiology, University of Western Australia,
     Nedlands 6009, Australia
RODRIGUES-DAGAEFF, CATHERINE
     Inst. of Physiology, Bugnon Str. 7, 1005 Lausanne, Switzerland
ROUILLER, ERICK
     Institute of Physiology, Bugnon Street 7, 1005 Lausanne,
     Switzerland
RUGGERO, MARIO A.
     Dept. of Otolaryngology, Medical Research East, 2630 University
     Avenue S. E., Minneapolis, MN 55414, USA
RUTTGERS, KARIN
     Department of Zoophysiology, University Tübingen, D-7400
     Tübingen 1, FRG
RYBAK, LEONARD
     Department of Surgery, SIV School of Medicine, P.O.Box 3926,
     Springfield, Illinois, 62708, USA
SALDANA, ENRIQUE
     Department of Biology, Medical Faculty, University of Salamanca,
     37007 Salamanca, Spain
SCHACHT, JOCHEN
     Kresge Hearing Institute, University of Michigan, 1301 E. Ann
     Street, Ann Arbor, MI 48109, USA
SCHÄFER, JÜRGEN W.
     ENT Department, University of Ulm, P.O.Box 4066, D-79oo Ulm, FRG
SCHERMULY, LOTHAR
     Zentrum der Physiologie, J. W. Goethe Universität, D-6000 Frankfurt
     70, FRG
SCHOONHOVEN, RUURD
     ENT Department, Leiden University Hospital, 2333 AA Leiden,
     Netherland
SCHREINER, CHRISTOPH
     Coleman Laboratory HSE 871, University of California, San
     Francisco, CA 94143, USA
SEDLÁČEK, KAREL
     Department of Phoniatry Medical Faculty, Charles University,
     Žitná 24, 12000 Prague 2, ČSSR
ŠEJNA, IVAN
     ENT Department, Institute for Postgrad. Med. Educ., Budínova 2,
     18000 Prague 8, ČSSR
SELDON, LEE H.
     ENT University Clinic, Josef Stelzmannstr. 9, 500 Köln 41, FRG
SHIMADA, HITOSHI
     Dept. of Otolaryngology, Tokkyo University, Koshigaya Hospital 343,
     Japan
SIMM, GERIT M.
     Institute of Physiology, 1005 Lausanne, Switzerland
ŠPÁTOVÁ, MILADA
     Institute of Experimental Medicine, Czechoslovak Academy of
     Sciences, Lidových milicí 61, 12000 Prague 2, ČSSR
STEFFEN, H.
     Inst. of Zoology, Techn. University, Schnittspahnstr. 3, D-6100,
     Darmstadt, FRG
STOKKUM van, IVO H. M.
     Dept. Med. Physics-Biophysics, University of Nijmegen, P.O.Box
     9101 6500 HB,Nijmegen, Netherland
SUGIHARA, IZUMI I.S.
     Dept. of Physiology, Tokyo Med.-Dent. University, 1-5-45 Yushima,
     Bunkyo-ku, Tokyo 113, Japan

SYKA, JOSEF
       Institute of Experimental Medicine, Czechoslovak Academy of
       Sciences, Lidovych milici 61, 12000 Prague 2, ČSSR
SYKOVÁ, EVA
       Institute of Physiological Regulations, Czechoslovak Academy of
       Sciences, Bulovka - Pavilon 11, 18000 Prague 8, ČSSR
TANAKA, YASUO
       Department of Otolaryngology, Koshigaya Hospital, Dokkyo University,
       2-1-50 Minami-Koshigaya, Koshigaya, Saitama 343, Japan
TOMLINSON, WARD R. V.
       2211 Wesbrook Mall, Vancouver, B. C., Columbia, V6T 2B5, Canada
ÚLEHLOVÁ, LIBUŠE
       Institute of Experimental Medicine Czechoslovak Academy of
       Sciences, Lidových milici 61, 12000 Prague 2, ČSSR
VLKOVÁ, ALEXANDRA
       Institute of Experimental Medicine, Czechoslovak Academy of
       Sciences, Lidovych milici 61, 12000 Prague 2, ČSSR
VOIGT, HERBERT F.
       Dept. of Biomed. Engn., Boston University, 110 Cummington St,
       Boston, MA 02215, USA
ZELENÝ, MOJMÍR
       Dept. of Otolaryngology, Central Military Hospital, Střešovická,
       16000 Prague 6, ČSSR
ZURITA, PATRICIA
       Institute of Physiology, Rue du Bugnon 7, 1005 Lausanne, Switzerland
ZWISLOCKI, JOZEF J.
       Institute for Sensory Research, Syracuse University, Syracuse,
       NY 13244, USA

Lateralization, 349-353
Lizard, 6, 7
Localization of sound, 335-345,
    349-353

Medial geniculate body, 280
  amplitude modulation, 327-333
  functional organization, 191-195
  impulse noise effect, 217-220
  moving sound, 349-353
  unit responses, 319-321
Melanin, 29-32
Middle ear reflex, 306, 307
Monkey, 123, 245, 323, 327
Moving sound, 349-353
Myosin, 23

Neuronal activity
  in auditory cortex, 245-249, 349
  in auditory nerve, 4, 5, 57, 101-
      105
  in cochlear nucleus, 141-146, 149-
      153, 207
  in inferior colliculus, 213, 285,
      289, 335, 349
  in medial geniculate body, 191-
      195, 319-321, 349
Noise
  hair cell loss, 45-49
  and evoked potentials, 217-220
Noradrenalin
  in cochlear nucleus, 112, 123-129
  in superior olivary complex, 112

Octopus cells, 67-71, 91, 92, 136
Olivocochlear bundle, 253-265, 269-
    277, 279-280, 301, 306
Omnidirectional units, 337, 338
Onset units, 145
Otoacoustic emission, 51-55
  and OCB stimulation, 260
Outer ear, 13-16

Pauser units, 145
Periodicity analysis
  neuronal model, 207-211
Phase sensitivity
  in cochlear nucleus, 213-215
  in inferior colliculus, 213-215
Pinna, 344
Positron emission tomography, 313-
    318
Prostacyclin, 35-38

Rabbit, 217
Rat, 41, 45, 89, 95, 203, 279, 293,
    305
Reptiles, 9, 10

Species-specific sound, 14, 319-321
Spherical cells, 67-71, 77-79

Spiral ganglion, 101-105, 270-277
Stereocilia
  deflection, 17
  stiffness, 21
  ultrastructure, 27
Stria vascularis, 37
Superior olivary complex, 279-281
  neurotransmitters, 107-118
  projections
      from inferior colliculus, 281,
          282, 299-303, 306, 307
      to inferior colliculus, 299-303
Superior temporal gyrus, 323-326
Synthetic syllables, 315, 316
Spontaneous activity, 102-104
  auditory fibers, 262-264
  efferents, 269-277
  in nonmammals, 4, 5

Tectorial membrane, 17-21
Thalamus
  and vocalization, 324-326
Threshold shift, 261-265
Tonotopic organization
  auditory cortex, 223-227
  auditory thalamus, 197-201
  efferents in cochlea, 271, 275
  human cortex, 315
Transfer function
  of outer ear, 13-16
Thromboxane, 35-38

Ventral cochlear nucleus, 67-71
  cytoarchitecture, 77-81
  electrical stimulation, 149-153
  immunostaining, 135-136
Vocalization, 323-326